PRINCIPLES
OF SOIL AND PLANT
WATER RELATIONS

PRINCIPLES
OF SOIL AND PLANT
WATER RELATIONS

M.B. KIRKHAM

Kansas State University

ELSEVIER
ACADEMIC
PRESS

AMSTERDAM • BOSTON • HEIDELBERG • LONDON
NEW YORK • OXFORD • PARIS • SAN DIEGO
SAN FRANCISCO • SINGAPORE • SYDNEY • TOKYO

Publisher: Dana Dreibelbis
Editorial Coordinator: Kelly Sonnack
Project Manager: Kristin Macek
Marketing Manager: Linda Beattie
Cover Design: Eric DeCicco
Composition: Kolam
Text Printer: Vail Ballou
Cover Printer: Phoenix Color

Elsevier Academic Press
200 Wheeler Road, 6th Floor, Burlington, MA 01803, USA
525 B Street, Suite 1900, San Diego, California 92101-4495, USA
84 Theobald's Road, London WC1X 8RR, UK

This book is printed on acid-free paper. ∞

Library of Congress Cataloging-in-Publication Data
Kirkham, M. B.
    Principles of soil and plant water relations / M. B. Kirkham.
        p. cm.
    Includes bibliographical references and index.
    1.   Plant-water relationships. 2.   Plant-soil relationships. 3.   Ground water flow.
    I.   Title
    QK870.K57 2004
572'.5392—dc22
2004019470

British Library Cataloguing in Publication Data
A catalogue record for this book is available from the British Library

ISBN: 0-12-409751-0

For all information on all Elsevier Academic Press publications
visit our Web site at www.books.elsevier.com

Printed in the United States of America
04   05   06   07   08   09   9   8   7   6   5   4   3   2   1

To my family

Contents

17 Measurement of Water Potential with Pressure Chambers

18 Stem Anatomy and Measurement of Osmotic Potential and Turgor Potential Using Pressure-Volume Curves

19 The Ascent of Water in Plants

Preface

This textbook is developed from lectures for a graduate class in soil-plant-water relations taught at Kansas State University. Students in the class are from a number of departments, including agronomy, biology, horticulture, forestry and recreational resources, biochemistry, and biological and agricultural engineering. The book can be used as a text for a graduate- or upper-level undergraduate courses or as a self-study guide for interested scientists. The book follows water as it moves through the soil-plant-atmosphere continuum. The text deals with principles and is not a review of recent literature. The principles covered in the book, such as Ohm's law and Poiseuille's law, are ageless. The book has equations, but no knowledge of calculus is required. Because plant anatomy is often no longer taught at universities, chapters review root, stem, leaf, and stomatal anatomy. Instrumentation to measure status of water in the soil and plant also is covered. Many instruments could have been described, but the ones chosen focus on traditional methods such as tensiometry and psychrometry and newer methods that are being widely applied such as tension infiltrometry and time domain reflectometry. Because the humanistic side of science is usually overlooked in textbooks, each chapter ends with biographies that tell about the people who developed the concepts discussed in that chapter.

Although a textbook on water relations might logically include developments in molecular biology, this topic is not covered. Rather, the text focuses on water in the soil and whole plant and combines knowledge of soil physics, plant physiology, and microclimatology. Chapter 1 reviews population and growth curves and provides a rationale for studying water in the soil-plant-atmosphere continuum. Chapter 2, which defines physical units, at first may appear elementary, but many students have not had a class in physics. The definitions in this chapter lay the foundation for understanding future chapters. Chapter 3 goes over the unique structure and properties of water, which makes life possible. Chapter 4, on tensiometry, is the first instrumentation lecture. Other instrumentation lectures include Chapter 9 on penetrometer measurements; Chapter 10 on measurement of the oxygen diffusion rate in the soil; part of Chapter 11 on applications of tension infiltrometry to determine soil hydraulic conductivity, sorptivity, repellency, and solute mobility; Chapter 13 on time domain reflectometry; Chapter 16 on psychrometry; Chapter 17 on pressure chambers; Chapter 22, which includes ways to measure stomatal opening and resistance; and Chapter 24 on infrared thermometers. Chapters 4 through 13 focus on water in the soil; Chapters 14 through 22, on water in the plant; and Chapters 23 through 27, on water as it leaves the plant and moves into the atmosphere.

Within any one chapter, the notation is consistent and abbreviations are defined when first introduced. When the same letter stands for different parameters, such as A for "ampere" or "area" and g for "acceleration due to gravity" or "grams," these differences are pointed out.

I express my appreciation to the following people who have helped make this book possible: my sister, Victoria E. Kirkham, professor of romance languages at the University of Pennsylvania, who first suggested on March 12, 1999, that I write this book; Dr. Kimberly A. Williams, Associate Professor of Horticulture, who audited my class in 2000 and then nominated me for the College of Agriculture Graduate Faculty Teaching Award, of which I was the inaugural recipient in 2001; Mr. Martin Volkmann in the laboratory of Prof. Dr. Rienk van der Ploeg of the University of Hannover in Germany, for converting the first drafts of the electronic files of the chapters, which I had typed in MS-DOS using WordPerfect 5.1 (my favorite word-processing software), to Microsoft Word documents, to ensure useable back-up copies in case my 1995 Compaq computer broke (it did not); my students, who have enthusiastically supported the development of this book; anonymous reviewers who had helpful suggestions for revisions and supported publication of the book; publishers and authors who allowed me to use material for the figures; and Mr. Eldon J. Hardy, my long-time professional draftsman at

Oklahoma State University, now retired. His drafting ability is unparalleled. He has redrawn the figures from the original and ensured that they are uniform, clear, and precisely done.

I am grateful to the publishers, Elsevier, including Michael J. Sugarman, Director, for accepting my book for publication, and Kelly D. Sonnack, Editorial Coordinator, for her help during the production of this book. Through my late father, Don Kirkham, former professor of soils and physics at Iowa State University, I have known of the venerable scientific publications of Elsevier since I was a child and truly am "non solus" with a book.

M.B. Kirkham
Manhattan, Kansas
April 6, 2004

Introduction

I. WHY STUDY SOIL-PLANT-WATER RELATIONS?

A. Population

Of the four soil physical factors that affect plant growth (mechanical impedance, water, aeration, and temperature) (Shaw, 1952; Kirkham, 1973), water is the most important. Drought causes 40.8% of crop losses in the United States, and excess water causes 16.4%; insects and diseases amount to 7.2% of the losses (Boyer, 1982). In the United States, 25.3% of the soils are affected by drought, and 15.7% limit crop production by being too wet (Boyer, 1982).

People depend upon plants for food. Because water is the major environmental factor limiting plant growth, we need to study soil-plant-water relations to provide food for a growing population. What is our challenge?

The earth's population is growing exponentially. The universe is now considered to be 13 billion years old (Zimmer, 2001). The earth is thought to be 4.45 billion years old (Allègre and Schneider, 1994). The earth's oldest rock is 4.03 billion years old (Zimmer, 2001). Primitive life existed on earth 3.7 billion years ago, according to scientists studying ancient rock formations harboring living cells (Simpson, 2003). Human-like animals have existed on earth only in the last few (less than 8) million years. In Chad, Central Africa, six hominid specimens, including a nearly complete cranium and fragmentary lower jaws, have been found that are 6 to 7 million years old (Brunet et al., 2002; Wood, 2002). In 8000 B.C., at the dawn of agriculture, the world's population was 5 million (Wilford, 1982). At the birth of Christ in 1 A.D., it was 200 million. In 1000, the population was 250 million (*National Geographic*, 1998a) (Fig. 1.1). By 1300, it had grown larger (Wilford, 1982). But by 1400, the population had dropped dramatically due to the Black Death, also called the bubonic plague (McEvedy, 1988), which is caused by a bacillus spread by fleas on rats. The Black Death raged in Europe between 1347 and 1351 and killed at least half of its population. It caused the depopulation or total disappearance of about 1,000 villages. Starting in coastal areas, where rats were on ships, and spreading inland, it was the greatest disaster in western European history (Renouard, 1971). People fled to the country to avoid the rampant spread of the disease in cities. The great piece of literature, *The Decameron*, published in Italian in 1353 and written by Giovanni Boccaccio (1313–1375), tells of 10 people who in 1348 went to a castle outside of Florence, Italy, to escape the plague. To pass time, they each told a tale a day for 10 days (Bernardo, 1982).

By 1500, the world's population was about 250,000,000 again. In 1650, it was 470,000,000; in 1750, it was 694,000,000; in 1850, it was 1,091,000,000. At the beginning of the nuclear age in 1945, it was 2.3 billion. In 1950, it was 2,501,000,000; in 1970, 3,677,837,000; in 1980,

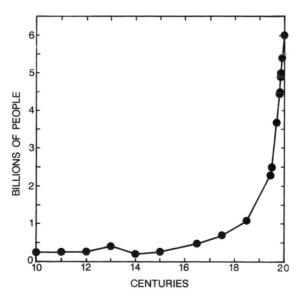

FIG. 1.1   The human population growth curve. (Drawn by author from data found in literature.)

4,469,934,000. In 1985, it was 4.9 billion, and in 1987 it was 5.0 billion (*New York Times*, 1987). In 1999, the world's population reached 6 billion (*National Geographic*, 1999). In 2002, the population of the USA was 284,796,887 (*Chronicle of Higher Education*, 2002).

Note that it took more than six million years for humans to reach the first billion; 120 years to reach the second billion; 32 years to reach the third billion; and 15 years to reach the fourth billion (*New York Times*, 1980). It took 12 years to add the last billion (fifth to sixth billion, 1987 to 1999). The United Nations now estimates that world population will be between 3.6 and 27 billion by 2150, and the difference between the two projections is only one child per woman (*National Geographic*, 1998b). If fertility rates continue to drop until women have about two children each—the medium-range projection—the population will stabilize at 10.8 billion. If the average becomes 2.6 children, the population will more than quadruple to 27 billion; if it falls to 1.6, the total will drop to 3.6 billion.

The population may also fall due to plagues (Weiss, 2002) such as the one that devastated Europe in the fourteenth century. Current potential plagues may result from AIDS (acquired immune deficiency syndrome), generally thought to be caused by a virus; influenza, another viral infection—for example, there may be a recurrence of the 1918 pandemic (Gladwell, 1997); sudden acute respiratory syndrome (SARS), a deadly infectious disease caused by a coronavirus (Lemonick and Park, 2003); and mad-cow disease, which is formally called bovine spongiform encephalopathy (BSE). BSE is called Creutzfeldt-Jakob (also spelled as Creutzfeldt-Jacob) disease (CJD), when it occurs in humans (Hueston and Voss, 2000). It is thought to be caused by *prions*, which were discovered by Stanley Prusiner (1942–) of the University of California School of Medicine in San Francisco; the discovery won Prusiner both the Wolf Prize (1996) and the Nobel Prize (1997) in medicine. Prions are a new class of protein, which, in an altered state, can be pathogenic and cause important neurodegenerative disease by inducing changes in protein structure. Prions are designed to protect the brain from the oxidizing properties of chemicals activated by dangerous agents such as ultraviolet light.

B. The "Two-Square-Yard Rule"

The population is limited by the productivity of the land. There is a space limitation that our population is up against. Many of us already have heard of this limitation, which is a space of two square yards per person. The sun's energy that falls on two square yards is the minimum required to provide enough energy for a human being's daily ration. Ultimately, our food and our life come from the sun's energy. The falling of the sun's energy on soil and plants is basic. We want to make as many plants grow on those two square yards per person as possible, to make sure we have enough to eat.

Let us do a simple calculation to determine how much food can be produced from two square yards, using the following steps:

1. Two square yards is 3 feet by 6 feet or 91 cm by 183 cm.

    91 cm × 183 cm = 16,653 cm^2 *or*, rounding, 16,700 cm^2.

2. The solar constant is 2.00 cal cm^{-2} min^{-1}, or, because 1 langley = 1 cal cm^{-2}, it is 2.00 langleys min^{-1}. The langley is named after Samuel Pierpoint Langley (1834–1906), who was a US astronomer and physicist who studied the sun. He was a pioneer in aviation.

    The solar constant is defined as the rate at which energy is received upon a unit surface, perpendicular to the sun's direction in free space at the earth's mean distance from the sun (latitude is not important) (Johnson, 1954). The brightness of the sun varies during the 11-year solar cycle, but typically by less than 0.1% (Lockwood et al., 1992).

3. 16,700 cm^2 × 2.00 cal cm^{-2} min^{-1} = 33,400 cal min^{-1}.

4. 33,400 cal min^{-1} × 60 min h^{-1} × 12 h d^{-1} = 24,048,000 cal d^{-1}, or, rounding, 24,000,000 cal d^{-1}. We multiply by 12 h d^{-1}, because we assume that the sun shines 12 hours a day. Of course, the length the sun shines each day depends on the day of the year, cloudiness, and location.

5. There is 6% conversion of absorbed solar energy into chemical energy in plants (Kok, 1967). This 6% is for the best crop yields achieved; 20% (Kok, 1967) to 30% (Kok, 1976) conversion is thought possible, but it has not been achieved; 2% is the conversion for normal yields; under natural conditions, ≤1% is converted (Kok, 1976). The solar energy reaching the earth's surface that plants do not capture to support life is wasted as heat (Kok, 1976). Let us assume a 6% conversion:

    24,000,000 cal d^{-1} × 0.06 = 1,440,000 cal d^{-1}.

6. The food "calories" we see listed in calorie charts are in kilocalories. So, dividing 1,440,000 cal d^{-1} by 1,000, we get 1,440 kcal d^{-1}, which is not very much. The following list gives examples of calories consumed per day in different countries (Peck, 2003):

Location	Kilocalories d^{-1}
USA, France	>3,500
Argentina	3,000–3,500
Morocco	2,500–2,999
India	2,000–2,499
Tanzania	<2,000

We recognize that the above calculation of productivity from two square yards is simplified, and more complex and thorough calculations of productivity, which consider geographic location, sky conditions, leaf display, and other factors, have been carried out (e.g., de Wit, 1967). Nevertheless, the 1,440 kcal d^{-1} is a useful number to know. It would be a starvation diet. One could live on it, but the calories probably would not provide enough for active physical work, creative intellectual activity, and reproduction. Women below a minimum weight cannot reproduce (Frisch, 1988). Civilization would advance slowly with this daily ration. People begin to die of starvation when they lose roughly a third of their normal body weight. When the loss reaches 40%, death is almost inevitable.

Triage is a system developed in World War I. It is the medical practice of dividing the wounded into survival categories to concentrate medical resources on those who could truly benefit from them and to ignore those who would die, even with treatment, or survive even without it. This practice has been advocated to allocate scarce food supplies. Wealthy countries should help only the most promising of the poorer countries since spreading precious resources too thin could jeopardize chances for survival of the strong as well as the weak. If we can grow more food, then this system does not need to be put into effect. In this book, we seek a better understanding of movement of water through the soil-plant-atmosphere continuum, or SPAC (Philip, 1966), because of the prime importance of water in plant growth.

We focus on principles rather than review the literature. Many references are given, but no attempt is made to cite the most recent papers. Articles explaining the principles are cited. They often are in the older literature, but we need to know them to learn the principles. No knowledge of calculus is required to understand the equations presented.

In this book, we divide the movement of water through the SPAC into three parts: 1) water movement in the soil and to the plant root; 2) water movement through the plant, from the root to the stem to the leaf; and 3) water movement from the plant into the atmosphere. However, before we turn to principles of water movement in the SPAC, let us first consider plant growth curves.

II. PLANT GROWTH CURVES

A. The Importance of Measuring Plant Growth and Exponential Growth

The world-population growth curve (Fig. 1-1) is an exponential curve. What do plant growth curves look like? Because water is the most important soil physical factor affecting plant growth, it is important to quantify plant growth to determine effects of water stress. In any experiment dealing with plant-water relations, some measure of plant growth (e.g., height, biomass) should be obtained. Plant growth curves also exemplify quantitative relationships that we seek to understand basic principles of plant-water relations. If we can develop equations to show relationships, then we can predict what is going to happen. Equations describing plant-growth curves demonstrate how we can quantify, and thus predict, plant growth.

We first consider the growth of the bacterium *Escherichia coli*. In the early nineteenth century, when plants and animals were being classified, the bacteria were arbitrarily included in the plant kingdom, and botanists first studied them (Stanier et al., 1963, p. 55–56). Even though bacteria are not plants or animals, we can follow their growth to understand plant growth curves.

Under ideal conditions, a cell of *E. coli* divides into two cells approximately every 20 minutes; for the sake of simplicity we assume that it is exactly 20 minutes. Let us consider the propagation of a single cell. Our purpose is to find a relation between the number N of cells at some moment in the future and the time t that has elapsed. At the start of our observations, at the time 0 min, there is 1 cell. When 20 min have elapsed there are 2 cells. When 40 min have elapsed there are $2 \times 2 = 2^2$ cells. When 60 min have elapsed there are $2 \times 2^2 = 2^3$ cells; that is, when 3 time intervals of 20 min each have passed, there are 2^3 cells. We observe a pattern developing: when m time intervals each of 20 min have passed, at the time $t = 20m$ min, there are $2^m = 2^{t/20}$ cells. Thus, if N denotes the number of cells present at the moment when t minutes have elapsed, then the relation we seek is given by the equation

$$N = 2^{t/20}. \tag{1.1}$$

Because the time t appears in the exponent of the expression $2^{t/20}$, this equation is said to describe *exponential growth* of the number N of cells (De Sapio, 1978, p. 21–23).

A famous book called *On Growth and Form* by D'Arcy Wentworth Thompson contains the following statement (Thompson, 1959, Vol. 1, p. 144): "Linnaeus shewed that an annual plant would have a million offspring in twenty years, if only two seeds grew up to maturity in a year." Linnaeus is, of course, Carolus Linnaeus (born Karl von Linné) (1707–1778), the great Swedish botanist. We can show that what Linnaeus said is true by adapting the preceding equation, as follows:

$$X = 2^{20}, \qquad (1.2)$$

where X is the number of offspring from the plant in twenty years.

To solve this equation, we need to use logarithms. John Napier (1550–1617), a distinguished Scottish mathematician, was the inventor of logarithms. (See the Appendix, Section III, for his biography.) To solve equations using logarithms, we need to know the fundamental laws of logarithms, which are as follows (Ayres, 1958, p. 83):

1. The logarithm of the product of two or more positive number is equal to the sum of the logarithms of the several numbers. For example,

$$\log_b (P \cdot Q \cdot R) = \log_b P + \log_b Q + \log_b R \qquad (1.3)$$

2. The logarithm of the quotient of two positive numbers is equal to the logarithm of the dividend minus the logarithm of the divisor. For example,

$$\log_b (P/Q) = \log_b P - \log_b Q \qquad (1.4)$$

3. The logarithm of a power of a positive number is equal to the logarithm of the number, multiplied by the exponent of the power. For example,

$$\log_b (P^n) = n \log_b P \qquad (1.5)$$

4. The logarithm of a root of a positive number is equal to the logarithm of the number, divided by the index of the root. For example,

$$\log_b P^{(1/n)} = (1/n) \log_b P. \qquad (1.6)$$

In calculus, the most useful system of logarithms is the *natural system* in which the base is a certain irrational number $e = 2.71828$, approximately (Ayres, 1958, p. 86). The natural logarithm of N, $\ln N$, and the common logarithm of N, $\log N$, are related by the formula

$$\ln N = 2.3026 \log N. \qquad (1.7)$$

To solve our equation, we take the logarithm of each side:

$$\log (2^{20}) = \log X$$

Using logarithm Rule No. 3, we get

$$20 \log 2 = \log X$$

Solving (and reading out all the digits on our hand calculator):

$$\log X = 20 \, (0.30103) = 6.0205999$$

$$X = 1,048,576.$$

Linnaeus was right.

B. Sigmoid Growth Curve

The S-shaped, or sigmoid, curve is typical of the growth pattern of individual organs, or a whole plant, and of populations of plants (Fig. 1.2). It can be shown to consist of at least five distinct phases: 1) an initial lag period during which internal changes occur that are preparatory to growth; 2) a phase of ever-increasing rate of growth. (Because the logarithm of growth rate, when plotted against time, gives a straight line during this period, this phase is frequently referred to as the log period of growth or "the grand period of growth."); 3) a phase in which growth rate gradually diminishes; 4) a point at which the organism reaches maturity and growth ceases. If the curve is prolonged further, a time will arrive when 5) senescence and death of the organism set in, giving rise to another component of the growth curve (Mitchell, 1970, p. 95).

C. Blackman Growth Curve

Since about 1900, people have used growth curves to analyze growth. Significant relationships of a mathematical nature, however, are difficult to

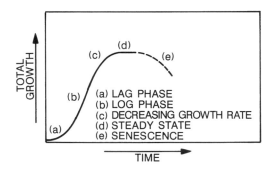

FIG. 1.2 Five phases in the sigmoid growth curve. (From Mitchell R.L., 1970, p. 95. Reprinted by permission of Roger L. Mitchell.)

apply to such a complex thing as growth (Hammond and Kirkham, 1949). One well-known theory of plant growth is the compound interest law of Blackman (1919). He related plant growth to money in a bank. When money accumulates at compound interest, the final amount reached depends on:

1. The capital originally used;
2. The rate of interest;
3. The time during which the money accumulates.

Comparing these factors to plants,

1 = the weight of the seed;
2 = the rate at which the seed material is used to produce new material;
3 = the time during which the plant increases in weight.

Blackman related the three factors into one exponential equation,

$$W_1 = W_o e^{rt}, \tag{1.8}$$

where
  W_1 = the final weight
  W_o = the initial weight
   r = the rate of interest
   t = time
   e = the base of natural logarithms (2.718 . . .).

The Blackman equation works best for early phases of growth (the log phase of growth in the sigmoid growth curve). In later growth stages, the decreasing relative growth rate has appeared to make impossible the application of this theory to the entire growth curve. Blackman attempted to do this, nevertheless, by using the average of all the different relative growth rate values as the r term in Equation 1.8. He called this term the "efficiency index" of plant growth.

Hammond and Kirkham found that the growth curves (dry weight versus time) of soybeans [*Glycine max* (L.) Merr.] and corn (*Zea mays* L.) were characterized by a series of exponential segments, which were related to the growth stages of the plants. The exponential equation for all segments had the form:

$$w = w_o \, e^{r(t-to)}, \tag{1.9}$$

where
  w = weight of the plant at time t

w_o = weight of the plant at an arbitrary time t_o
r = relative growth rate
e = base of natural logarithms (2.718 . . .).

Taking the natural logarithm of each side, we get

$$\ln w = \ln w_o + \ln e^{r(t-to)}$$
$$\ln w_o + [r(t-to)] \times 1]$$
$$\ln w_o + r(t-to).$$

Converting to common logarithms by dividing each term by 2.303, we get

$$\log w = \log w_o + [r(t-to)]/2.303.$$

Now let
$y = \log w$
$a = \log w_o$
$b = r/2.303$
$x = t-to.$

We get $y = a + bx$, which is the equation of a straight line.
The differential form of Equation 1.9, $w = w_o e^{r(t-to)}$, is

$$dw/(wdt) = r \qquad\qquad (1.10)$$

where r, the relative growth rate, is the increase in weight per unit weight per unit time. It is obtained by multiplying the slope, b, of the line by 2.303.

Hammond and Kirkham (1949) plotted the common logarithm of dry weight versus time and found that soybeans have three growth stages, I, II, and III. The analysis showed that the plants produce dry matter at the greatest relative rate during period I; at a smaller rate during period II; and at a still smaller rate during period III. That is, the slopes declined with age (slope = $r/2.303$). They saw that the dates of change in the growth curves from period I to period II were also the dates when the plants began to bloom. The dates of the second change in the growth curve, or the change from period II to period III, were the dates when the plants reached maximum height. The soybeans grew on two different soils, a Clarion loam and a Webster silt loam. The soybean plants in the Clarion soil bloomed and reached maximum height about a week earlier than the soybeans on the Webster silt loam soil. The growth curves clearly showed this difference (Fig. 1.3). Growth curves, therefore, can be used to see the effect of the soil environment on plant growth. Hammond and Kirkham (1949) did not give a reason for the difference in rate of growth on the Webster and Clarion

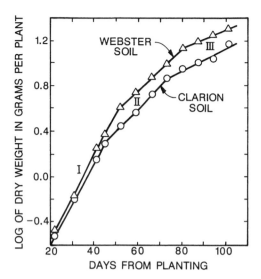

FIG. 1.3 Logarithmic dry matter accumulation curves of soybeans grown in the field in Iowa on Webster and Clarion soils. (From Hammond L.C., and Kirkham D. ©1949, American Society of Agronomy, Madison, Wisconsin. Reprinted by permission of the American Society of Agronomy.)

soils, but it must have been related to one of the four soil physical factors that affect plant growth: water, temperature, aeration, or mechanical impedance. For corn, they found four periods of growth (Fig. 1.4). The additional period in corn apparently was related to the difference in time of appearance of male and female flowers in corn. The physiological changes associated with the breaks in the curves were associated with tasseling, silking, and cessation of vegetative growth. The last break occurred after the corn plants had reached maximum height. In sum, the data for soybeans and corn showed that a quantitative analysis of the complete growth curve can be accomplished if the overall growth is partitioned into segments based on the growth stages of the plants.

The equations for plant growth show that we can develop significant mathematical relationships for a quantitative analysis of plant growth. This is probably because plant growth is governed by basic chemical and physical laws. From these relationships, we can predict plant growth.

III. APPENDIX: BIOGRAPHY OF JOHN NAPIER

John Napier (1550–1617), a distinguished Scottish mathematician, was the inventor of logarithms. The son of Scottish nobility, Napier's life was spent

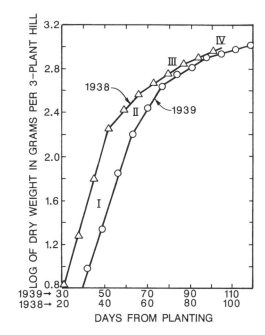

FIG. 1.4 Logarithmic dry matter accumulation curves of Iowa 939 corn in 1938 and 1939. (From Hammond L.C., and Kirkham D. ©1949, American Society of Agronomy, Madison, Wisconsin. Reprinted by permission of the American Society of Agronomy.)

amid bitter religious dissensions. He was a passionate Protestant. His great work, *A Plaine Discouery of the Whole Reuelation of Saint John* (1594), has a prominent place in Scottish ecclesiastical history as the earliest Scottish work on the interpretation of the scriptures. He then occupied himself by inventing instruments of war, including two kinds of burning mirrors, a piece of artillery, and a metal chariot from which shot could be discharged through small holes. Napier devoted most of his leisure to the study of mathematics, particularly to developing methods of facilitating computation. His name is associated with his greatest method, logarithms. His contributions to this mathematical invention are contained in two treatises: *Mirifici logarithmorum canonis descriptio* (1614; translated into English in 1857) and *Mirifici logarithmorum canonis constructio*, which was published two years after his death (1619) and translated into English in 1889. Although Napier's invention of logarithms overshadows all his other mathematical work, he has other mathematical contributions to his credit. In 1617, he published his *Rabdologiae, seu numerationis per virgulas libri duo* (English translation, 1667). In this work, he describes ingenious methods of performing the fundamental operations of multiplication and division with small rods (Napier's bones). He also made important contributions to spherical trigonometry (Scott, 1971).

REFERENCES

Allègre, C.J., and Schneider, S.H. (1994). The evolution of the earth. *Scientific American* 271(4); 66–75.

Ayres, F., Jr. (1958). *Theory and Problems of First Year College Mathematics*. Schaum Publishing: New York.

Bernardo, A.S. (1982). The plague as key to meaning in Boccaccio's "Decameron." In *The Black Death. The Impact of the Fourteenth-Century Plague* (Williman D., Ed.), pp. 39–64. Center for Medieval and Early Renaissance Studies, State University of New York at Binghamton, Binghamton, New York.

Blackman, V.H. (1919). The compound interest law and plant growth. *Annals of Botany* (old series) 33; 353–360.

Boyer, J.S. (1982). Plant productivity and environment. *Science* 218; 443–448 + cover.

Brunet, M., Guy, F., Pilbeam, D., Mackaye, H.T., Likius, A., Ahounta, D., Beauvilain, A., Blondel, C., Bocherens H. Boisserie, J.-R., De Bonis, L., Coppens, Y., Dejax, J., Denys, C., Duringer, P., Elsenmann, V., Fanone, G., Fronty, P., Geraads, D., Lehmann, T., Lihoreau, F., Louchart, A., Mahamat, A., Merceron, G., Mouchelin, G., Otero, O., Pelaez Campomanes, P., Ponce De Leon, M., Rage, J.-C., Sapanet, M., Schuster, M., Sudre, J., Tassy, P., Valentin, X., Vignaud, P., Viriot, V., Zazzo, A., and Zollikofer, C. (2002). A new hominid from the Upper Miocene of Chad, Central Africa. *Nature* 418; 145–151.

Chronicle of Higher Education. (2002). Almanac 2002-3. August 30, 2002, 49(1); 12.

De Sapio, R. (1978). *Calculus for the Life Sciences*. WH Freeman and Co: San Francisco.

De Wit, C.T. (1967). Photosynthesis: Its relationship to overpopulation. In *Harvesting the Sun* (San Pietro A, Greer FA, and Army TJ, Eds.), pp. 315–320. Academic Press: New York.

Frisch, R.E. (1988). Fatness and fertility. *Sci Amer* 258(3); 88–95 + cover.

Gladwell, M. (1997). The dead zone. *New Yorker* September 29, 1997; 52–65.

Hammond, L.C., and Kirkham, D. (1949). Growth curves of soybean and corn. *Agronomy J* 41; 23–29.

Hueston, W.D., and Voss, J.L. (2000). "Transmissible Spongiform Encephalopathies in the United States." CAST Report No. 136. Council for Agricultural Science and Technology, Ames, Iowa.

Johnson, F.S. (1954). The solar constant. *J Meteorol* 11; 431–439.

Kirkham, D. (1973). Soil physics and soil fertility. *Bulletin des Recherches Agronomiques de Gembloux Faculté des Sciences Agronomiques de l'État* (new series) 8(2); 60–88.

Kok, B. (1967). Photosynthesis—physical aspects. In *Harvesting the Sun* (San Pietro, A., Greer, F.A., and Army, T.J., Eds.), pp. 29–48. Academic Press: New York.

Kok, B. (1976). Photosynthesis: The path of energy. In *Plant Biochemistry* (Bonner, J., and Varner, J.E., Eds.), pp. 845–885. Academic Press: New York.

Lemonick, M.D., and Park, A. (2003). The truth about SARS. *Time*, May 5, 2000; 48–53.

Lockwood, G.W., Skiff, B.A., Baliunas, S.L., and Radick, R.R. (1992). Long-term solar brightness changes estimated from a survey of Sun-like stars. *Nature* 360; 653–655.

McEvedy, C. (1988). The bubonic plague. *Sci Amer* 258(2); 118–123.

Mitchell, R.L. (1970). *Crop Growth and Culture*. Iowa State University Press, Ames.

National Geographic. (1998a). Population. 193(1); 6–8.

National Geographic. (1998b). Millennium in maps. Population. Map insert for 194(4).

National Geographic. (1999). Global population reaches a milestone. 196(4); (unpaged).

New York Times. (1980). Population of the earth is said to be 4.5 billion. Section I, page 2. March 16, 1980.

New York Times. (1987). Zagreb baby hailed as 5 billionth person. Section I, page 5. July 12, 1987.

Peck, D. (2003). The world in numbers. The weight of the world. *Atlantic Monthly* 291(5); 38–39.

Philip, J.R. (1966). Plant water relatons: Some physical aspects. *Annu Rev Plant Physiol* 17; 245–268.

Renouard, Y. 1971. Black Death. *Encyclopaedia Britannica* 3; 742–743.

Scott, J.F. (1971). Napier (Neper), John. *Encyclopaedia Britannica* 15; 1174–1175.

Shaw, B.T. (1952). *Soil Physical Conditions and Plant Growth.* Academic Press: New York.

Simpson, S. (2003). Questioning the oldest signs of life. *Sci Amer* 288(4); 70–77.

Stanier, R.Y., Doudoroff, M., and Adelberg, E.A. (1963). *The Microbial World.* 2nd ed. Prentice-Hall: Englewood Cliffs, New Jersey.

Thompson, D'A.W. (1959). *On Growth and Form.* 2 vols. 2nd ed. Cambridge University Press: Cambridge. (First published in 1917.)

Weiss, R. (2002). War on disease. *Nat Geog* 201(2); 2–31.

Wilford, J.N. (1982). The people boom. Off the chart! Condensed from the *New York Times*, October 6, 1981. *Reader's Digest* 120(717); 114–115.

Wood, B. (2002). Hominid revelations from Chad. *Nature* 418; 133–135.

Zimmer, C. (2001). How old is... *Nat Geog* 200(3); 78–101.

2

Definitions
of Physical Units
and the
International System

I. DEFINITIONS

In plant-water relations, we will be using units based on physical definitions. Therefore, we need to review the definitions. We will define seven different units: force, weight, work, energy, power, pressure, and heat. The definitions come from Schaum (1961), but they can be found in any physics textbook.

A. Force

A *force* is a push or pull exerted on a body. If an unbalanced force acts on a body, the body accelerates in the direction of the force. Conversely, if a body is accelerating, there must be an unbalanced force acting on it in the direction of the acceleration. The unbalanced force acting on a body is proportional to the product of the mass and of the acceleration produced by the unbalanced force.

Newton's Laws of Motion.

For completeness, we now review these three laws, even though the second law is the one we are interested in for the definition of force. (See the Appendix, Section IV, for a biography of Newton.)

1. A body will maintain its state of rest or of uniform motion (at constant speed) along a straight line unless compelled by some unbalanced force to change that state. In other words, a body accelerates only if an unbalanced force acts on it.
2. An unbalanced force F acting on a body produces in it an acceleration a which is in the direction of the force and directly proportional to the force, and inversely proportional to the mass m of the body.

    In mathematical terms, this law states that $ka = F/m$ or $F = kma$, where k is a proportionality constant. If suitable units are chosen so that $k = 1$, then $F = ma$.
3. To every action, or force, there is an equal and opposite reaction, or force. In other words, if a body exerts a force on a second body, then the second body exerts a numerically equal and oppositely directed force on the first body. These two forces, although equal and oppositely directly, do not balance each other, because both are not exerted on the same body.

Units of Force.

In the equation $F = ma$, it is desirable to make $k = 1$; that is, to have units of mass, acceleration, and force such that $F = ma$. To do this, we specify two fundamental units and derive the third unit from these two.

1. In the *meter-kilogram-second* or *mks absolute system*, the fundamental mass unit chosen is the kilogram and the acceleration unit is the m/s^2. The corresponding derived force unit, called the *newton* (nt or N), is the unbalanced force that will produce an acceleration of 1 m/s^2 in a mass of 1 kg.
2. In the *centimeter-gram-second* or *cgs absolute system*, the fundamental mass unit is the gram and the acceleration unit is the cm/s^2. The corresponding derived force unit, called the *dyne*, is that unbalanced force that will produce an acceleration of 1 cm/s^2 in a mass of 1 gram.
3. In the *English gravitational system*, the fundamental force unit is the *pound* and the acceleration unit is the ft/s^2. The corresponding derived mass unit, called the *slug*, is the mass that when acted on by a 1 lb force acquires an acceleration of 1 ft/s^2.

Thus the following indicate three consistent sets of units that may be used with the equation $F = ma$ ($F = kma$ with $k = 1$):

- **mks system:** F (newtons) $= m$ (kilograms) $\times a$ (m/s^2)
- **cgs system:** F (newtons) $= m$ (grams) $\times a$ (cm/s^2)
- **English system:** F (pounds) $= m$ (slugs) $\times a$ (ft/s^2)

B. Mass and Weight

The mass m of a body refers to its inertia, and the weight w of a body is the pull or force due to gravity acting on the body which varies with location. (Inertia is the tendency of matter to remain at rest if at rest, or, if moving, to keep moving in the same direction, unless affected by some outside force.) Weight w is a force with a direction approximately toward the center of the earth.

If a body of mass m is allowed to fall freely, the resultant force acting on it is its weight, w, and its acceleration is that due to gravity, g. Then in any consistent system of units the equation $F = ma$ becomes

$$w = mg.$$

Thus

w (newtons) $= m$ (kilograms) $\times g$ (m/s^2)
w (dynes) $= m$ (grams) $\times g$ (cm/s^2)
w (pounds) $= m$ (slugs) $\times g$ (ft/s^2).

It follows that $m = w/g$. For example, if a body weighs 64 lb at a place where $g = 32$ ft/s^2, its mass is $m = w/g = 64$ lb/(32 ft/s^2) = 2 slugs. If a body weighs 49 newtons at a place where $g = 9.8$ m/s^2, its mass $m = w/g = 49$ newtons/(9.8 m/s^2) = 5 kg.

C. Work

A force does work on a body when it acts against a resisting force to produce motion in the body. Consider that a constant external force F acts on a body at an angle θ with the direction of motion and causes it to be displaced a distance d (Fig. 2.1). Then the work W done by the force F on the body is the product of the displacement d and the component of F in the direction of d. Thus

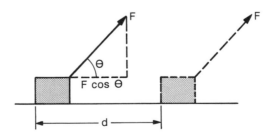

FIG. 2.1   Illustration for definition of work. (Adapted from Schaum D., *Theory and Problems of College Physics*, p. 49, ©1961, Schaum Publishing Co., New York. This material is reproduced with permission of The McGraw-Hill Companies.)

$$W = (F \cos \theta)d.$$

If d and F are in the same direction, $\cos \theta = \cos 0° = 1$ and $W = Fd$.

Units of Work.

Any unit of work equals a unit of force × a unit of length.

- One *foot-pound* (ft-lb) of work is done when a constant force of 1 lb moves a body a distance of 1 ft in the direction of the force.
- One *newton-meter* (nt-m), called 1 *joule*, is the work done when a constant force of 1 nt moves a body a distance of 1 meter in the direction of the force. Because 1 newton = 0.2248 lb and 1 meter = 3.281 ft,

> 1 joule = 1 newton-meter = 0.7376 *ft-lb*
> 1 ft-lb = 1.356 joules.

One *dyne-cm*, called 1 *erg*, is the work done when a constant force of 1 dyne moves a body a distance of 1 cm in the direction of the force. Since 1 nt = 10^5 dynes and 1 m = 10^2 cm,

> 1 joule = 10^7 ergs.

D. Energy

The *energy* of a body is its ability to do work. Because the energy of a body is measured in terms of the work it can do, it has the same units as work.

The *potential energy (PE)* of a body is its ability to do work because of its position or state. The potential energy of a mass m lifted a vertical distance h, where g is the acceleration due to gravity, is

$$PE = mgh.$$

In the mks system: PE (joules) = m (kg) × g (m/s^2) × h (m). In the cgs system: PE (ergs) = m (grams) × g (cm/s^2) × h (cm).

Because $mg = w$, we may also write: $PE = mgh = wh$.

E. Power

Power is the time rate of doing work. Average power = (work done)/(time taken to do this work) = force applied × velocity of body to which force is applied, where the velocity of the body is in the direction of the applied force.

Units of Power.

The unit of power in any system is found by dividing the unit of work in that system by the unit of time. Thus two units of power are the joule/s (or watt) and the ft-lb/s. Other practical units of power are the kilowatt and the horsepower.

- 1 watt = 1 joule/s
- 1 kilowatt (kw) = 1000 watts = 1.34 horsepower
- 1 horsepower (hp) = 550 ft-lb/s = 33,000 ft-lb/min = 746 watts

Work done = power × time taken. Hence the total work done in 1 hour, when the rate of doing work is 1 kilowatt, is 1 kilowatt-hour (kw-hr). The total work done in 1 hour, when the rate of doing work is 1 horsepower, is 1 horsepower-hour (hp-hr). Every month, I receive a bill for electricity, and the meter readings on the bill, taken from the meter for my apartment, are in units of kilowatt-hours. This is a unit of work.

F. Pressure

Pressure or p is a force per unit area.

$$p = \frac{\text{(force F acting perpendicular to an area)}}{\text{(area A over which the force is distributed)}}$$

or

$$p = F/A.$$

Some units of pressure are the lb/ft^2, lb/in^2, nt/m^2, and $dyne/cm^2$. The $dyne/cm^2$ is a unit that we will use often in plant-water relations.

$$1 \times 10^6 \text{ dyne/cm}^2 = 1 \text{ bar.}$$

However, a bar is not an SI unit. The SI unit is the Pascal, and 10 bars = 1 megaPascal or 10 bars = 1 MPa. We will talk about the SI system of units in Section II.

G. Heat

Heat is a form of energy. The three units most commonly used in measuring the quantity of heat are defined as follows.

1. **One calorie** (cal) = the quantity of heat required to raise the temperature of 1 gram of water by 1 centigrade degree.

Because the calorie was originally defined as stated above, it has been recognized that the energy requirement for raising the temperature of 1 gram of water by 1 degree depends slightly on the temperature, with a variation of about half a percent over the interval from 0° to 100°C. For work requiring an accuracy no greater than 1 percent, the above definition is satisfactory. For the most precise work, it has been agreed to define the calorie in terms of electrical units of energy, so that 1 calorie = 4.1840 joules. This is very close to the amount of energy required to raise the temperature of 1 gram of water from 16.5° to 17.5°C.

2. **1 kilocalorie** or **kilogram-calorie** (kcal or kg-cal) = 1000 calories.
3. **1 British thermal unit** (Btu) = the quantity of heat required to raise the temperature of 1 pound of water by 1 fahrenheit degree. 1 Btu = 453.6 × 5/9 cal = 252 cal.

II. LE SYSTÈME INTERNATIONAL D'UNITÉS

Le Système International d'Unités, which is French for "The International System of Units" or SI units, is a listing of decisions promulgated since 1889 on units of measurement. The General Conference on Weights and Measures (CGPM) meets regularly to update the units. The document was originally written in French, and consequently the name of the system has a French name. CGPM stands for the French *La Conférence Générale de Poids et Mesures*. The goal of the General Conference on Weights and Measures is "to make recommendations on the establishment of a *practical system of units of measurement* suitable for adoption by all signatories to the Meter Convention" (U.S. Department of Commerce, 1977, p 1). Another goal, in my opinion, is to have standard units worldwide, so that anybody reading an article in the world now or in the future will know exactly what the unit is. This allows replication of experiments. It is important that people understand published works after the works' authors die. Several years ago I investigated buying a chlorophyll meter, also called a SPAD meter (I do not know what SPAD stands for), but decided against it, because it does not read out in SI units. Someone 100 years from now could not replicate or understand experiments with a SPAD meter, unless the meter was used then. Graphs of absorption spectra of chlorophyll *a* and chlorophyll *b* (Steward, 1964, p 51) can be replicated 100 years from now, because they are expressed in standard units for absorption coefficients and wavelengths.

The tenth CGPM (1954), by its Resolution 6, and the fourteenth CGPM (1971), by its Resolution 3, adopted as base units of this "practical system of units," the units of the following seven quantities: length, mass, time, electric current, thermodynamic temperature, amount of substance, and luminous intensity. The eleventh CGPM (1960), by its Resolution 12, adopted the name *International System of Units*, with the international abbreviation SI, for this practical system of units of measurements and laid down rules for the prefixes and the derived and supplementary units.

SI units are divided into three classes: base units, derived units, and supplementary units (Taylor, 1991, p. 1). The base units and their abbreviations are (U.S. Department of Commerce, 1977, p. 6; Taylor, 1991, pp. 3–5) (Table 2.1):

- length = meter (m)
- mass = kilogram (kg)
- time = second (s)
- electric current = ampere (A)
- thermodynamic temperature = kelvin (K)
- amount of substance = mole (mol)
- luminous intensity = candela (cd)

The derived units are combinations of the base units. Examples of SI derived units expressed in terms of base units are (U.S. Department of Commerce, 1977, p. 6; Taylor, 1991, p. 6) (Table 2.2):

- area = square meter (m^2)
- volume = cubic meter (m^3)
- speed, velocity = meter per second (m/s)
- acceleration = meter per second squared (m/s^2)
- wave number = 1 per meter (m^{-1})
- density, mass density = kilogram per cubic meter (kg/m^3)

TABLE 2.1 SI base units[a]

Quantity	SI Unit	
	Name	Symbol
Length	meter	m
Mass	kilogram	kg
Time	second	s
Electric current	ampere	A
Thermodynamic temperature	kelvin	K
Amount of substance	mole	mol
Luminous intensity	candela	cd

[a]From Taylor (1991, p. 5)

TABLE 2.2   Example of SI derived units expressed in terms of base units[a]

	SI Unit	
Quantity	Name	Symbol
Area	square meter	m^2
Volume	cubic meter	m^3
Speed, velocity	meter per second	m/s
Acceleration	meter per second squared	m/s^2
Wave number	reciprocal meter	m^{-1}
Density, mass density	kilogram per cubic meter	kg/m^3
Specific volume	cubic meter per kilogram	m^3/kg
Current density	ampere per square meter	A/m^2
Magnetic field strength	ampere per meter	A/m
Concentration (of amount of substance)	mole per cubic meter	mol/m^3
Luminance	candela per square meter	cd/m^2

[a]From Taylor (1991, p. 6)

Other derived units with special names are given by the U.S. Department of Commerce (1977) and Taylor (1991).

The class of supplementary units contains two purely geometrical units: the SI unit of plane angle, the *radian*, and the SI unit of solid angle, the *steradian*. (A radian is an arc of a circle equal in length to the radius.) Another supplementary unit is the *astronomical unit*. This unit does not have an international symbol; abbreviations used, for example, are *AU* in English, *UA* in French, and *AE* in German. The astronomical unit of distance is "the length of the radius of the unperturbed circular orbit of a body of negligible mass moving round the Sun with a sidereal angular velocity of 0.017 202 098 950 radian per day of 86 400 ephemeris seconds" (US Department of Commerce, 1977, p. 11). In the system of astronomical constants of the International Astronomical Union, the value adopted is

$$1 \text{ AU} = 149\ 597.870 \times 10^6 \text{ m.}$$

Note about units: The appropriate unit should be used for each situation. For example, it would be inappropriate to use astronomical units to report the height of a corn plant, just as it would be inappropriate to use nanometers to report the distance from the earth to the sun. On graphs, sometimes data are reported on an axis as, for example, "$\times\ 10^3$." This notation is ambiguous, because the reader is not sure whether to multiply the number by 1,000 or divide the number by 1,000. It is best to use the prefixes to define a unit. The SI prefixes range from a factor of 10^{18} (exa) to 10^{-13} (atto) (US Department of Commerce, 1977, p. 10).

III. EXAMPLE: APPLYING UNITS OF WORK AND PRESSURE TO A ROOT

Let us use our knowledge of the definitions of work and pressure to quantify the work done by a root as it pushes through the soil (Kirkham, 1973). When roots open up the soil, they expend energy. We can obtain a simple mathematical expression for the amount of work a root does as it grows. To make matters easy, let us first assume that, as the root moves along, only its end exerts forces, to push the root through the soil; and, let us assume that the end of the root is blunt rather than rounded. By "blunt," let us mean that the end of the root is flat like the end of a solid right circular cylinder. Let us see what happens as we push such a solid cylinder through the soil. As we push the cylinder through the soil with a force, F, the resisting force of the soil will also have the value F, and, if we push the cylinder a distance, d, the work, W, done will be the product of the force and the distance, d:

$$W = Fd. \tag{2.1}$$

We now wish to get Equation 2.1 in terms of the soil pressure. If we take r to be the radius of the cylinder that we are pushing in the soil, then the soil pressure P (force per unit area) on the end of the cylinder in contact with the soil will be

$$P = F/(\pi r^2). \tag{2.2}$$

So we divide both sides of Equation 2.1 by the factor πr^2 and in the result substitute P for $F/(\pi r^2)$ to find that Equation 2.1 becomes

$$W/(\pi r^2) = Pd \tag{2.3}$$

which may be written in the form

$$W/(\pi r^2 d) = P. \tag{2.4}$$

But $\pi r^2 d$ is the volume, say V, of soil displaced. So we may write Equation 2.4 as

$$W/V = P \tag{2.5}$$

or

$$W = PV. \tag{2.6}$$

Equation 2.6 says that the work done by the root end as it pushes its way through the soil is equal to the product of root pressure and the volume of soil displaced by the root. Equation 2.6 gives us useful information. The equation says that if the pressure, P, encountered by the roots in a soil of poor structure is twice as great as for a soil in good structure that the roots must do twice as much work to establish the same size root system in the poor soil as in the good soil.

IV. APPENDIX: BIOGRAPHY OF ISAAC NEWTON

The following biographical material on Newton comes from Tannenbaum and Stillman (1959).

Newton was born on December 25, 1642, in Woolsthorpe, Lincolnshire, England. He was premature, and the midwife thought he would not live the night. He did not have the advantage of a happy, loving family. His father was said to be extravagant and wild, but he had no influence on Isaac, because he died more than two months before Isaac was born. His mother, Hannah, remarried to a Reverend Mr. Smith to avoid a life of poverty. After the marriage, Isaac lived in a separate house with his grandmother, but near his mother's home.

Because Isaac turned out to be a hopeless farmer, his family sent him to Cambridge in 1661. He became engaged to his childhood friend Catherine Storey, but they were never married. After Newton got his undergraduate degree, he wanted to get his master's degree with the support of a teaching assignment, but the rules of the University of Cambridge were strict and no members of the faculty could marry. Isaac and Catherine remained lifelong friends.

During the time of the great plague, Cambridge University was closed (the doors closed on August 8, 1665), and Newton went home and studied the movement of the planets. This is when the famous "apple" story occurred. There he reasoned that the moon does not fly off into space because the earth and the moon attract each other with a force that is directly proportional to the product of their masses. The same law holds the planets in orbit around the sun. Newton returned to Cambridge after the plague and started to study telescopes. His home headquarters at Cambridge served as his laboratory, because scientists did not have laboratories then.

On October 22, 1669, Isaac was appointed Lucasian Professor of Mathematics after his tutor, Dr. Isaac Barrow (1630–1677), stepped down so Newton could have the chair. As a professor he was expected to give lectures, but few students came to Newton's lectures. After his death, many people read his written versions of the lectures.

In 1671, Isaac was asked to build a telescope for the Royal Society, which had been formed in 1645. Members called themselves the "Invisible

College" and they met in bars. The group welcomed Isaac's work on the telescope, and on January 11, 1672, Newton was elected to membership in the Royal Society and was invited to present more of his work, which he did. He also presented his theory of light and color, which some members, especially Robert Hooke (1635–1703), criticized because they had their own theories. Shy and studious Newton was deeply hurt by the criticism. He defended his work in several papers, but then gave up, saying he did not want to "become a slave to defend it." He submitted no more papers for publication to the Society. Newton then wrote the *Principia* (*Philosophiae Naturalis Principia Mathematica*; begun in 1684 and first published in 1687), which was an instant success and made him one of the most famous men at Cambridge.

About 1693, he became involved in an argument with Gottfried Wilhelm Leibniz (1646–1716), a philosopher and mathematician, who had independently discovered calculus in Germany. Leibniz's method of writing calculus was eventually the one that was universally accepted.

While the arguments between Leibniz and Newton were raging over who discovered calculus first, Newton's good friend Charles Montague was elected president of the Royal Society. On March 19, 1696, Montague got the king to make Newton Warden of the Mint. People used to chip pieces off of coins for the valuable metal, so coins got smaller and smaller as they circulated. During Newton's tenure, new coins were made and recoinage was complete in 1699. He was promoted to Master of the Mint, a position that provided a good salary that enabled him to help his poorer relatives. He believed that "they who gave away nothing till they died, never gave." He also gave freely to many worthy causes such as the fund for building a new library at Trinity College, which was designed by Newton's friend Sir Christopher Wren. The library is still in use today. Newton also contributed money toward the purchase of a permanent building for the Royal Society.

Newton became president of the Royal Society on November 30, 1703 and remained president until his death. On April 16, 1705, Queen Anne knighted him. Of all his nieces and nephews, only one, Catherine, the daughter of his half-sister Hannah, was intellectually related to him. Newton gave Catherine the best education available to a woman at that time, and she presided over his London home in his later years. To their home she attracted men such as Jonathan Swift (1667–1745) and Alexander Pope (1688–1744). John Dryden (1631–1700) wrote poems about her. Voltaire (1694–1778) also visited their house and was the first to publish the falling apple story. Newton had an excellent library in his study with over 1800 books, each with his bookplate "Philosophimur," meaning "let us seek knowledge." In the quiet of his study he could pursue the religious studies that were so important to him.

Newton's last years were peaceful. His old enemies, Hooke and Leibniz, had died. Catherine had married, and her husband, John Conduit, took over the management of the Mint. The Conduits and their daughter lived with Newton, and this daughter inherited all of Newton's papers. Her son was the Earl of Portsmouth, and Newton's papers are known as the Portsmouth Collection. Newton developed gout, and, on the way back from a meeting of the Royal Society, became ill and died on March 20, 1727. The inscription on his tomb reads, "Let Mortals rejoice/That there has existed such and so great/AN ORNAMENT OF THE HUMAN RACE." Of his own place in history, Newton simply said, "If I have seen farther . . ., it is by standing on the shoulders of giants."

REFERENCES

Kirkham, D. (1973). Soil physics and soil fertility. *Bulletin des Recherches Agronomiques de Gembloux Faculté des Sciences Agronomiques de l'État* (new series) 8(2); 60–88.

Schaum, D. (1961). *Theory and Problems of College Physics*. 6 ed. Schaum Publishing: New York.

Steward, F.C. (1964). *Plants at Work*. Addison-Wesley: Reading, Massachusetts.

Tannenbaum, B., and Stillman, M. (1959). *Isaac Newton: Pioneer of Space Mathematics*. McGraw-Hill: New York.

Taylor, B.N. (United States of America Ed.). (1991). *The International System of Units (SI). Approved translation of the sixth edition (1991) of the International Bureau of Weights and Measures publication* Le Système International d'Unités (SI). NIST Special Publication 330, 1991 Ed. US Department of Commerce, National Institute of Standards and Technology: Washington, DC.

United States Department of Commerce. (1977). *The International System of Units (SI)*. NBS Special Publication 330. US Department of Commerce, National Bureau of Standards: Washington, DC.

3

Structure and Properties of Water

I. STRUCTURE OF WATER

To understand the nature of water in soil and plants, we need a mental picture of the water molecule. The water molecule (Fig. 3.1) is composed of two hydrogen atoms and one oxygen atom. The water molecule is positively charged on one side and negatively charged on the other and is, thus, a dipole. Two hydrogen atoms each share a pair of electrons with a single oxygen atom. The two hydrogen atoms of the water molecule are separated at an angle of 103 to 106 degrees, measured with the oxygen atom as the apex of the angle and with the two hydrogen protons as points on the angle sides. The electron pairs shared between the oxygen nucleus and the two hydrogen protons only partially screen (neutralize) the positive charge of the protons. The result is that the proton side of the molecule becomes the positive side of the water molecule (Kirkham and Powers, 1972, p. 2).

There are two concentrations of negative electricity, one concentration above and one concentration below a plane defined by the two hydrogen protons and the oxygen atom. They are called the *lone-pair electrons*. One pair is above the plane and one pair is below. These two lone pairs of electrons do not take part directly in bond formation, as do the electrons shared between the hydrogen and oxygen atoms of the water molecule. The electric charge structure of the water molecule resembles a tetrahedron with the oxygen near the center, two of its corners positively charged due to the partially screened protons of the hydrogen, and the remaining two corners

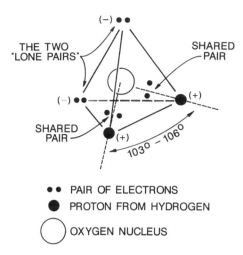

FIG. 3.1 Tetrahedral charge structure of a water molecule. (From *Advanced Soil Physics* by Kirkham D., and Powers W.L, p 2, ©1972, John Wiley & Sons, Inc., New York. This material is used by permission of John Wiley & Sons, Inc.)

of the tetrahedron negatively charged due to the two pairs of lone-pair electrons. (The word *tetrahedron* comes from the late Greek *tetraedros*, which means four-sided, and is a solid figure with four triangular surfaces.) This arrangement makes the water molecule a dipole, that is, one end of the molecule tends to be positive and the other end tends to be negative (Kirkham and Powers, 1972, p. 3). *Dipole* is a term used in physics and physical chemistry and is anything having two equal but opposite electric charges or magnetic poles, as in a hydrogen atom with its positive nucleus and negative electron.

II. FORCES THAT BIND WATER MOLECULES TOGETHER

A. Hydrogen Bonding

There are two attractive forces between water molecules: hydrogen bonding and the van der Waals-London force. Hydrogen bonding results from the electrical structure of water molecules that makes them group together in a special way. The negative lone-pair electrons of one water molecule are attracted to a positive partially screened proton of another water molecule (Kirkham and Powers, 1972, p. 3). Thus each corner of the four corners of the water tetrahedron can be attached, by electrostatic attraction, to four other water tetrahedron molecules in solution (Fig. 3.2). This type of bonding is called hydrogen bonding.

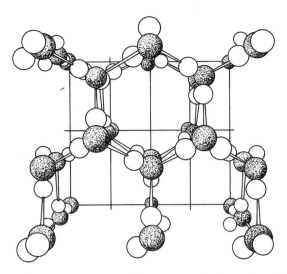

FIG. 3.2   Diagram showing approximately how water molecules are bound together in a lattice structure in ice by hydrogen bonds. The dark spheres are oxygen atoms, and the light spheres are hydrogen atoms. (From *Water Relations of Plants* by Kramer P.J., p. 10, ©1983, Academic Press, New York. Reprinted by permission of Academic Press.)

Hydrogen bonding is important in binding water molecules together. Hydrogen bonds have a binding force of about 1.3 to 4.5 kilocalories per mole in water (Kramer, 1983, p. 11). Nobel (1974, p. 46) puts the value of hydrogen bonding at about 4.8 kilocalories per mole. Only part of the structure of water due to hydrogen bonding is destroyed by heating, and about 70% of the hydrogen bonds found in ice remain intact in liquid water at 100°C (Kramer, 1983, p. 12). Postorino et al. (1993) found that at 400°C almost all hydrogen bonding is broken down.

B. van der Waals-London Force

A van der Waals-London force is one that exists between neutral nonpolar molecules, and, therefore, does not depend on a net electrical charge. The force was first described by van der Waals (1837–1923), a Dutch physicist. (See the Appendix, Section IV, for his biography.) London (1930) used quantum mechanics to obtain a quantitative expression for the van der Waals attractive force. This attractive force occurs because the electrons of one atom oscillate in such a way as to make it a rapidly fluctuating (about 10^{15} or 10^{16} Hertz) dipolar atom, which in turn polarizes an adjacent atom, making it, too, a rapidly fluctuating dipole atom such that the two atoms attract each other. The generated force varies inversely as the seventh power of the distance between the atoms (Kirkham and Powers, 1972, p. 4). An article in *Nature* puts the value as the sixth power (Maddox, 1985). Because this attractive force varies inversely as the sixth or seventh power, it is short ranged. Quantum mechanics predicts that at distances greater than 100 Ångstroms one atom cannot polarize another. (1 Å = 10^{-10} m. The unit Ångstrom is named after Anders Jöns Ångström, 1814–1874, a Swedish physicist.) The exact proportion of attraction that we can attribute to the van der Waals-London force is not known. Kramer (1983, p. 11) puts the attractive force at about 1 kcal/mole. It is generally felt that this force contributes little to the attraction of water to itself.

III. PROPERTIES OF WATER

We now look at the unique physical properties of water. These properties permit life. If life exists on other planets, it probably is based on water rather than on any other molecule, like ammonia. In 1999, when scientists discovered a solar system 44 AU (astronomical units) from earth, which was the first planetary system ever found around a normal star, aside from our solar system, it was postulated that some of those planets could be located at the right distance from their host star to have liquid water and, hence, life (Showstack, 1999). Life depends on water.

We will list 15 properties of water. The list is extracted from the book by Kramer (1983, pp. 8–9 and 14), and the definitions of the properties come from the *Handbook of Chemistry and Physics* (Weast, 1964).

1. Specific Heat.

Water has the highest specific heat of any known substance except liquid ammonia, which is about 13 percent higher. The handbook states (Weast, 1964, p. F-57) that if a quantity of heat H calories is necessary to raise the temperature of m grams of a substance from t_1 to t_2 °C, the specific heat, s, is

$$s = H/[m(t_2 - t_1)]$$

The units of specific heat are cal gram^{-1} C^{-1}. Table 3.1, adapted from van Wijk and de Vries (1966, p. 41), gives the specific heat of water at different temperatures.

From Table 3.1, one sees that the specific heat of water decreases with an increase of temperature up to 35°C, and then the specific heat increases with further increase in temperature. Paul (1986) noted that the majority of

TABLE 3.1 Physical properties of liquid water[a]

Temperature (°C)	Density (g cm^{-3})	Surface tension (g s^{-2})	Dynamic viscosity (g cm^{-1} s^{-1}) $\times 10^{-2}$	Heat of vaporization (cal g^{-1})	Specific heat (cal g^{-1}°C^{-1})	Thermal conductivity (cal cm^{-1} s^{-1}°C^{-1}) x 10^{-3}
−10	0.99794	⋯	⋯	603.0	1.02	⋯
−5	0.99918	76.4	⋯	⋯	⋯	⋯
0	0.99987	75.6	1.7921	597.3	1.0074	1.34
4	1.00000	⋯	⋯	⋯	⋯	⋯
5	0.99999	74.8	1.5188	594.5	1.0037	1.37
10	0.99973	74.2	1.3077	591.7	1.0013	1.40
15	0.99913	73.4	1.1404	588.9	0.9998	1.42
20	0.99823	72.7	1.0050	586.0	0.9988	1.44
25	0.99708	71.9	0.8937	583.2	0.9983	1.46
30	0.99568	71.1	0.8007	580.4	0.9980	1.48
35	0.99406	70.3	0.7225	577.6	0.9979	1.50
40	0.99225	69.5	0.6560	574.7	0.9980	1.51
45	0.99024	68.7	0.5988	571.9	0.9982	1.53
50	0.98807	67.9	0.5494	569.0	0.9985	1.54

[a]From van Wijk, W.R., and de Vries, D.A. 1966. The atmosphere and the soil. In *Physics of Plant Environment*, 2nd ed. (W.R. van Wijk, Ed.), p. 41, ©1966, North Holland Publishing: Amsterdam. (Reprinted by permission of Prof. Daniel A de Vries.)

isothermic animals maintain their body temperatures, during non-hibernation, within a few degrees of 36°C (human temperature = 98.6°F or 37°C), because the specific heat of water is a minimum at 35°C. He said:

> As usual the key to much of life's mystery lies in the extraordinary behaviour of water. . . . [T]he relationship between the specific heat of water and temperature reveals that at 35°C the specific heat of water is at its minimum value of 4.1779 J g^{-1} °C^{-1}. . . . An organism functioning at this temperature will find it necessary to generate or dissipate the minimum amount of heat energy in order to maintain its temperature constant. From the point of view of the organism's energy economy this temperature is clearly the most efficient at which to function. It seems likely that since the environmental temperatures on Earth are, with few exceptions, lower than 35°C, organisms which have been able to set their working temperatures at a point just slightly above the temperature at which the specific heat of water is at its minimum value have thrived.

The idea supposes that the evolution of warm-blooded animals and the chemical properties of water are related (Kadler and Prockop, 1987). Paul's suggestion created a series of responses (Calder, 1986; Dunitz and Benner, 1986; Bird, 1987; Kadler and Prockop, 1987; McArthur and Clark, 1987; Stevenson, 1987).

The high specific heat of water stabilizes temperatures and results in the relatively uniform temperature of islands and land near large bodies of water (Kramer, 1983, p. 8). This is important for the growth of crops and natural vegetation.

2. Heat of Vaporization.

The heat of vaporization of water is the highest known. The heat of vaporization is defined as the amount of heat needed to turn one gram of a liquid into a vapor, without a rise in the temperature of the liquid. This term is not in the list of definitions given by Weast (1964), so the definition comes from *Webster's New World Dictionary of the American Language* (1959). The units are cal/gram and values for the heat of vaporization of water at different temperatures are given in Table 3.1. The heat of vaporization is a latent heat. *Latent* comes from the Latin *latere*, which means to lie hidden or concealed. Latent heat is the additional heat required to change the state of a substance from solid to liquid at its melting point, or from liquid to gas at its boiling point, after the temperature of the substance has reached either of these points. Note that a latent heat is associated with no change in temperature, but a change of state. Because of the high heat of vaporization, evaporation of water has a pronounced cooling effect and condensation has a warming effect (Kramer, 1983, p. 8). The cooling effect from evaporation is important in semi-arid regions, such as Kansas (see Chapter 26).

3. Heat of Fusion.

The heat of fusion of water is unusually high. The heat of fusion is the quantity of heat necessary to change one gram of a solid to a liquid with no temperature change (Weast, 1964, p. F-44). It is also a latent heat and is sometimes called the latent heat of fusion. It has only one value for water, because water freezes at one value (0°C), and it is 79.71 cal/gram or the rounded number 80 cal/gram.

The high heat of fusion of water is used in frost control. Irrigation water drawn from the ground is often at a uniform temperature above freezing. In Nebraska, for example, groundwater is generally at about 12°C (Rosenberg, 1974, p. 276), and each gram can supply about 12 cal to the air with which it comes in contact. This thermal effect is small, however, when compared to the liberation of heat that occurs when water freezes (80 cal/gram). Irrigation water may contribute more than 90 cal/gram (12 cal/gram + 80 cal/gram) in the process of cooling and freezing (Rosenberg, 1974). Fields may be flooded as a means of protecting crops from frost. Such a measure is extreme and is likely to be ineffective in an advective frost, because winds would remove the liberated heat from the flooded fields rapidly and freeze the unprotected vegetation. (In simple terms, advection means that wind is blowing and bringing hot or cold air into an area. We shall discuss advection when we talk about evapotranspiration in Chapter 26.)

Sprinkling is a more effective use of irrigation water in frost protection (Rosenberg, 1974, p. 276). Plants are sprinkled at the onset of freezing temperatures. As water freezes onto the plant parts, the heat of fusion is liberated. As long as the freezing continues, the temperature of the ice will remain at 0°C. Sprinkling must be continued after the sun comes up the next day or until the temperatures have risen to melt the ice. If sprinkling is discontinued prematurely, heat will be drawn from the plant parts to melt the ice and frost damage may occur. Care must be taken when sprinkling tall plants. Ice loads of great weight will break them (Rosenberg, 1974, p. 276).

4. Heat Conduction.

Water is a good conductor of heat compared with other liquids and nonmetallic solids, although it is poor compared to metals (Kramer, 1983,

p. 8). Thermal or heat conductivity is not defined by Weast (1964), so we use the definition from an earlier edition of the *Handbook of Chemistry and Physics* (Hodgman, 1959, p. 2431). Heat conductivity is "the quantity of heat in calories which is transmitted per second through a plate one centimeter thick across an area of one square centimeter when the temperature difference is one degree Centigrade." The units, therefore, are cal s^{-1} cm^{-2} (°C/cm)$^{-1}$ or cal cm^{-1} s^{-1}°C^{-1}. Table 3.1 gives values of thermal conductivity of water at different temperatures. The table shows that at 20°C, water has a thermal conductivity of 0.00144 cal s^{-1} cm^{-1}°C^{-1}. For comparison, copper, a metal, at 18°C has a thermal conductivity of 0.918 cal s^{-1} cm^{-1} °C^{-1} (Hodgman, 1959, p. 2431), or about a thousand times greater than water.

Survival of crops in the spring can depend on thermal conductivity. There are cases in which soil surface management can prevent a radiation freeze from occurring in row crops. In the spring of 1993, a number of Kansas farmers reported freeze damage in corn fields that had been cultivated. Uncultivated fields had no damage. In some cases the damaged and undamaged areas were adjacent, apparently in locations where the farmer had stopped cultivating for the day. Disturbing the soil by cultivation reduced the thermal conductivity and prevented stored heat from being released back toward the surface (Gerard J. Kluitenberg, Department of Agronomy, Kansas State University, personal communication, June 18, 1993). Both the solid soil and the water channels in the soil were disturbed by cultivation. If continuous water channels were broken, then the heat conducted by water upward from lower in the soil was reduced. Lower levels of the soil profile store heat, and, in the early cool spring months in Kansas, this heat can move upward to the cooler soil surface. Remember that in Nebraska groundwater is generally at about 12°C. As we shall see later, when we discuss linear flow laws (Chapter 7), Fourier's heat flow law (Kirkham and Powers, 1972, p. 75) is a linear flow law and shows that heat is transported according to a temperature gradient.

5. Transparency to Visible Radiation.

Water is transparent to visible radiation. This allows light to penetrate bodies of water and makes it possible for algae to photosynthesize at considerable depths (Kramer, 1983, p. 8).

6. Opaqueness to Infrared Radiation.

Water is nearly opaque to longer wavelengths in the infrared range. Thus, water filters are good heat absorbers (Kramer, 1983, p. 8). A practical use of water to absorb heat was observed in Israel by Kirkham (1985), who said, "An unusual use of water is illustrated . . . at the Dead Sea, in Israel, where experiments are underway to get energy from the sun's heat. The salt water in these ponds is heated very hot by the sun, transferred by pipe to a nearby building; exposure to very cold water produces steam, which becomes a potential source of energy." The good absorption of heat by water helps to make this energy production in Israel possible.

We now define the visible and infrared regions of the spectrum. The visible region extends from 390–7800 Å. The spectrum goes from violet (shortest wavelength), through blue, green, yellow, orange, and to red (longest wavelength). Chlorophyll absorbs in the blue and red regions and reflects in the green. That is why it looks green. The maximum (peak) absorptions in the blue and red regions for chlorophyll a and b are (Stewart, 1964, p. 51):

Chlorophyll a
4300 Å (blue)
6600 Å (red)

Chlorophyll b
4700 Å (blue)
6500 Å (red)

The infrared region goes from 7800 to 4×10^6 Å (Giese, 1962, p. 164). The near infrared extends from 7800 to 2×10^4 Å, and the far infrared extends from 2×10^4 to 4×10^6 Å.

Figure 3.3 illustrates that water is transparent in visible wavelengths, but absorbs at wavelengths in the infrared region. Note that the abscissa in Fig. 3.3 extends to 1400 nm (1 nm = 1 mμ), which is in the near infrared region. The near infrared extends to 2000 nm.

7. Surface Tension.

Water has a much higher surface tension than most other liquids because of the high internal cohesive forces between molecules (Kramer, 1983, p. 8). In Chapter 6 we will discuss surface tension in detail, including LaPlace's surface tension theory. The high surface tension of water provides the tensile strength required for the cohesion theory for the ascent of sap (water in the xylem). The cohesion theory is only a theory, but appears to be the best explanation for the rise of water in plants. We shall go into it in detail in Chapter 19 and calculate the tensile strength of water. Surface tension has units of force per unit length or dyne/cm. Table 3.1 gives values of

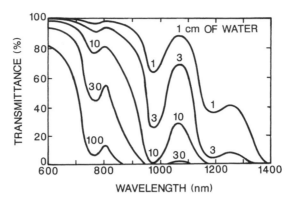

FIG. 3.3 Transmission of radiation of various wavelengths through layers of water of different thicknesses. The numbers on the curves refer to the thickness of the layers in centimeters. Transmission is much greater at short than at long wavelengths. (From *Water Relations of Plants* by Kramer P.J., p. 8, ©1983, Academic Press, New York. Reprinted by permission of Academic Press.)

surface tension at different temperatures. Note that the unit given for surface tension in the table is g s^{-2}. This is equivalent to dyne/cm, because, as we saw in Chapter 2,

$$F = ma$$
$$1 \text{ dyne} = 1 \text{ gram} \times 1 \text{ cm/s}^2.$$

Rearranging, we get

$$\text{gram/s}^2 = \text{dyne/cm}.$$

8. Density.

Water has a high density and is remarkable in having its maximum density at 4°C instead of at the freezing point (Kramer, 1983, pp. 8–9).

9. Expansion Upon Freezing.

Water expands on freezing, so that ice has a volume about 9% greater than the liquid water from which it was formed (Fig. 3.4). This explains why ice floats and pipes and radiators burst when the water in them freezes. If ice sank, bodies of water in the cooler parts of the world would be filled permanently with ice, and aquatic organisms could not survive (Kramer, 1983, p. 9).

10. Ionization.

Water is very slightly ionized. Only one molecule in 55.5×10^7 is dissociated (Kramer, 1983, p. 9).

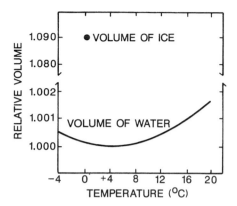

FIG. 3.4 Change in volume of water with change in temperature. The minimum volume is at 4°C, and below that temperature there is a slight increase in volume as more molecules are incorporated into the lattice structure. The volume increases suddenly when water freezes, because all molecules are incorporated into a widely spaced lattice. Above 4°C, there is an increase in volume caused by increasing thermal agitation of the molecules. (From *Water Relations of Plants* by Kramer P.J., p. 9, ©1983, Academic Press, New York. Reprinted by permission of Academic Press.)

11. Dielectric Constant.

Water has a high dielectric constant. We will not give the equation defining a dielectric, because it requires knowledge of capacitance, which is studied in physics courses. An interested reader can read a physics textbook such as Shortley and Williams (1971, pp. 517–520) to learn more about dielectrics. We will understand a dielectric by using the definition for dielectric in *Webster's New World Dictionary of the American Language* (1959): "[*dia*= through, across + *electric*: so called because it permits the passage of the lines of force of an electrostatic field, but does not conduct the current]; a material, as rubber or glass, that does not conduct electricity; insulator."

Water, therefore, is a good insulator. This might seem contradictory to the fact that we know we shall get electrocuted if we stand in water and put our finger in an electrical outlet. This is because we are standing in tap water, and tap water has salts in it. The electrical conductivity of tap water in Manhattan, Kansas, is 0.54 mmho cm^{-1} (= 0.54 dS/m where dS is a deciSiemen) (Kirkham, 1982). So when we say water is a good dielectric, we mean pure water. One should never touch any electrical appliance while taking a bath, including a radio.

12. Solvent for Electrolytes.

Water is a good solvent for electrolytes, because the attraction of ions to the partially positive and negative charge on water molecules results in each ion being surrounded by a shell of water molecules, which keeps ions of opposite charge separated (Fig. 3.5).

13. Solvent for Nonelectrolytes.

Water is a good solvent for many nonelectrolytes, because it can form hydrogen bonds with amino and carbonyl groups (Kramer, 1983, p. 9). An amino group has one hydrogen atom in the ammonia molecule replaced by an alkyl or other nonacid radical. A carbonyl is the bivalent radical CO.

14. Adsorption.

Water tends to be adsorbed, or bound strongly, to the surfaces of clay micelles, cellulose, protein molecules, and many other substances. This characteristic is of great importance in soil and plant water relations (Kramer, 1983, p. 9). We will see that it is important in the cohesion theory.

15. Viscosity.

Water has a high viscosity. The *Handbook of Chemistry and Physics* defines viscosity as follows (Weast, 1964, p. F-62): "All fluids possess a definite resistance to change of form and many solids show a gradual yielding to forces tending to change their form. This property, a sort of internal friction, is called viscosity; it is expressed in dyne-seconds per cm^2 or poises." The poise is named after Poiseuille, and we will study Poiseuille's law for flow of liquids through capillary tubes in Chapter 14. Table 3.1 gives values for the viscosity of water at different temperatures. The units in the

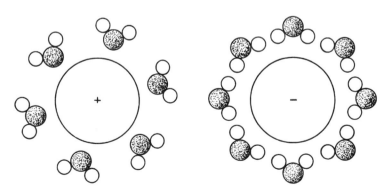

FIG. 3.5 Diagram showing approximate arrangement of water molecules in shells oriented around ions. These shells tend to separate ions of opposite charge and enable them to exist in solution. They also disrupt the normal structure of water and slightly increase the volume. (From *Water Relations of Plants* by Kramer P.J., p. 12, ©1983, Academic Press, New York. Reprinted by permission of Academic Press.)

table are g cm^{-1} s^{-1}. Again, using $F = ma$, we can see that these units are equivalent to dyne-seconds per cm^2, as follows. We substitute (gram-cm)/s^2 for dyne, and we get

$$(dyne-s)/cm^2 = [(gram-cm)/s^2] (s/cm^2) = gram\ cm^{-1}\ s^{-1}.$$

IV. APPENDIX: BIOGRAPHY OF JOHANNES VAN DER WAALS

Johannes Diderik van der Waals (1837–1923), a Dutch physicist, was born in Leyden, The Netherlands, November 23, 1837. He was a self-taught man, who took advantage of the opportunities offered by the University of Leyden. He first attracted notice in 1873 with his treatise "On the Continuity of the Liquid and Gaseous State," the basis for his doctoral degree. He taught physics at various schools, and, in 1877, he was appointed professor of physics at the University of Amsterdam, a post he kept until 1907 (Preece, 1971).

Van der Waals combined the determination of cohesion in the theory of capillarity by Laplace (1749–1827) with the kinetic theory of gases, and this led to the conception of the continuity of the liquid and gaseous states. He arrived at an equation that was the same for all substances by using the values of the volume, temperature, and pressure divided by their critical values (Preece, 1971). His work enabled the liquefaction of gases, which had important practical application during World War I (1914–1918). Although others had studied liquefaction of gases, it was van der Waals who was the first to treat the subject of the continuity of gases and liquids from the standpoint of the mathematical theory of gases (Cajori, 1929, pp. 210–211). Van der Waals was awarded the 1910 Nobel Prize in physics for his research on the equations of state for gases and fluids. He died in Amsterdam on March 9, 1923.

REFERENCES

Bird, R.J. (1987). Body temperature and the thermodynamics of water. *Nature* 325; 488–489.
Cajori, F. (1929). *A History of Physics.* Macmillan: New York.
Calder, W.A. III. (1986). Body temperature and the specific heat of water. *Nature* 324; 418.
Dunitz, J.D., and Benner, S.A. (1986). Body temperature and the specific heat of water. *Nature* 324; 418.
Giese, A.C. (1962). *Cell Physiology.* W.B. Saunders: Philadelphia.
Hodgman, C.D. (1959). *Handbook of Chemistry and Physics.* 40th ed. Chemical Rubber Publishing: Cleveland, Ohio.
Kadler, K., and Prockop, D.J. (1987). Protein structure and the specific heat of water. *Nature* 325; 395.

Kirkham, D. (1985). Soil, water, and hungry man. Lecture given at the Institute of Water Conservancy and Hydroelectric Power Research, Beijing, China, May, 1985. (Copy available from M.B. Kirkham, Department of Agronomy, Kansas State University, Manhattan, KS 66506.)

Kirkham, D., and Powers, W.L. (1972). *Advanced Soil Physics*. Wiley-Interscience: New York.

Kirkham, M.B. (1982). Water and air conductance in soil with earthworms: An electrical-analogue study. *Pedobiologia* 23; 367–371.

Kramer, P.J. (1983). *Water Relations of Plants*. Academic Press: New York.

London, F. (1930). Zur Theorie und Systematik der Molekularkrafte. *Zeitschrift für Physik* 63; 245–279.

Maddox, J. (1985). Recalculating interatomic forces. *Nature* 314, 315 (one page only).

McArthur, A.J., and Clark, J.A. (1987). Body temperature and heat and water balance. *Nature* 326: 647–648.

Nobel, P.S. (1974). *Introduction to Biophysical Plant Physiology*. W.H. Freeman and Company: San Francisco.

Paul, J. (1986). Body temperature and the specific heat of water. *Nature* 323; 300.

Postorino, P., Tromp, R.H., Ricci, M.-A., Soper, A.K., and Neilson G.W. (1993). The interatomic structure of water at supercritical temperatures. *Nature* 366; 668–670.

Preece, W.E. (Gen. Ed.). (1971). Waals, Johannes Diderik van der. *Encyclopaedia Britannica*, 23; 133.

Rosenberg, N.J. (1974). *Microclimate: The Biological Environment*. Wiley-Interscience: New York.

Shortley, G., and Williams, D. (1971). *Elements of Physics*. 5th ed. Prentice-Hall: Englewood Cliffs, New Jersey.

Showstack, R. (1999). Scientists discover solar system 44 AU from earth. *EOS Trans Amer Geophy Union* 80; 193–194.

Stevenson, R.D. (1987). Body temperature and the thermodynamics of water. *Nature* 324; 489.

Stewart, F.C. (1964). *Plants at Work*. Addison-Wesley: Reading, Massachusetts.

van Wijk, W.R., and de Vries, D.A. 1966. The atmosphere and the soil. In *Physics of Plant Environment* (van Wijk, W.R., Ed.). 2nd ed., pp. 17–61. North-Holland Pub: Amsterdam.

Weast, R.C. (1964). *Handbook of Chemistry and Physics*. 45th ed. Chemical Rubber Company: Cleveland, Ohio (pages not numbered sequentially).

Webster's New World Dictionary of the American Language. (1959). College ed. World Publishing: Cleveland and New York.

4

Tensiometers

As we noted in Chapter 1, in this book we shall study water as it moves through the soil-plant-atmosphere continuum (SPAC) and ways to measure this water. Now that we have learned basic physical definitions and the structure and properties of water, we turn to the main topic of water in the SPAC. We first focus on water in soil. We begin by learning how to measure the status of water in the soil using a tensiometer.

I. DESCRIPTION OF A TENSIOMETER

A tensiometer is a device for measuring, when the soil is not too dry, the soil matric potential. In old terminology, the matric potential was called the soil moisture tension (see Chapter 5 for terminology used in soil-water relations). Because the instrument measures tension, it was called a tensiometer. For a review of the early literature on tensiometers, see Richards (1949). However, the instrument could have been called an "ergmeter" (Don Kirkham, Departments of Physics and Agronomy, Iowa State University, personal communication, February 10, 1994). As we shall see later (Chapter 5), we can express tension of water in soil in terms of tension head (using units of length) or in terms of potential energy per unit mass (e.g., ergs/gram) or potential energy per unit volume (e.g., joule/m^3 or dyne/cm^2; remember 1 bar = 1×10^6 dyne/cm^2).

A tensiometer consists of a porous, permeable ceramic cup connected through a water-filled tube to a manometer, vacuum gauge, pressure transducer, or other pressure measuring device (Soil Science Society of America, 1997). As noted, we use the tensiometer to measure matric potential. A matric potential exists in soil when the soil is unsaturated and the water in the soil is under tension. We use a piezometer to measure water in saturated soil. A piezometer is an instrument used to measure pressure. Some soil physicists call the matric potential the pressure potential or the pressure head. In our work, we shall confine the terms *pressure potential* and *pressure head* to saturated soil (see Chapter 5). Tensiometers do not measure osmotic potential, because they are not sensitive to the osmotic effects of dissolved salts in the soil solution (Richards, 1965).

Because much soil-water theory and experimentation deals with the matric potential, we need to know how a tensiometer works (Kirkham and Powers, 1972, p. 29). Two key relationships that are necessary before soil physicists can model water in unsaturated soil are: 1) the relationship between soil matric potential and soil water content and 2) the relationship between soil hydraulic conductivity and soil matric potential (van Genuchten, 1980).

Let us consider an impractical, but instructive, type of tensiometer in which a tension height h_t is developed by means of the tensiometer porous cup in contact with moist soil (Fig. 4.1). The small pores in a tensiometer cup serve to make connections between the soil water held in soil pores and a tension column. The pores of the cup must be smaller than the soil pores in which the tension is to be measured; otherwise, air may enter the cup. If air enters the cup, we have reached the air-entry value for that porous cup. Each porous cup has an air-entry value. Air comes out of solution. If air enters, we have cavitation, which is the formation of partial vacuums in a flowing liquid, as a result of the separation of its parts. Cavitation comes from the Latin *cavitas*, which means a hollow or a cavity.

In Fig. 4.1, we have dug a pit into the soil to accommodate the tensiometer. But it is usually impractical to dig a pit, so in Fig. 4.2 a column of water of height d_1 extending into the ground is replaced by an equivalent, and much shorter, column of mercury of height d_2 above ground. With distances h_t, H, d_1, and d_2 as shown in the figure, the tension height h_t is given by $h_t = d_1 - H = 13.6d_2 - H$, where d_1 is replaced by $13.6d_2$ because 13.6 gram/cm^3 is the density of mercury, making 1 cm of mercury column give the same pressure as 13.6 cm of water column. In our equation we consider that 13.6 is the specific gravity of mercury taken to be numerically equal to its density in g/cm^3. The units of specific gravity equal unity. Specific gravity is the ratio of the weight or mass of a given volume of a substance to that of an equal volume of another substance (water for liquids and solids; air or hydrogen for gases) used as a standard, and its dimensions are unity.

The pores in the porous cup and the reservoirs in the two figures (Figs. 4.1 and 4.2) have been drawn large for instructive purposes. In practice, the reservoirs are made as small as practical to limit water flow to the soil from the cup. In the laboratory, one can use a porous plate apparatus (Fig. 4.3)

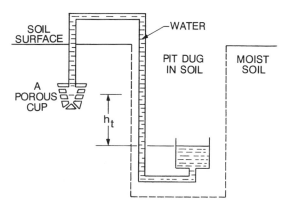

FIG. 4.1 A form of tensiometer. (From *Advanced Soil Physics* by Kirkham, D., and Powers, W.L., p. 29, ©1972, John Wiley & Sons, Inc.: New York. This material is used by permission of John Wiley & Sons, Inc. and William L. Powers.)

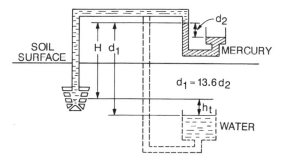

FIG. 4.2 Equivalent water and mercury tensiometers. (From *Advanced Soil Physics* by Kirkham, D., and Powers, W.L., p. 30, ©1972, John Wiley & Sons, Inc.: New York. This material is used by permission of John Wiley & Sons, Inc. and William L. Powers.)

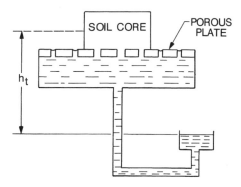

FIG. 4.3 Porous plate apparatus. (From *Advanced Soil Physics* by Kirkham, D., and Powers, W.L., p. 31, ©1972, John Wiley & Sons, Inc.: New York. This material is used by permission of John Wiley & Sons, Inc. and William L. Powers.)

to measure soil moisture tension in soil cores. In Fig. 4.3, the height h_t is the tension height.

The tension height h_t of a tensiometer cannot in practice exceed about 3/4 bar or about 750 cm of water column. (In Chapter 5, we shall show that 1020 cm water = 1 bar.) This is due to air coming out of solution at the reduced pressure to break the continuity of the water column. The 750-cm value does not mean that water will not have a tension greater than 750 cm in a soil, but that 750 cm of tension is all the tension a tensiometer can measure (Kirkham and Powers, 1972, p. 30).

Dr. L.A. Richards was one of the early developers of the tensiometer (see the Appendix, Section V, for his biography). Before he moved to the U.S. Salinity Laboratory in Riverside, California, he worked at Iowa State University (then called Iowa State College) in Ames, Iowa. In his laboratory in Curtiss Hall the early tensiometers he used were water-filled tensiometers. To have a long enough water tube, he drilled a hole through the ceiling of one of the floors in Curtiss Hall so the tube could span the length of two floors. When one of his successors, Don Kirkham, moved to Iowa State in 1946, this hole was still in Curtiss Hall. Mercury manometers obviate the need to drill holes through ceilings to take measurements with tensiometers.

For values of h_t greater than 750 cm, pressure apparatus may be used. To see how a pressure apparatus works, we consider Fig. 4.4. In this figure three capillary tubes are shown with the heights of the water rise as h_{t1}, h_{t2}, and h_{t3}. The bottoms of the capillary tubes rest either on a porous ceramic plate or a porous membrane. A bell jar covers the capillary tubes and pressure may be introduced in the jar. The figure is drawn for the case of initially atmospheric pressure. If now a pressure equal to an h_{t1} cm of water column is

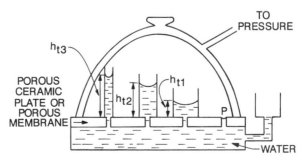

FIG. 4.4 Pressure apparatus. (From *Advanced Soil Physics* by Kirkham, D., and Powers, W.L., p. 31, ©1972, John Wiley & Sons, Inc.: New York. This material is used by permission of John Wiley & Sons, Inc. and William L. Powers.)

applied, the column h_{t1} will drain to the top level of the membrane. If h_{t2} cm of pressure is applied, column h_{t2} will drain to the top level of the membrane, and if h_{t3} cm of pressure is applied, all three will drain. If a porous membrane is used, pores in the membrane such as the one at P will not drain unless pressures of about 30 or more bars are applied. At high pressures a bell jar cannot be used. Instead, pressure equipment made of heavy steel plate, called pressure plate or pressure membrane apparatus, is used (Fig. 4.5). It works up to a pressure of 30 or even 100 bars (Kirkham and Powers, 1972, p. 32).

II. TYPES OF TENSIOMETERS

Tensiometers have three types of read-outs: mercury manometer assemblies, vacuum dial gauges, and current transducers. Instruments that have mercury manometers are attached to tubes of varying lengths with porous cups at the base that are inserted into the soil (Fig. 4.6). Soilmoisture Equipment Corporation (Santa Barbara, California) provides tubes that are 6, 12, 24, 36, 48, and 60 inches long (15, 30, 61, 91, 122, 152 cm, respectively).

Soilmoisture also sells tensiometers with vacuum dial gauge read-outs, the "Jet Fill" (Fig. 4.7) for fixed installation and the "Quick Draw" probe (Fig. 4.8), which is a portable probe designed for rugged field use. Vacuum dial gauge tensiometers also can be obtained for different depths from Soilmoisture (6-, 12-, 18-, 24-, 36-, 48-, and 60-inch depths or 15, 30, 46, 61, 91, 122, and 152 cm, respectively). For greenhouse work with pots, Soilmoisture's Model 2100F with a vacuum dial gauge (ceramic cup: 0.6 cm diameter; 2.4 cm long) can be used, because it is a miniature tensiometer (Zhang and Kirkham, 1995).

The Tensimeter™ (Fig. 4.9) sold by Soil Measurement Systems (Tucson, Arizona) is a fast, simple, and portable method to read tensiometers

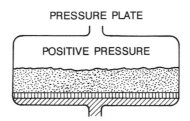

FIG. 4.5 Pressure plate. Inside the plate, soil is on a porous membrane. The heavy steel plate allows high pressures to be applied. (From *Soil Physics* by Baver, L.D., Gardner, W.H., and Gardner, W.R., p. 294, ©1972, John Wiley & Sons, Inc.: New York. This material is used by permission of John Wiley & Sons, Inc.)

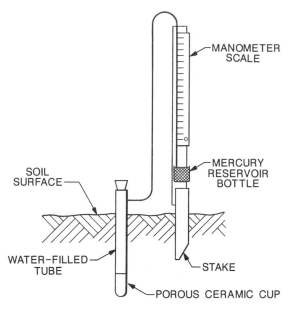

FIG. 4.6 A mercury manometer type of tensiometer. (Redrawn from a figure in a Soilmoisture Equipment Corp., Santa Barbara, California, brochure. Reproduced by permission of Soilmoisture Equipment Corp.)

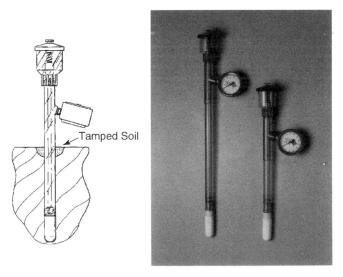

FIG. 4.7 The "Jet Fill" tensiometer of Soilmoisture Equipment Corporation. (Courtesy of Soilmoisture Equipment Corp., Santa Barbara, California.)

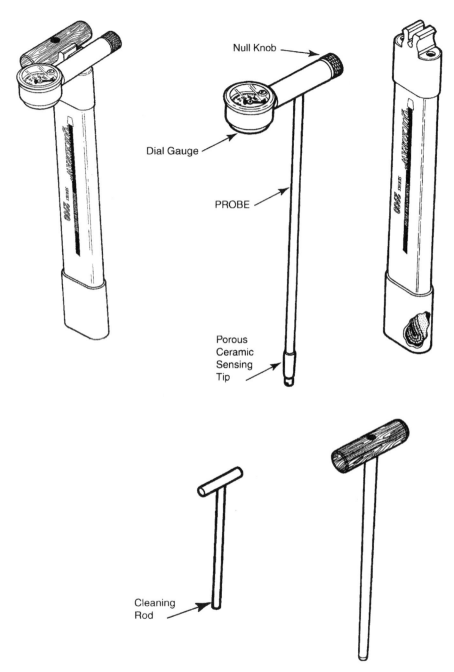

FIG. 4.8 The "Quick Draw" tensiometer of Soilmoisture Equipment Corporation. One side holds a soil corer and the other side holds the tensiometer. (Courtesy of Soilmoisture Equipment Corp., Santa Barbara, California.)

FIG. 4.9 Photograph of pressure transducer on tensiometer and digital read-out. (From Marthaler, H.P. et al. A pressure transducer for field tensiometers. Soil Sci. Soc. Amer. J. 47; p. 626, ©1983. Soil Science Society of America: Madison, Wisconsin. Reprinted by permission of the Soil Science Society of America.)

with 1 mbar sensitivity using a pressure transducer. This method was originally described by Marthaler et al. (1983), and a diagram of the tensiometer is shown in Fig. 4.10. Any ordinary tensiometer can be used. The tubing is closed off with a septum stopper, which forms an airtight seal during and after insertion of a syringe needle through the stopper. The air pressure in the upper end of the tubing is measured by inserting a syringe needle attached to a pressure transducer through the septum (Fig. 4.11). A guide tube keeps the transducer system in a vertical position when placed on the tensiometer and centers the needle in the septum. The inside diameter of the guide tube fits the outside diameter of the stopper and plexiglass tube. The transducer consists of a steel enclosure with a steel transducer membrane

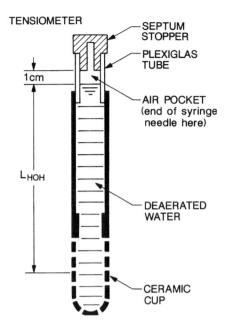

FIG. 4.10 Diagram of a tensiometer with septum stopper. (From Marthaler, H.P. et al. A pressure transducer for field tensiometers. Soil Sci. Soc. Amer. J. 47; p. 625, ©1983. Soil Science Society of America: Madison, Wisconsin. Reprinted by permission of the Soil Science Society of America.)

separating the enclosure into an upper chamber and a lower chamber. The upper chamber is at atmospheric pressure. Through the syringe needle, the air pressure in the lower chamber equilibrates with the pressure inside the tube, causing a small deflection of the steel membrane. This deflection changes the resistance of silicon semiconductors embedded into the membrane. A shielded four-lead wire connects the silicon element with a resistivity meter. The meter is calibrated to read directly in millibars or centimeters of water.

In field use, the tensiometers are inserted into the soil for permanent use during a season. To operate the Tensimeter™, one simply places the transducer over a tensiometer. The needle probe penetrates the septum stopper of the tensiometer. The tension inside the tensiometer is measured and digitally displayed (in mb or cm). One can get as many as 45 readings per septum (45 needle insertions) before the rubber septum needs to be replaced (Loyd R. Stone, Department of Agronomy, Kansas State University, personal communication, January 23, 1992). One can take about 75 readings (read 75 tensiometers) per hour; it takes about 15 seconds per reading. Readings can be hand-recorded, because this is a cheap

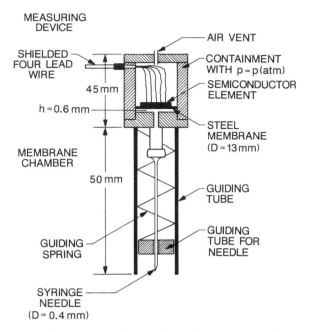

FIG. 4.11 Diagram of pressure transducer with attached syringe needle. (From Marthaler, H.P. et al. A pressure transducer for field tensiometers. Soil Sci. Soc. Amer. J. 47; p. 625, ©1983. Soil Science Society of America: Madison, Wisconsin. Reprinted by permission of the Soil Science Society of America.

and accurate method. In a subsurface drip irrigation study of corn (*Zea mays* L.) in the western part of Kansas, dozens of tensiometers were installed for several seasons (Darusman et al., 1997a, 1997b; Lamm et al., 1997), and speed in taking readings was important when measuring numerous tensiometers.

III. TEMPERATURE EFFECTS ON TENSIOMETERS

Temperature affects readings with tensiometers in two ways:

1. Effects of temperature on water in soil, and
2. Effects of temperature on the instrument.

Temperature affects the physical properties of water, including density and surface tension (Table 3.1). Therefore, the matric potential (tension) of water in the soil is affected by temperature changes. The major effect of temperature occurs at the soil surface, where temperature changes are greatest. Temperature effects on soil tension account for measurable amounts of water flow (Loyd R. Stone, personal communication, February 1, 1989).

Temperature affects the instrument directly because the mercury is heated (the mercury expands with heating). The metal on the part of the tensiometer that is inserted into the soil can conduct heat from the air to the soil. Plastic instead of metal in the construction of the tensiometer minimizes temperature effects. Tensiometers can be insulated by using shade boxes or temperature effects can be minimized by taking readings early in the morning when the sun is not far up in the sky. The instruments cannot be used in freezing weather, because the water in the tensiometer will freeze and break the tensiometer. But one can bury tensiometers deeply or use methanol-water solutions, which can protect the tensiometers as low as −18°C (Cassel and Klute, 1986, p. 586). One would need to do a controlled laboratory experiment to determine gradients caused by temperature in the soil and the magnitude of change in soil-water tension with temperature (Loyd R. Stone, personal communication, February 1, 1989). With the dual-probe heat-pulse method, we can measure soil temperature with resolution as fine as measurements 1.5 cm apart (Song et al., 1999). However, tensiometers have not yet been miniaturized enough to measure matric potential at such fine resolution. In the field, one just recognizes that temperature does affect readings with tensiometers and tries to minimize this effect (Loyd R. Stone, personal communication, February 1, 1989), especially in hot, semi-arid regions such as Kansas.

IV. APPLICATIONS OF TENSIOMETERS

Tensiometers have five applications (Richards, 1965).

1. They are used to determine rooting depth. One can follow readings with time, and the rate of increase in soil tension at any given depth can be related to the density of the active roots.
2. They are used for timing of field irrigations. It is time to irrigate when tensiometer readings reach a prescribed value for a soil depth where feeder root concentration is greatest. The duration of an irrigation is judged with tensiometers measuring soil tension at a second or greater depth. If tension readings at this second depth are low (high matric potential), an irrigation of short duration is indicated. Conversely, if they are high, irrigation water should be allowed to run until readings are low.
3. They are used to determine timing of greenhouse irrigations for potted plants and greenhouse beds. Under these conditions, only one depth is read.

4. The water table level is determined using tensiometers. When a reading on a tensiometer in the field is below zero, this is evidence that a water table occurs above the depth of the cup. The negative reading in millibars is equivalent to the distance in centimeters from the water table to the cup depth. The water table may rise and fall, and tensiometers will show this (negative versus positive readings).

5. The hydraulic gradient is determined from measurements using two tensiometers. (See Chapter 7 for the definition of the hydraulic gradient.) If one knows the hydraulic gradient, one knows which way water is moving in the soil. The use of two tensiometers at two different depths is the only way to determine the direction of movement of water in the soil (up or down). This knowledge is necessary if one wants to leach out salts or determine the water balance. Two typical depths of installation of tensiometers are at 4.5 and 5.5 feet (1.4 and 1.7 m), because these depths are below the root zone (Darusman et al., 1997a). By reading the tensiometers, one can determine if the soil is wetting up or draining. This knowledge is critical in water-balance studies because one needs to know the drainage component. In semi-arid Kansas, the goal is to minimize drainage to improve irrigation efficiency (Darusman et al., 1997b).

V. APPENDIX: BIOGRAPHY OF L.A. RICHARDS

Lorenzo Adolph Richards, known as "L.A." or "Ren," was born April 24, 1904, in Fielding, Utah. At Utah State College he received his B.S. degree in 1926 and his M.A. in 1927; in 1931 he received his Ph.D. in physics from Cornell University. He was an assistant at Utah State from 1924 to 1927, an assistant at Cornell from 1927 to 1929, an instructor at Cornell from 1929 to 1935, and a research physicist at Battelle Memorial Institute in Ohio in 1935. He then moved to Iowa State College, where he was an assistant and associate professor of physics and a research assistant and associate professor of soil between 1935 and 1939. He left Iowa State to become senior soil physicist at the U.S. Salinity Laboratory in Riverside, California, where he stayed from 1939 to 1942. He was a National Defense Research Fellow and Group Leader at the California Institute of Technology from 1942 to 1945 (Cattell, 1961). He then returned to the Salinity Laboratory, where he was chief physicist from 1945 to 1966. He married Zilla Linford of Logan, Utah, in 1930, and they had three children: L. Willard Richards, a partner in Sonoma Technology, Santa Rosa, California; Paul L. Richards, professor of physics at the University of California, Berkeley; and Mary Armstead of Carmel, California (American Society of Agronomy, 1993).

He was the recipient of many awards, including honorary Doctor of Science degrees from the Israel Institute of Technology in Haifa in 1952, and later in life, from his alma mater, Utah State University. He was Fellow of the American Society of Agronomy and won its Stevenson Award in 1949. He was president of the Soil Science Society of America (1952) and the American Society of Agronomy (1965). In recognition of his military contributions during World War II, he received the U.S. Department of Navy Ordnance Development Award in 1945 and the Presidential Certificate of Merit in 1948. He received the USDA Superior Service Award in 1959 and honorary membership in the International Society of Soil Science in 1968. In 1981, the American Geophysical Union organized a symposium on the impact of the Richards's (1931) equation, to honor the fiftieth anniversary of his influential publication in *Physics* (American Society of Agronomy, 1993). He was a long-time member of the American Geophysical Union.

His research interests were in soil physics, retention and flow of water in soil, vacuum tube circuits, rocket ordnance, diagnosis and improvement of saline and alkali soils, the relation of soil water to plant growth, and measurement of aqueous vapor pressure at high humidity (Cattell, 1961). His brother, Sterling Jacob Richards (born 1909; Cattell, 1961), also made contributions in soil physics, but was not as famous as his older brother.

L.A. Richards made continuous improvements in the design and operation of the instruments used to provide a quantitative understanding of the energy status of water in the soil. He initiated and edited the influential 1954 USDA Agricultural Handbook No. 60 entitled, *Diagnosis and Improvement of Saline and Alkali Soils*, a publication still in use today (United States Salinity Laboratory Staff, 1954). He died of Alzheimer's disease on March 12, 1993, in Carmel, California at the age of 88 (American Society of Agronomy, 1993).

REFERENCES

American Society of Agronomy. (1993). Lorenzo A. Richards. *Agronomy News*, May, 1993, p. 26 (one page only). American Society of Agronomy, Madison, Wisconsin.

Baver, L.D., Gardner, W.H., and Gardner, W.R. (1972). *Soil Physics*. 4th ed. Wiley: New York.

Cassel, D.K., and Klute, A. (1986). Water potential: Tensiometry. In *Methods of Soil Analysis. Part 1. Physical and Mineralogical Methods* (Klute, A., Ed.), 2nd ed., pp. 563–596. American Society of Agronomy and Soil Science Society of America, Madison, Wisconsin.

Cattell, J. (Ed.). (1961). *American Men of Science. A Biographical Directory. The Physical and Biological Sciences*. 10th ed., p. 3362. The Jacques Cattell Press: Arizona State University, Tempe, Arizona.

Darusman, Khan, A.H., Stone, L.R., and Lamm, F.R. (1997a). Water flux below the root zone vs. drip-line spacing in drip-irrigated corn. *Soil Sci Soc Amer J* 61; 1755–1760.

Darusman, Khan, A.H., Stone, L.R., Spurgeon, W.E., and Lamm, F.R. (1997b). Water flux below the root zone vs. irrigation amount in drip-irrigated corn. *Agronomy J* 89; 375–379.

Kirkham, D., and Powers, W.L. (1972). *Advanced Soil Physics*. Wiley-Interscience: New York.

Lamm, F.R., Stone, L.R., Manges, H.L., and O'Brien, D.M. (1997). Optimum lateral spacing for subsurface drip-irrigated corn. *Trans Amer Soc Agr Eng* 40: 1021–1027.

Marthaler, H.P., Vogelsanger, W., Richard, F., and Wierenga, P.J. (1983). A pressure transducer for field tensiometers. *Soil Sci Soc Amer J* 47; 624–627.

Richards, L.A. (1931). Capillary conduction of liquids through porous mediums. *Physics* 1; 318–333.

Richards, L.A. (1949). Methods of measuring soil moisture tension. *Soil Sci* 88; 95–112.

Richards, S.J. (1965). Soil suction measurements with tensiometers. In *Methods of Soil Analysis. Part 1. Physical and Mineralogical Properties, Including Statistics of Measurement and Sampling* (Black, C.A., Evans, D.D., Ensminger, L.E., White, J.L., and Clark, F.E., Eds.), pp. 153–163. American Society of Agronomy: Madison, Wisconsin.

Soil Science Society of America. (1997). *Glossary of Soil Science Terms*. Soil Science Society of America: Madison, Wisconsin.

Song, Y., Kirkham, M.B., Ham, J.M., and Kluitenberg, G.J. (1999). Measurement resolution of the dual-probe heat-pulse technique. In *Characterization and Measurement of the Hydraulic Properties of Unsaturated Porous Media* (van Genuchten, M.T., Leij, F.J., and Wu, L., Eds.), pp. 381–386. University of California: Riverside.

United States Salinity Laboratory Staff. (1954). *Diagnosis and Improvement of Saline and Alkali Soils* (Richards, L.A., Ed.). Agricultural Handbook No. 60. United States Department of Agriculture: Washington, DC.

van Genuchten, M.T. (1980). A closed-form equation for predicting the hydraulic conductivity of unsaturated soils. *Soil Sci Soc Amer J* 44; 892–898.

Zhang, J., and Kirkham, M.B. (1995). Sap flow in a dicotyledon (sunflower) and a monocotyledon (sorghum) by the heat-balance method. *Agronomy J* 87; 1106–1114.

5

Soil-Water Terminology and Applications

Two important expressions used to describe the state of water in the soil are *water content* and *water potential*.

I. WATER CONTENT

Water content is a measurement of the amount of water in the soil either by weight or volume and is defined as the water lost from the soil upon drying to constant mass at $105°C$ (Soil Science Society of America, 1997). It is expressed in units of either mass of water per unit mass of dry soil (kg/kg) or in units of volume of water per unit bulk volume of soil (m^3/m^3).

II. WATER POTENTIAL

The second expression utilizes the potential energy status of a small parcel of water (say a milligram) in the soil. The expression applies also to a small parcel of water in a plant. All water in the soil (or plant) is subjected to force fields originating from four main factors: the presence of the solid phase (the matrix); the gravitational field; any dissolved salts; and the action of external gas or water pressure. If the force fields in the soil are compared to a reference point, then they can be expressed on a potential energy basis, and each of the four factors can be assigned a separate potential energy value. The sum of these four potential energy values is called the *water potential* of the soil or the *total water potential* to emphasize that it is comprised of several factors. The water potential is abbreviated using the Greek letter psi, Ψ. Subscripts are sometimes added to the letter: Ψ_w stands for water potential and Ψ_T stands for total water potential.

The reference point for these potential energies is taken as pure free water at some specified height or elevation. Because water is held in the soil by forces of adsorption, absorption, cohesion, and solution, soil water is usually not capable of doing as much work as pure free water. Hence the soil water potential is normally negative. The old term for soil water potential, no longer used, is the *total soil moisture stress* and it is defined using positive values.

We now describe each component of the water potential.

1. Matric (Capillary) Potential, Ψ_m.

The *matric potential energy* or the *matric potential* is the portion of the water potential that can be attributed to the attraction of the soil matrix for water. The matric potential used to be called the *capillary potential*, because, over a large part of its range, the matric potential is due to capillary action akin to the rise of water in small, cylindrical capillary tubes (Baver et al., 1972, p. 293). (In the next chapter, we study in detail the rise of water in soil pores.) However, as the water content decreases in a porous material, water that is held in pores due to capillarity becomes negligibly small, when compared to the water held directly on particle surfaces. The term *matric potential*, therefore, covers phenomena beyond those for which a capillary analogy is appropriate.

The matric potential may be determined with a tensiometer, which measures matric potential of water in situ. As we saw in Chapter 4, the word *tensiometer* refers to the fact that it measures the *soil moisture tension*, a term no longer used in defining the components of the water potential. Other old terms used to describe the matric potential are the *soil moisture suction* or the *matric suction*. As we know, a tensiometer consists of a porous, permeable ceramic cup connected through a water-filled tube to a manometer, vacuum gauge, or other pressure measuring device. Water pressure in the manometer comes into equilibrium with the adjacent soil through flow across the ceramic cup. The height of the liquid column at that time is an index of matric potential. Soil moisture tension has been represented often with a positive sign, in which case it can be considered to be numerically equivalent to, but opposite in sign to, the matric potential.

The units used to measure matric potential, and other potentials, become evident when we consider measurement of matric potential with a tensiometer. The force per unit area, or negative pressure of the water in the porous cup, is the weight per unit cross section of the hanging column. This is the volume of the column divided by the area multiplied by the density of the liquid water and the acceleration due to gravity (Baver et al., 1972, pp. 294–295):

$$P = F/A = mg/A = (V)\,(\rho_w)\,(g/A) = (hA\rho_w g)/A = h\rho_w g \qquad (5.1)$$

where P = pressure; F = force; m = mass; g = acceleration due to gravity; A = area; V = volume; ρ_w = density of water; h = height, and the potential (negative pressure) is in units of potential energy or work per unit volume. In the centimeter-gram-second (cgs) system of units, 1020 cm of water, would exert a negative pressure of

$$(1020 \text{ cm}) \, (1 \text{ g/cm}^3) \, (980 \text{ cm/sec}^2) = 999,600 \text{ dyne/cm}^2 \qquad (5.2)$$

or 1×10^6 dyne/cm^2 = 1 bar, because, in cgs pressure units, 1 bar = 1×10^6 dyne/cm^2. The SI (Système International) unit for pressure is the Pascal, which is one newton per square meter, and thus 10 bars = 1 MPa or 1 MegaPascal. The unit (dyne/cm^2) is the same as potential energy/volume, because if we multiply the top and bottom of the fraction, F/A, in Equation 5.1 by 1 cm (= cm/cm = unity), we get work/volume = potential energy/volume:

$$[(\text{dyne})(\text{cm})]/[(\text{cm}^2)(\text{cm})] = \text{potential energy/volume} = \text{erg/cm}^3,$$

because 1 dyne-cm = 1 erg.

Units of potential energy per unit volume can be converted to units of potential energy per unit mass by dividing by the density of water, which we shall take to be 1 g/cm^3:

$$(1 \times 10^6 \text{ dyne/cm}^2)/1 \text{ gram/cm}^3 = 1 \times 10^6 \text{ dyne-cm/gram}$$
$$= 1 \times 10^6 \text{ erg/gram} = 100 \text{ joule/kg,}$$

because 1 joule = 1×10^7 ergs. Or, 1 bar can be considered to be the equivalent of 100 joules/kg. Note that the units of matric potential are not equal to potential energy units (ergs; joules), but can be given in units of potential energy/vol or potential energy/mass.

2. Gravitational Potential, ψ_g or ψ_z.

The *gravitational potential energy* or the *gravitational potential* is the potential energy associated with vertical position. The reference height or datum assigned can vary according to need and is often based on utility. It is generally convenient to keep the reference level sufficiently low so that one does not get negative values. Solutions to problems are prone to error when negative numbers are used. Land surveyors take their datum at a level below the lowest level that they expect to encounter on their survey to ensure that all of their levels will be positive. Soil scientists often take either the soil surface or the groundwater level as the reference level. The reference level usually depends on the direction of water movement: rising or infiltration. If the reference level is below the point in question, work must be done on the water and the gravity potential is positive; if the level is above the point in question, work is done by the water and the gravity potential is negative (Baver et al., 1972, p. 296).

3. Solute Potential, ψ_s.

The *solute potential energy* or *solute potential* is the portion of the water potential that can be attributed to the attraction of solutes for water. If pure water and solution are separated by a membrane, pressure will build up on the solution side of the membrane that is equivalent to the energy difference in the water on the two sides of the membrane. This pressure, which is usually called the *osmotic pressure*, is numerically equivalent, but opposite in sign, to the solute potential. The solute potential in soil is often called the *osmotic potential*, ψ_o, even if no membranes are present. The osmotic potential is usually ignored in determining water movement in the soil, unless the soil is saline. Osmotic potential is important in plants, and is discussed in Chapter 18.

An osmometer is used to measure osmotic pressure (Fig. 5.1). It consists of a U-tube that contains dissolved substances in solution on one side, which is separated from pure water in the other arm by a semipermeable membrane at the bottom. The osmotic pressure of the solution is equal to the pressure that must be applied to prevent movement of water into it (Kramer, 1983, p. 18). The membrane must have openings that are large enough to permit passage of water molecules but too small for ions of the dissolved substance to pass through (Baver et al., 1972, p. 298).

4. Pressure Potential, ψ_p.

The *pressure potential energy*, or *pressure potential*, is the potential energy due to the weight of water at a point under consideration, or to gas pressure that is different from the pressure that exists at a reference position (Baver et al., 1972, p. 297). Sometimes this pressure potential energy is divided into two separate components: the *air pressure potential*, which occurs under unsaturated conditions when the soil has an air phase, and the *hydrostatic pressure potential*, which occurs when the soil is saturated and

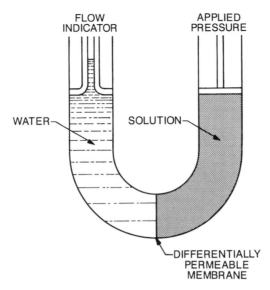

FIG. 5.1 Diagram of an osmometer in which a membrane permeable to water but impermable to a solution separates pure water from the solution. The osmotic pressure of the solution is equal to the pressure that must be applied to prevent movement of water into it. Water movement is observed by a change in the level of water in the capillary tube on the left. (From *Water Relations of Plants* by Kramer, P.J., p. 18, ©1983, Academic Press: New York. Reprinted by permission of Academic Press.)

there is a hydrostatic pressure from an overlying water phase (Jury et al., 1991, p. 51). In saturated soil, the pressure potential is sometimes called the *piezometric potential* (Baver et al., 1972, p. 297), because it can be measured with a *piezometer*. As noted in Chapter 4, a piezometer is an instrument used to measure pressure. The word comes from the Greek *piezein*, which means "to press," and *meter*, which comes from *metron*, a measure (*Webster's New World Dictionary of the American Language*, 1959). It is a tube placed in soil with its top end open to the atmosphere. It also may have openings in the wall at the point where the pressure measurement is to be taken. The level of water in the tube, measured from a suitable reference, is the piezometer reading. Piezometers are used to measure groundwater depth. Pressure potentials due to gas may be measured with manometers.

5. Other Potentials Defined.

Occasionally, a *tensiometer pressure potential*, which is the potential measured with a tensiometer, is defined (Jury et al., 1991, p. 50). The matric potential differs from the potential measured with a tensiometer, because the soil air pressure is maintained at the reference pressure. The reference pressure can be atmospheric pressure. However, the difference between atmospheric pressure and air pressure in the soil is usually ignored, and the potential measured with a tensiometer is considered to be the matric potential. But if one were comparing measurements of matric potential made with a tensiometer on top of a mountain and at sea level, then one would have to consider air pressure differences.

Other potentials may be defined according to need, such as an *overburden potential*, which occurs when the soil is free to move and some part of its weight becomes involved as a force acting upon water at the point in question (Baver et al., 1972, p. 299). Such an overburden potential might occur in soil under a ridge in a ridge-till system. But the pressure exerted by the weight of the soil in the ridge would be small. When a potential that is not zero is neglected, it must be assumed that it is implicitly included in one of those that is explicit in the definition. For example, when overburden potential is neglected, it becomes implicit in the pressure potential or matric potential.

In both soil and plant systems the water potential is usually considered to be the sum of the four potentials described in the preceding sections: matric potential, gravitational potential, solute potential (or osmotic potential), and pressure potential, or

$$\psi = \psi_m + \psi_g + \psi_s \ (or \ \psi_o) + \psi_p \tag{5.3}$$

Water moves in response to differences in water potential. The difference is called the *water potential difference*. The *water potential gradient* is the potential difference per unit distance of flow. Water moves from high potential energy to low potential energy. Under nonsaline, unsaturated conditions, the two most important potentials in the soil are the matric potential and the gravitational potential, and both must be considered in determining the direction of flow of water. Under nonsaline, saturated conditions, the two most important potentials in the soil are the (hydrostatic) pressure potential and the gravitational potential, and the difference in the sum of these two potentials, called the *hydraulic head difference*, governs the soil water flow. In plants, the two most important components of the water potential are the osmotic potential and the pressure potential, also called the turgor potential (see Chapter 18).

III. HEADS IN A COLUMN OF SOIL

A head is a source of water kept at some height to supply, for example, a mill. Hence, it is a pressure, as in a *head* of steam (*Webster's New World*

Dictionary of the American Language, 1959). Instead of using potential terminology, we can express potential in terms of head (a length). As we saw in Equation 5.1, the tension, or negative pressure, that develops in a tensiometer is

$$F/A = h\rho_w g,$$

where F = force; A = area; h = height; ρ_w is the density of water; and g is acceleration due to gravity. Because the density of water and acceleration due to gravity are constant, we can relate a pressure or negative pressure, in terms of potential energy per unit volume, to a length. Engineers prefer to work with lengths (heads) rather than potentials because they are easier to measure and keep track of.

So our equation for water potential in soil, Equation 5.3, $\psi = \psi_m + \psi_g + \psi_s$ (or ψ_o) + ψ_p, becomes in head terminology the following for a non-saline soil (ψ_s or ψ_o is negligible):

$$h = h_t + h_g + h_p \tag{5.4}$$

where h = total head; h_t = tension head (we call it the tension head instead of the matric head); h_g = gravitational head; and h_p = pressure head.

Under saturated conditions when there is no tension head, the total head is

$$h = h_p + h_g.$$

When the soil is under tension (but the pores can be filled with water, i.e., saturated, yet the water is under tension, which we discuss in the next chapter), we have

$$h = h_t + h_g.$$

or

$$h = (h_p \text{ or } h_t) + h_g.$$

Let us apply head terminology to determine heads in a column of soil. We follow the analysis of Kirkham and Powers (1972, pp. 36–37) and plot the four heads h, h_g, h_p, and h_t, in a column of wet soil. (For a biography of W.L. Powers, see the Appendix, Section V.) In Figure 5.2 we have plotted the various heads against distance z above the base of a soil column standing in 5 cm of water. The water in the column is assumed to be saturated to the top above the 5-cm level by capillarity and is at equilibrium. Evaporation from the top of the soil is prevented. The reference level for heads is the level of the base of the column. The line AE is a graph of height

z versus gravitational head h_g; the line BC is of z versus pressure head h_p; CD is of z versus tension head h_t; and BF is of z versus total head h. The line BF is vertical because at all heights z in the column the abscissas h_g and h_t (or h_p) must add up to the same value h, the total head (= 5 cm). Otherwise, the water in the column could not be in equilibrium and the water would move. The equilibrium conditions for the column may be written symbolically as

$$h_g + h_p = h = 5 \text{ cm, for } z \text{ between 0 cm and 5 cm}$$

and

$$h_g + h_t = h = 5 \text{ cm, for } z \text{ between 5 cm and 20 cm.}$$

The line BC in the graph shows that the pressure head decreases linearly from 5 cm of head to 0 cm of head as the height z increases from 0 to 5 cm. From $z = 5$ cm to $z = 20$ cm the pressure becomes negative; that is, tension develops. The tension at level E is 15 cm of water, a negative pressure. If evaporation were permitted at the soil surface, the tension head there would be of greater magnitude than 15 cm and curves CD and BF would change from their shown positions. In Figure 5.2 the tension head h_t is that resulting from water curvature in the pores, which we shall discuss in the next chapter. "Equivalent" tension heads obtained by indirect methods such as freezing-point depression do not reflect film curvature, and when such equivalent tension heads are plotted on a graph with tensions as measured by a tensiometer, the freezing-point-depression values may be three times as great in magnitude as the tensiometer values, and confusion may

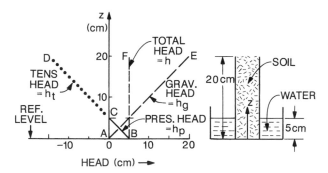

FIG. 5.2   Heads in a column of wet soil. (From *Advanced Soil Physics* by Kirkham, D., and Powers, W.L., p. 36, ©1972, John Wiley & Sons, Inc.: New York. This material is used by permission of John Wiley & Sons, Inc. and William L. Powers.)

result (Kirkham and Powers, 1972, p. 37). The freezing-point method to determine the free energy of water in soil is described by Richards (1965, pp. 137–139).

In Figure 5.2, tension could be determined with a tensiometer for the distance of z between 5 and 20 cm, and pressure could be determined with a piezometer for the distance of z between 0 and 5 cm.

IV. MOVEMENT OF WATER BETWEEN TENSIOMETERS

Because we can equate pressure or negative pressure or tension with height, let us use height to determine the direction of movement of water in soil with two tensiometers. All we need is a ruler. Figure 5.3, as modified from Richards (1940) and reproduced by Kirkham and Powers (1984, p. 240), gives a physical picture of how the tension head h_t and the gravitational head z combine to give the total head h. Tensiometers are shown with water manometers inside of the soil. In actuality, the water manometers would be replaced by mercury manometers above the soil. The physical principles are clearer if the water manometers are represented as shown. The reference level for head is taken at a depth z_D below the center of the uppermost tensiometer cup.

In Figure 5.3 we notice that for each location, A, B, C, and D, the total head is given by $h = h_t + z$ (where h_t is negative). For location A and B, h_{tB}

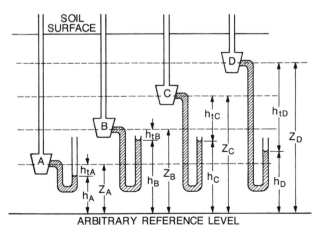

FIG. 5.3 Diagram to illustrate water movement between tensiometers. (From *Advanced Soil Physics* by Kirkham, D., and Powers, W.L., p. 240, 1984, Reprint edition, Robert E. Krieger Pub. Co.: Malabar, Florida; ©1972, John Wiley & Sons, Inc.: New York. This material is used by permission of John Wiley & Sons, Inc. and William L. Powers.)

has been taken equal to h_{tA}. Thus, the soil moisture is of equal dryness at A and B; but this does not mean that the soil moisture will be statically held between A and B. On the contrary, since the total head h_B is seen to be greater than h_A, moisture will move from B to A. At locations C and B we see that h_{tC} has been taken to be more negative than h_{tB}. Thus, the soil is drier at C than at B, but moisture will not move from the wetter soil at B to the drier soil at C because we have chosen conditions such that the total head h_B is equal to h_C. Therefore, moisture neither moves from B to C nor from C to B; the moisture is in equilibrium. Finally, we note that moisture will not move upward from C to D because our moisture conditions at points C and D have been chosen such that h_C is greater than h_D. However, we notice that because h_{tD} is more negative than h_{tC} (tension is greater for D than C), the soil at D has to be considerably drier than at C to cause the upward movement. In Figure 5.3, if the reference level were at the top of the figure, the values h_A, h_B, and so on would always be negative; h_t, of course, is negative for unsaturated soil regardless of the location of the reference level (Kirkham and Powers, 1984, pp. 241–242).

V. APPENDIX: BIOGRAPHY OF WILLIAM L. POWERS

William LeRoy Powers is the son of LeRoy Powers, a geneticist and breeder of agronomic and horticultural crops, who worked for the Agricultural Research Service of the United States Department of Agriculture on the campus of Colorado State University in Fort Collins (Cattell, 1961). LeRoy Powers was a famous geneticist and excellent statistician (George L. Liang, personal communication, July 26, 2002), who published a classic paper in plant genetics (Powers, 1963). William Powers received his B.S. degree at Colorado State and joined the Department of Agronomy at Iowa State University in the fall of 1960 to pursue a master's degree and Ph.D. under Don Kirkham. He obtained his M.S. degree in 1962 and the Ph.D. in 1966; the title of his Ph.D. dissertation was "Solution of Some Theoretical Soil Drainage Problems by Generalized Orthonormal Functions." After he obtained his Ph.D. he moved to Kansas State University, where he rose through the academic ranks to professor and became director of the Evapotranspiration Laboratory and director of the Kansas Water Resources Research Institute. In 1980, he accepted the position of Director of the Water Resources Center at the University of Nebraska in Lincoln.

He wrote, with Don Kirkham, *Advanced Soil Physics* (Wiley: New York, 1972). His research included developing formulas for applying organic wastes to soils; water-balance studies; pore-size distribution as an index of atrazine movement; tillage effects on soil water release curves; estimating soil water content from soil strength; TNT sorption in soil; spatial series analysis of horizontal soil cores to characterize tracer patterns; physical and chemical characteristics of aging golf greens; and spatial analysis of machine-wheel traffic and its effects on soil physical properties.

He married in 1958, and he and his wife, Marty, a registered nurse, have two daughters, Jenny and Susan. He retired from the University of Nebraska on July 31, 2001, and spends his retirement volunteering at the Veteran's Administration Hospital and St. Elizabeth Medical Center in Lincoln, Nebraska, and delivering meals on wheels.

REFERENCES

Baver, L.D., Gardner, W.H., and Gardner, W.R. (1972). *Soil Physics*. 4th ed. Wiley: New York.

Cattell, J. (Ed.). (1961). *American Men of Science. A Biographical Directory. The Physical and Biological Sciences*. 10th ed., p. 3234. Jaques Cattell Press: Tempe, Arizona.

Jury, W.A., Gardner, W.R., and Gardner, W.H. (1991). *Soil Physics*. 5th ed. Wiley: New York.

Kirkham, D., and Powers, W.L. (1972). *Advanced Soil Physics*. Wiley: New York.

Kirkham, D., and Powers, W.L. (1984). *Advanced Soil Physics*. Reprint ed. Robert E. Krieger: Malabar, Florida.

Kramer, P.J. (1983). *Water Relations of Plants*. Academic Press: New York.

Powers, L. (1963). The partitioning method of genetic analysis and some aspects of its application to plant breeding. In *Statistical Genetics and Plant Breeding* (Hanson, W.D., and Robinson, H.F., Eds.), pp. 280–318. Pub. No. 982. National Academy of Sciences-National Research Council: Washington, DC.

Richards, L.A. (1940). Hydraulics of water in unsaturated soil. *Agr Eng* 22; 325–326.

Richards, L.A. (1965). Physical condition of water in soil. In *Methods of Soil Analysis. Part 1. Physical and Mineralogical Properties, Including Statistics of Measurement and Sampling* (Black, C.A., Evans, D.D., White, J.L., Ensminger, L.E., and Clark, F.E., Eds.), pp. 128–152. American Society of Agronomy: Madison, Wisconsin.

Soil Science Society of America. (1997). *Glossary of Soil Science Terms*. Soil Science Society of America: Madison, Wisconsin.

Webster's New World Dictionary of the American Language. (1959). College ed. World Publishing: Cleveland.

6

Static Water in Soil

We now look at how water interacts with the solid system of the soil. In particular, we shall study surface tension, and then see how it is related to the rise and fall of water in soil pores, which, in turn, explains hysteresis.

I. SURFACE TENSION

We first recall the definitions of some terms from elementary physics (Kirkham and Powers, 1972, p. 11). An object is under tension if a pull is being exerted on it. In Fig. 6.1, the cross section A of the cylinder is under tension due to the forces F. Tension is a pull or stretching force per unit area. Pressure implies a push and is a compression force per unit area. If we reverse the directions of the arrows in Fig. 6.1, the cylinder will be under pressure. In talking about soil water and plant water, we sometimes say that the water is under stress. Stress may be either a pull or push, tension, or compression. So stress may properly be expressed as a pull or push per unit area.

The term *surface tension* should not be confused with tension (Kirkham and Powers, 1972, pp. 11–12). Surface tension, or more specifically, the surface tension coefficient, an energy per unit area, is equivalently a force per unit length, whereas tension is a force per unit area. We abbreviate the surface tension coefficient using the Greek letter sigma (σ).

$$\sigma = \text{energy/area} = (\text{force}) (\text{distance})/\text{area} \qquad (6.1)$$

or

$$\sigma = \text{force/length.} \qquad (6.2)$$

Surface tension may be compared with the force that develops in a sheet of paper when we pull it on opposite edges. In Figure 6.2 the force F when

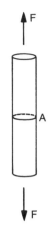

FIG. 6.1 Illustration of tension. (From *Advanced Soil Physics* by Kirkham, D., and Powers, W.L., p. 11, ©1972, John Wiley & Sons, Inc., New York. This material is used by permission of John Wiley & Sons, Inc. and William L. Powers.)

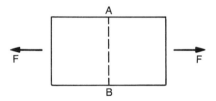

FIG. 6.2 Surface tension in a sheet. (From *Advanced Soil Physics* by Kirkham, D., and Powers, W.L., p. 12, ©1972, John Wiley & Sons, Inc.: New York. This material is used by permission of John Wiley & Sons, Inc. and William L. Powers.)

divided by the length AB gives a surface tension coefficient σ, which may be denoted by

$$\sigma = F/(2AB), \tag{6.3}$$

where the 2 in the denominator is used because the sheet of paper has an upper and a lower surface even though the paper is thin.

Laplace (1749–1827), a French mathematician and astronomer, explained surface tension. (See the Appendix, Section IV, for a history of surface tension and the Appendix, Section V, for a biography of Laplace.) A molecule in the body of a fluid (Fig. 6.3) is attracted equally from all sides. But a molecule at the surface undergoes a resultant inward

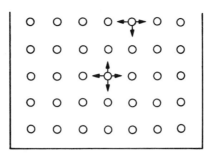

FIG. 6.3 Laplace's surface tension theory. (From *Advanced Soil Physics* by Kirkham, D., and Powers, W.L., p. 12, ©1972, John Wiley & Sons, Inc.: New York. This material is used by permission of John Wiley & Sons, Inc. and William L. Powers.)

pull, because there are no molecules outside the liquid causing attraction. Hence, molecules in the surface have a stronger tendency to move to the interior of the liquid than molecules in the interior have to move to the surface. What results is a tendency for any body of liquid to minimize its surface area. A molecule at the surface of a liquid is acted on by a net inward cohesive force that is perpendicular to the surface. So it requires work to move molecules to the surface against this opposing force, and the surface molecules have more energy than interior ones (Schaum, 1961, p. 108).

This tendency to minimize surface area is often opposed by external forces acting on the body of liquid, as gravity acting on a water drop resting on a flat surface, or as adhesive forces between water and other materials. Thus, the actual surface may not be an absolute minimum, but rather a minimum depending on the conditions in which the body of liquid is found (Kirkham and Powers, 1972, pp. 12–13).

The surface tension coefficient has been expressed as a force per unit length. If a wire is pulled horizontally from beneath a liquid, as illustrated in Fig. 6.4, the force required to pull it out depends on the length of wire. Let the symbols in Fig. 6.4 be defined as follows:

- F = upward pull required to balance surface tension forces (gravitational forces neglected)
- L = length of wire
- σ = surface tension (units of force per unit length)
- d = distance wire is raised.

Then we have

$$F = 2(\sigma L), \tag{6.4}$$

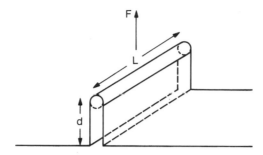

FIG. 6.4 Wire being pulled from water with adhering water film. (From *Advanced Soil Physics* by Kirkham, D., and Powers, W.L., p. 13, ©1972, John Wiley & Sons, Inc.: New York. This material is used by permission of John Wiley & Sons, Inc. and William L. Powers.)

where the 2 is used because the force to be overcome by surface tension is developed on the two sides of the wire. (Note that the wire is circular, but water adheres to two "sides.") Now the work W required to pull the wire against surface tension forces through the distance d is

$$W = Fd. \tag{6.5}$$

That is, using the relation $F = 2(\sigma L)$, we have

$$W = \sigma(2Ld) \tag{6.6}$$

or

$$\sigma = W/(2Ld) \tag{6.7}$$

or

$$\sigma = W/(\text{increased area of surface}). \tag{6.8}$$

That is, σ is the energy stored in the surface per unit increase in its area. So by pulling the wire out of a liquid, we can see that the coefficient of surface tension may be expressed as the energy stored per unit area of increase in the surface.

Surface tension causes the rise or fall of a liquid in a capillary tube. We are going to relate the rise of water in soil to the rise of water in capillary tubes, so we need to understand the rise of water in capillary tubes. The equation for the height of rise in a capillary tube, h, is (Schaum, 1961, p. 108)

$$h = (2\sigma \cos \alpha)/(r\rho g), \tag{6.9}$$

where

- σ = surface tension of the liquid
- r = radius of the tube
- ρ = density of the liquid
- α = contact angle between the liquid and tube (called the wetting angle in terminology used in soil science)
- g = acceleration due to gravity.

We shall now prove this equation in a simple manner, recognizing that more complicated proofs exist using calculus (Porter, 1971).

Let us look at Fig. 6.5 (Schaum, 1961, p. 109). Consider the body of liquid inside the tube and above the outside level. The vertical (downward and upward) forces acting on it must balance. The downward force is its weight. Remember from Chapter 2 that weight, w, is a force and $w = mg$, where m is mass and g is defined above.

weight of liquid inside tube = volume × weight per unit volume

$$= \pi r^2 h \times mg/V = \pi r^2 h \times (m/V)g \qquad (6.10)$$

$$= \pi r^2 h \rho g \text{ acting downward.} \qquad (6.11)$$

The upward force is due to surface tension. Remember that surface tension, σ, is a force/length and the length of the tube is its circumference, $2\pi r$. The force is the perpendicular force, so to get the normal component we must

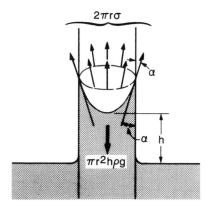

FIG. 6.5 Height of rise of a liquid in a capillary tube. (From Schaum, D., *Theory and Problems of College Physics*, p. 109, ©1961, Schaum Publishing Co.: New York. This material is reproduced with permission of The McGraw-Hill Companies.)

multiply $2\pi r$ by cos α. So the upward force is $2\pi r\sigma\cos\alpha$, and this is the force due to surface tension. For vertical equilibrium:

upward force = downward force

$$2\pi r\sigma \cos \alpha = \pi r^2 h\rho g \qquad (6.12)$$

or

$$h = (2\sigma \cos)/(r\rho g). \qquad (6.13)$$

The meniscus in the capillary tube can be either convex or concave. (In physics, *meniscus* is defined at the curved upper surface of a column of liquid. It comes from the Greek word *meniskos*, which is a diminutive of *mene*, "the moon.") A convex meniscus is illustrated by mercury on glass and a concave meniscus is illustrated by water on glass (Fig. 6.6). Mercury in contact with glass has an angle of contact of 130 to 140 degrees (Porter, 1971, p. 447). For water in contact with most soil minerals the wetting angle, α (also abbreviated θ), is close to zero (Linford, 1930), so in the equation for the height of rise in a capillary tube, we can take cos 0 = 1 or cos α = 1. However, in highly repellent soils, the contact angle is large.

Leon Linford (1930) developed a clever way to measure the wetting angle in soil by using mirrors and the well-known laws of reflection in physics. The angle of incidence is the angle between the incident ray and the normal to the reflecting surface at the point of incidence (Schaum, 1961, p. 214) (Fig. 6.7). The angle of reflection is the angle between the reflected ray and the normal to the surface. The laws of reflection are: 1) The incident ray, reflected ray, and normal to the reflecting surface lie in the same plane. 2) The angle of incidence equals the angle of reflection. Concave mirrors form real and inverted images of objects located outside of the principal focus; if the object is between the principal focus and the

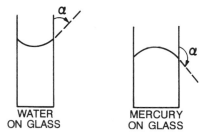

FIG. 6.6 Angles of contact. (From *Advanced Soil Physics* by Kirkham, D., and Powers, W.L., p. 22, ©1972, John Wiley & Sons, Inc.: New York. This material is used by permission of John Wiley & Sons, Inc. and William L. Powers.)

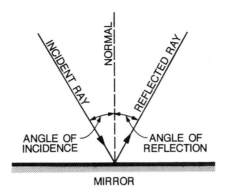

FIG. 6.7 The angles of incidence and reflection. (From Schaum, D., *Theory and Problems of College Physics,* p. 214, ©1961, Schaum Publishing Co.: New York. This material is reproduced with permission of The McGraw-Hill Companies.)

mirror, the imagine is virtual, erect, and enlarged. Convex mirrors produce only virtual, erect, and smaller images. Linford (1930) beamed light down into soil and said,

"Any point of the [water] meniscus, acting as a cylindrical concave mirror would reflect the light back and upwards at an angle such that it is twice the angle between the tangent to the meniscus at the point in question and the vertical. If the angle of contact is zero, the light reflected from the top of the meniscus would come back horizontally. . . . [B]y photographing the reflected light, the angle of contact was shown to be very small if not zero."

Leon Linford, born July 8, 1904 (Cattell, 1955), was a physicist who worked at Utah State in Logan, Utah, and joined the famous Radiation Laboratory at the Massachusetts Institute of Technology (MIT) in Cambridge, Massachusetts, during the Second World War (Buderi, 1996; Seitz, 1996). After the war, Linford became head of the Department of Physics at the University of Utah in Salt Lake City, but died of cancer in 1957 (Cattell, 1961), perhaps from exposure to radioactivity at the Radiation Laboratory.

II. EXAMPLES OF SURFACE TENSION

The importance of surface tension can be illustrated in five ways.

1. A water beetle or other small aquatic organisms can float on water because of surface tension (Porter, 1971, p. 442). The fact that small

insects can float on water shows the close relationship between the way they have evolved and water. (We remember from Chapter 2 that another example of the relationship between animal evolution and properties of water is the fact that water has a minimum specific heat at 35°C.) One can simulate a water beetle by floating a razor blade on water. The razor blade will float on an unbroken water surface, but it will not float if the surface tension is broken by soap.

2. We can kill mosquitoes by putting oil on water. Because surface tension is broken and the mosquitoes cannot float, they sink and die. Just a little oil will do (Don Kirkham, personal communication, January 30, 1992).

3. Ducks cannot float as easily on a farm pond with oil as on a pond with no oil (Don Kirkham, personal communication, January 30, 1992).

4. Oil is put on ocean waters to calm the waves (Don Kirkham, personal communication, January 30, 1992). For example, this might be done at shipwrecks. Familiar quotations allude to oil calming water. Plutarch (A.D. 46–120) said, "Why does pouring oil on the sea make it clear and calm? Is it for that the winds, slipping the smooth oil, have no force, nor cause any waves?" (Bartlett, 1955, p. 49). Pliny the Elder (A.D. 23–79) said, "Everything is soothed by oil, and this is the reason why divers send out small quantities of it from their mouths, because it smooths every part which is rough" (Barlett, 1955, p. 49).

5. Walnut shells clump together if floated on water. An experiment can be done in which half-shells of walnuts (with the nut meat removed) are used to make little boats floating on water. Put the walnuts in a pan of water—the walls of the pan must be clean (i.e., no soap or grease on the sides of the pan)—and the walnuts pull together by themselves in the small pan. They pull together to keep the surface energy to a minimum. The walnuts will form chains or trains as they clump together. Adding soap to the water would change the wetting angle and the walnuts would not clump together. As noted above, surface tension occurs in any body of liquid and causes the surface area to be minimized.

The walnuts in water simulate soil conditioners, which cause aggregation. There are many types of soil conditioners (Schamp et al., 1975; see their Fig. 2 for polymers used as soil conditioners). Some of them have charged ions in the molecule (e.g., sodiumpolyacrylate with Na^+ or K-polystyrene-sulfonate with a K^+), but many of them have no charged ions in their molecular structure. The popular polyacrylamide (PAM), widely used as a soil conditioner (see the cover of the 1998 November-December issue of the *Soil Science Society of America Journal*, and the accompanying article by

Sojka et al., 1998), comes in a nonionic form (Aly and Letey, 1998). For these nonionic polymers, hydrogen bonding may not be as important in binding the polymers to water and soil and helping in aggregation as it would be for ionic polymers. The nonionic soil conditioners probably have an effect like the walnuts; they minimize the surface energy of water in the soil. Marcel De Boodt, who got his Master's Degree at Iowa State University, said (personal communication, 1976) that his career was based on the simple walnut demonstration, which Don Kirkham showed in his soil physics class. Professor De Boodt built a world-famous laboratory focusing on soil conditioners at the University of Ghent, Belgium. The laboratory has projects around the world and has utilized soil conditioners to stabilize the sands around landing strips at airports in oil-wealthy desert countries in the Middle East and Africa. De Boodt (1975) reviews use of soil conditioners.

III. RISE AND FALL OF WATER IN SOIL PORES

Water is attracted into soil pores predominantly because of the attraction of water to other surfaces (adhesion) and because of capillarity. Surface tension controls the rise or fall of a liquid in a capillary tube. We have discussed surface tension and the equation to determine the height of rise in capillary tubes. We now discuss the rise and fall of water in soil pores (capillary tubes) and how the rise and fall determine the soil moisture characteristic curve. We follow the analysis of Kirkham (1961, pp. 24–29).

If one keeps track of the moisture withdrawn from an initially saturated soil core as greater tension is successively applied, and then plots on the x-axis (abscissa) water content (moisture percent by volume in the soil, not the water sucked out) and on the y-axis (ordinate), tension head (positive units) or matric potential (negative units), the curve so obtained will be the so-called *moisture characteristic* (ABCD in Fig. 6.8). The moisture percentage on such a curve may be based on oven-dry weight, but in drainage work, as in the figure, the soil moisture characteristic is most useful when the moisture is expressed on a volume basis, because then the surface centimeters (depth) of irrigation water needed to replenish moisture in the sample is obtained from the characteristic. For example, a moisture percentage of 30% by volume at saturation means that, for a 10-cm dry soil layer, 3 cm of water must be applied to the surface to bring the 10 cm to saturation.

In Fig. 6.8 one may think of the tension as being produced by a falling water table. One may verify the following on the figure: Initially (point A),

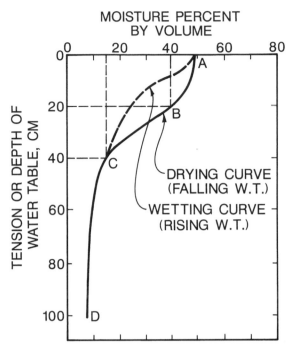

FIG. 6.8 A soil moisture characteristic curve for a loam soil. (From Kirkham, D., *Lectures on Agricultural Drainage*, p. 24, ©1961, Institute of Land Reclamation, College of Agriculture, Alexandria University: Alexandria, Egypt.)

the bulk volume of the soil has all of its pore space, that is, 50% of its bulk volume, filled with water. For a 20-cm depth of water table, the moisture percent at the soil surface is 40%; for a 40-cm depth of water table, 15%; and for a 100-cm depth, 8%. In Fig. 6.8, if the water table had fallen to a 40-cm depth and then risen slowly to the soil surface, the moisture percentages would be those corresponding to the dashed line. The failure of the curve to retrace itself in the reverse direction is called *hysteresis*. In Fig. 6.8, the soil moisture characteristic ABCD is that of a loam; for finer-textured soils, the curves would be higher. If, for Fig. 6.8, the water table for the dashed curve had not risen slowly, the moisture percent for zero depth of the water table would be, because of trapped air, less than 50%. Even if the water table rises slowly, there is usually a small amount of trapped air, and, when hysteresis loops are determined experimentally, they are not seen to return to the original point.

We pause here to say a few words about hysteresis. It comes from the Greek *hysteresis*, a coming short or deficiency. It is a word used in physics and its definition is "a lag in the effect in a body when the force acting on it is changed; especially, a lag in the change of a magnetization behind the varying magnetizing force" (*Webster's New World Dictionary of the American Language*, 1959). Because hysteresis relates to a physical system, a hysteretic curve can be repeated (e.g., the curves in Fig. 6.8 for a loam soil). Its use to describe a biological system should be approached with caution. For example, curves relating leaf water potential to evapotranspiration have been called hysteretic (Sharratt et al., 1983), but such curves are not repeatable and depend on physiological factors such as stomatal closure.

If the soil is saturated to the surface and covered by a thin layer of water, there will be no tension in the soil pores (voids). If the water table falls through the soil surface, tension will develop in the soil pores. If the pores are of the same diameter, they will start to drain and the water level in them will fall the same distance the water table falls. The maximum tension that the falling water table can exert on a soil pore at the soil surface is

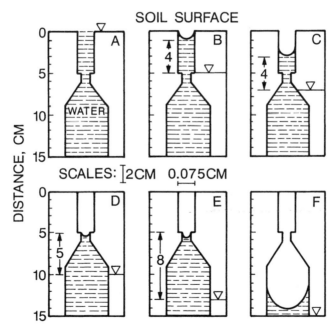

FIG. 6.9 The falling water table in a soil pore (channel) of variable diameter. Note the difference in the vertical and horizontal scales. The water table is indicated by the inverted Greek "delta." (From Kirkham, D., *Lectures on Agricultural Drainage*, p. 27, ©1961, Institute of Land Reclamation, College of Agriculture, Alexandria University: Alexandria, Egypt.)

$\rho_w gh$ dynes/cm², where h is the depth of the water table below the soil surface. If the diameter of the pore is too large to support this tension, the pore will not be subject to the maximum tension.

However, pores in the soil are not all the same diameter. Figure 6.9 illustrates what happens in a soil pore of variable diameter, when the water table falls for six different cases of water table fall. The depth of soil and the length of the pore channel for each case is taken as 15 cm, so that, for the heights of capillary rise shown, the diameter of tube nearest the surface is calculated to be 0.075. Thus the scale in the horizontal direction is, as seen in the figure (2/0.075 =), 27-fold that of the vertical direction. In part A of the figure the soil is shown saturated to the surface. In parts B, C, D, E, and F the water table is shown at successively greater depths. In parts B and C a 4 cm height of water column is held. In part D only sufficient water curvature has developed in the narrow neck to support about 5 cm height of water. In part E additional curvature has developed in the narrow neck, such that about 8 cm height of water is supported. In part E the water table is at 13 cm depth, and in part F it is at 15 cm depth, a drop of 2 cm. In dropping these 2 cm, the ability of the narrow neck to support the needed 2 cm is exceeded and the pore then empties suddenly and discontinuously to about the level of the water table. This example shows that the emptying of individual pores occurs discontinuously. When the water is removed from a large number of pores, as for any soil sample, a graph of moisture percentage versus tension (or matric potential) does not show the discontinuous nature of the pore-emptying process. The example also shows that soil pores can be filled with water (saturated), yet the water is under tension in the pores.

In Fig. 6.10, at the left, three shapes of pores are shown when the water table has fallen from level A to level B. The same three pores are shown at the right when the water table has risen from level C (say) to level B. At the left, the pores are filled up to the height h_c, the capillary height of rise. At the right, only one pore is filled up to the height h_c; one pore is empty; and one is partially filled. The soil at the left, for the water table falling, has a much higher moisture percentage than the soil at the right, for the water table rising.

Figure 6.10 also gives a physical picture for hysteresis shown in Fig. 6.8. A soil that is being wetted up from a rising water table holds less water than a soil that is being dried down. For the falling water table, water is held in tubes of supercapillary size, if there is a restriction of capillary size at or below the height of capillary lift. Water can be drawn up above a

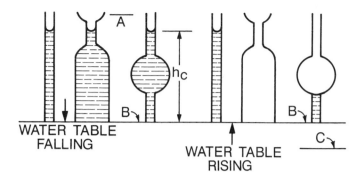

FIG. 6.10   Soil pore conditions for a falling and for a rising water table. (From Kirkham, D., *Lectures on Agricultural Drainage*, p. 28, ©1961, Institute of Land Reclamation, College of Agriculture, Alexandria University: Alexandria, Egypt.)

water table, however, only by a continuous capillary opening without supercapillary enlargements. Hence, more water is held in the *capillary fringe*, which is the thickness of saturated water held by capillarity above the water table, above a sinking water table than above a rising water table. These concepts were explained in 1937 by Cyrus Fisher Tolman of Stanford University in his classic book, *Ground Water*.

It is apparent from Fig. 6.10 that applications of subirrigation water to raise the water table will not result in the same amount of moisture in the capillary fringe as will applications of surface water. Subirrigation would provide more soil aeration than surface addition of water. This may be desirable in some cases.

IV. APPENDIX: HISTORY OF SURFACE TENSION

Atomic, nuclear, and high-energy physics is now emphasized in undergraduate physics courses, so the physics of earlier years is left out of textbooks, including the physics of surface tension (Don Kirkham, personal communication, 2 February 1992). Therefore, we need to look to other sources for an explanation of surface tension. The following history of surface tension comes from Porter (1971), who was a professor of physics at the University of London and died in 1939. Porter's historical summary is taken from the classic article by James Clerk Maxwell (1831–1879) in the ninth edition of the *Encyclopaedia Britannica*, as modified by the third Lord Rayleigh (1842–1919) in the tenth edition.

Leonardo de Vinci must be considered as the discoverer of capillary phenomena, but the first accurate observations of the capillary action of tubes and glass plates were made by Francis Hauksbee (*Physico-Mechanical Experiments*, London, 1709, pp. 139–169; and *Phil Trans* 1711 and 1712), who ascribed the action to an attraction between the glass and the liquid. Dr. James Jurin (*Phil Trans* 1718, p. 739; and 1719, p. 1083) showed that the height at which the liquid is suspended depends on the section of the tube at the position of the meniscus, and is independent of the form of the lower part. Sir Isaac Newton devoted the thirty-first query in the last edition of his *Opticks* to molecular forces and gave several examples of the cohesion of liquids, such as the suspension of mercury in a barometer tube at more than double the height at which it usually stands. This arises from its adhesion to the tube, and the upper part of the mercury sustains a considerable tension, or negative pressure, without the separation of its parts.

Alexis Claude Clairaut (*Théorie de la figure de la terre*, Paris, 1808, pp. 105, 128) appears to have been the first to show the necessity of taking account of the attraction between the parts of the fluid itself in order to explain the phenomena. He did not, however, recognize the fact that the distance at which the attraction occurs is not only small but immeasurable. J.A. von Segner (*Comment Soc Reg Götting* i. p. 301) introduced the important idea of the surface tension of liquids, which he ascribed to attractive forces, the sphere of action of which is so small that it cannot be perceived.

In 1756, J.G. Leidenfrost (*De aquae communis nonnullis qualitatibus tractatus*, Duisburg) showed that a soap bubble tends to contract, so that if the tube with which it was blown is left open the bubble will diminish in size and will expel through the tube the air that it contains. In 1802, John Leslie (*Phil Mag*, 1802, vol. xiv, p. 193) gave the first modern explanation of the rise of a liquid in a tube by considering the effect of the attraction of the solid on the thin stratum of the liquid in contact with it.

In 1804, Thomas Young (essay on the "Cohesion of Fluids," *Phil Trans*, 1805, p. 65) founded the theory of capillary phenomena on the principle of surface tension. (In Chapter 21, we return to Thomas Young and determine Young's modulus for plant leaves.) Young also observed the constancy of the angle of contact of a liquid surface with a solid, and showed, from these two principles, how to deduce the phenomena of capillary action. His essay contains the solution of a great number of cases, including most of those later solved by Laplace. Young supposed the particles to act on one another with two kinds of force, one of which, the attractive force of cohesion, extends to particles at a greater distance than those to which the repulsive force is confined.

The subject was next taken up by Pierre Simon Laplace (*Mécanique céleste*, supplement to the tenth book, published in 1806). His results are in many respects identical with those of Young, but his methods of arriving at them are different because they were conducted entirely by mathematical calculations. To study the molecular constitution of bodies, it is necessary to analyze the effect of forces that are understandable only at immeasurable distances. Laplace furnished us with an example of the method of this study that has never been surpassed.

The next great step in the treatment of the subject was made by J.C.F. Gauss (*Principia generalia Theoriae Figurae Fluidorum in statu Aequilibrii*, Göttingen, 1830; or *Werke*, v. 29, Göttingen, 1867). The principle that he adopted is that of virtual velocities, now known as the principle of the conservation of energy. Instead of calculating the direction and magnitude of the resultant force on each particle arising from the action of neighboring particles, he formed a single expression that is the aggregate of all the potentials arising from the mutual action between pairs of particles. This expression has been called the force function. With its sign reversed it is now called the potential energy of the system. It consists of three parts, the first depending on the action of gravity, the second on the mutual action between the particles of the fluid, and the third on the interaction of the particles of the fluid and the particles of a solid or fluid in contact with it. The condition of equilibrium is that this expression, which can be called the potential energy, shall be a minimum.

J.A.F. Plateau (*Statique expérimentale et théorique des liquides*, 1873), who made elaborate study of the phenomena of surface tension, adopted the following method to get rid of the effects of gravity. He formed a mixture of alcohol and water of the same density as olive oil, and then introduced a quantity of oil into the mixture, which assumed the form of a sphere under the action of surface tension alone. He then, by means of rings or iron wire, discs, and other contrivances, altered the form of certain parts of the surface of the oil. The free portions of the surface then assumed new forms depending on the equilibrium of surface tension. In this way he produced a great many of the forms of equilibrium of a liquid under the action of surface tension alone, and compared them with the results of mathematical calculations. The debt science owes to Plateau is augmented by the fact that, while investigating these beautiful phenomena, he never saw them, having lost his sight in about 1840. Plateau was a Belgian, and there is a street named after him in Ghent, Belgium (Don Kirkham, personal communication, 2 February 1992).

G.L. van der Mensbrugghe (*Mém de l'Acad Roy de Belgique*, xxxvii, 1873) devised a great number of illustrations of the phenomena of surface tension, and showed their connection to the experiments of Charles Tomlinson on the figures formed by oils dropped on the clean surface of water. Athanase Dupré in his fifth, sixth, and seventh memoirs on the mechanical theory of heat (*Ann. de Chimie et de Physique*, 1866–1868) applied the principles of thermodynamics to capillary phenomena, and his son Paul's experiments were ingenious and well devised, tracing the influence of surface tension in a great number of different circumstances and deducing from independent methods the numerical value of the surface tension.

V. APPENDIX: BIOGRAPHY OF MARQUIS DE LAPLACE

Pierre Simon Laplace (1749–1827), the great French mathematician and astronomer, was born at Beaumont-en-Auge in Normandy on March 28, 1749, where his father owned a small estate. At the age of 16 he went to the Univerity of Caen, where his mathematical genius was soon recognized. In 1767 he went to Paris and was appointed professor at the École Militaire. Shortly afterward Laplace discovered that any determinant is equal to the sum of all its minors that can be formed from any selected set of its rows, each minor being multiplied by its algebraic supplement. This theorem has been described as the most important in the subject and has been named after him (Whitrow, 1971).

Laplace next turned his attention to celestial mechanics. In 1773 he took up one of the outstanding problems that until then had resisted all attempts at solution in terms of Newtonian gravitation: the problem of why Jupiter's orbit appeared to be continually shrinking while Saturn's was continually expanding (Whitrow, 1971). In a memoir published in three parts (Academy of Science, 1784–1786), Laplace showed that this phenomenon has a period of 929 years. The phenomenon arises because the mean motions of the two planets are nearly commensurable. The main object of this memoir, however, was to establish the permanence of the solar system for all time. The mutual gravitation interactions of the component bodies of the solar system were so many and varied that Newton had come to the conclusion that divine intervention was required from time to time if the system were to be preserved in anything like its present state. Despite increasing knowledge of planetary perturbations, no advance beyond this position was made until Laplace finally succeeded in showing that, because all planets revolve around the sun in the same direction, the eccentricities and inclinations of their orbits to each other will always remain small, provided they are small at a particular epoch, as in fact they are at present.

Laplace's monumental *Mécanique céleste* appeared in five volumes beween 1799 and 1825. It summarized the work of three generations of mathematicians on gravitation. In 1796 he published *Exposition du système du monde*, a semipopular book which is a model of French prose. In a celebrated memoir on the gravitational fields of spheroids, published in 1785, he introduced the potential function and the equation named after him. Laplace is also famous for his *Théorie analytique des probabilités* published in 1812 and his *Essai philosophique* on the same subject published in 1814. The former introduced important new ideas in pure mathematics, in particular the theory of Laplace transforms (Whitrow, 1971).

In 1799 when Napoleon I became first consul, he appointed Laplace minister of the interior, but dismissed him after six weeks for bringing "the spirit of infinitesimals into administration" and elevated him to the senate. Later, Laplace was made a count of the empire, and after the restoration of the Bourbons he was created a marquis. He died in Paris on March 5, 1827 (Whitrow, 1971).

REFERENCES

Aly, Saleh M., and Letey, J. (1988). Polymer and water quality effects on flocculation of montmorillonite. *Soil Sci Soc Amer J* 52; 1453–1458.

Barlett, John. (1955). Bartlett's Familiar *Quotations*. 13th and Centennial ed. Little, Brown and Company: Boston.

Buderi, R. (1996). *The Weapon that Won the War*. Simon Shuster, Little, Brown: New York.

Cattell, J. (Ed.). (1955). *American Men of Science. A Biographical Directory. Volume I. Physical Sciences*. 9th ed., p. 1164. Science Press: Lancaster, PA; and R.K. Bowker: New York, NY.

Cattell, J. (Ed.). (1961). *American Men of Science. A Biographical Directory. The Physical and Biological Sciences*. 10th ed., p. 2437. Jaques Cattell Press: Tempe, Arizona.

De Boodt, M. (1975). Use of soil conditioners around the world. In *Soil Conditioners* (Gardner, W.R., Clapp, C.E., Gardner, W.H., Moldenhauer, W.C., Mortland, M.M., and Rich, C.I., Eds.), pp. 1–12. SSSA Special Publication No. 7. Soil Science Society of America: Madison, Wisconsin.

Kirkham, D. (1961). *Lectures on Agricultural Drainage*. Institute of Land Reclamation, College of Agriculture, Alexandria University: Alexandria, Egypt. (Copy in the Iowa State University Library, Ames, Iowa.)

Kirkham, D., and Powers, W.L. (1972). *Advanced Soil Physics*. Wiley-Interscience: New York.

Linford, L.B. (1930). Soil moisture phenomena in a saturated soil. *Soil Sci* 29; 227–235.

Porter, A.W. (1971). Surface tension. *Encyclopaedia Britannica* 21; 442–450.

Schamp, N., Huylebroeck, J., and Sadones, M. (1975). Adhesion and adsorption phenomena in soil conditioning. In *Soil Conditioners* (Gardner, W.R., Clapp, C.E., Gardner, W.H.,

Moldenhauer, W.C., Mortland, M.M., and Rich, C.I., Eds.), pp. 13–23. SSSA Special Publication No. 7. Soil Science Society of America: Madison, Wisconsin.

Schaum, D. (1961). *Theory and Problems of College Physics*. Schaum Publishing: New York.

Seitz, F. 1996. How allies shared radar secrets. Book review of *The Weapon That Won the War* by Robert Buderi, Simon and Schuster, Little, Brown: 1996. *Nature* 384; 424–425.

Sharratt, B.S., Reicosky, D.C., Idso, S.B. and Baker, D.G. (1983). Relationships between leaf water potential, canopy temperature, and evapotranspiration in irrigated and nonirrigated alflafa. *Agronomy J* 75; 891–894.

Sojka, R.E., Lentz, R.D., and Westermann, D.T. (1998). Water and erosion management with multiple applications of polyacrylamide in furrow irrigation. *Soil Sci Soc Amer J* 62; 1672–1680 + cover.

Tolman, C.F. (1937). *Ground Water*. McGraw-Hill: New York.

Webster's New World Dictionary of the American Language. (1959). College ed. World Publishing: Cleveland.

Whitrow, G.J. 1971. Laplace, Pierre Simon. *Encyclopaedia Britannica* 13; 716–717.

7

Water Movement in Saturated Soil

Understanding movement of water in saturated soil is important in drainage and groundwater studies. The French hydraulic engineer, Henry Darcy (1803–1858) determined experimentally the law that governs the flow of water through saturated soil (1856), which is called Darcy's law. (See the Appendix, Section VII, for a biography of Darcy.)

I. DARCY'S LAW

To illustrate Darcy's law, let us consider Fig. 7.1, which shows water flowing through a soil column of length L and cross-sectional area, A (Kirkham and Powers, 1972, p. 47). The law can be stated as follows:

$$Q = -KA(h_2 - h_1)/(z_2 - z_1) \tag{7.1}$$

where Q is the quantity of water per second such as in cubic centimeters per second, often called the *flux*; K, centimeters per second, is the *hydraulic conductivity* (the law defines K); heads h_1 and h_2 and distances z_1 and z_2 are as shown in Fig. 7.1. The reference level here is the x, y plane. The head h_1 is the hydraulic head for all points at the bottom of the soil column, that is, at $z = z_1$, and similarly the head h_2 applies to all points at the top of the soil column, $z = z_2$. The length of the column is $z_2 - z_1 = L$. The negative sign in the Darcy equation is used so that a positive value of Q will indicate flow in the positive z direction. The positive z direction is measured from z_1 to z_2 (Kirkham and Powers, 1972, p. 46-47).

In the Darcy law equation the quantity $(h_2 - h_1)/(z_2 - z_1)$ is called the *hydraulic gradient i*; the ratio Q/A is called the *flux per unit cross section* or

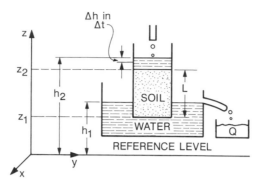

FIG. 7.1 Illustration of Darcy's law. (From *Advanced Soil Physics* by Kirkham, D., and Powers, W.L., p. 47, ©1972, John Wiley & Sons, Inc.: New York. This material is used by permission of John Wiley & Sons, Inc. and William L. Powers.)

flux density (cm³/sec)/cm². The ratio Q/A is also called the *Darcy velocity v* or, very often, just the *velocity v*. Therefore, Darcy's law may be written as $v = -Ki$. The *actual velocity* of the water in the soil is much greater than the Darcy velocity. The actual velocity is on the average v/f where f is the *porosity* (Kirkham and Powers, 1972, p. 47). The porosity is the volume of pores in a soil sample divided by the bulk volume of the sample (Soil Science Society of America, 1997). The pores can be filled with air and/or water. (Because Darcy's law is for saturated soil, the pores are filled with water when it applies.) The percent porosity in the soil can be determined from the following equation (Millar et al., 1965, p. 54):

$$\% \text{ porosity} = [1 - (\text{bulk density/particle density})] \times 100 \qquad (7.2)$$

The Darcy velocity v means more than flux per unit area Q/A. In Fig. 7.1, suppose that the supply of water shown dripping into the soil column is abruptly cut off during a short time interval Δt during which h_2 decreases by Δh. We let Δq be the volume of water flowing downward through the soil in Δt. Because Q is the flow per second, we may write Δq as $\Delta q = Q\Delta t$, and we also have by continuity of flow $\Delta q = A\Delta h$. Therefore, $Q\Delta t = A\Delta h$, and $Q/A = \Delta h/\Delta t$. Physically, $\Delta h/\Delta t$ is a velocity; therefore, so is $v = Q/A$. Thus the Darcy velocity v represents the rate $\Delta h/\Delta t$ approaches dh/dt of fall of surface water in Fig. 7.1. If the hydraulic gradient is unity (pressure potential is the same at the top and bottom of the soil column), then $v = -K$. Thus, it is determined that K is numerically equal to the rate of fall of a thin layer of ponded water into the soil, under only the force of the earth's gravitational pull. We also see that K is the velocity under a unit hydraulic gradient (Kirkham and Powers, 1972, p. 48).

Flow in a vertical soil column has been used to derive and illustrate Darcy's law. However, the law and principles developed in the preceding paragraphs apply for flow of water in any direction in the soil.

II. HYDRAULIC CONDUCTIVITY

The hydraulic conductivity should not be confused with the *intrinsic permeability*, sometimes just called the *permeability*, of the flow medium. The intrinsic permeability, symbolized by k by M. Muskat in his classic treatise (1946) (Muskat was a petroleum engineer in the United States well known for his studies in the 1930s and 1940s of fluid flow through porous media), is equal to $K\eta/\rho g$, where K is the Darcy hydraulic conductivity, η is the fluid viscosity, ρ is the fluid density, and g is the acceleration due to gravity. Dimensionally, k is an area (L^2). The units of K are m/day, which is the same as (m^3/m^2)/day. That is, K may be interpreted as the m^3 of water seeping through a m^2 of soil per day under a unit hydraulic gradient (Kirkham and Powers, 1972, p. 48–49).

Hydraulic conductivity in natural field soil is governed by factors such as cracks, root holes, worm holes, and stability of soil crumbs. Texture, that is, the percent of the primary particles of sand, silt, and clay, usually has a minor effect on hydraulic conductivity, except for disturbed soil materials. The hydraulic conductivity of natural soils in place varies from about 30 m/day for a silty clay loam to 0.05 m/day for a clay (Kirkham, 1961a, p. 46; Kirkham, 1961b). The hydraulic conductivity for disturbed soil materials varies from about 600 m/day for gravel to 0.02 m/day for silt and clay. The value of K can be made higher or lower by soil management. Roots of crops after decay increase K; compaction of soil by animals or machinery decreases K, at least in the surface soil.

Ordinarily one considers K in $v = -Ki$ to be a constant under saturated flow. It is a constant if 1) the physical condition of the soil and of the water does not change in space or time as the water moves through the soil (e.g., the soil is *isotropic*, that is hydraulic conductivity is the same regardless of the direction of measurement) and if 2) the type of flow is laminar, that is, not turbulent. In laminar flow, two particles of water seeping through the soil will describe paths (streamlines) that never cross each other. In turbulent flow, eddies and whirls develop. The possibility of turbulent flow is considered in soil only if the soil is a coarse sand or gravel, and then only if the hydraulic gradients are large (larger than those found in most problems of interest to agricultural soil scientists).

III. LAPLACE'S EQUATION

To solve groundwater seepage and drainage problems, it is desirable to have a general differential equation (Kirkham and Powers, 1972, p. 49), and Laplace's equation, which is a familiar equation occurring in nearly all branches of applied mathematics, applies. Laplace's equation is derived from Darcy's law and the *equation of continuity*. The equation of continuity states mathematically that mass can neither be created nor destroyed. We can state the equation of continuity in words, as follows: For a volume element x times y times z, the change in velocity of water in the x direction plus change in velocity of water in the y direction plus change in velocity of water in the z direction is equal to the total change in water content, θ, per unit time of the volume element under consideration. That is, inflow of water in the element minus outflow of water is equal to the water accumulated. Let us imagine a rectangular x, y, z system of coordinates that is established in a homogeneous porous medium of constant hydraulic conductivity, and let h be the hydraulic head referred to an arbitrary reference level for a point (x, y, z) and let time be t and v_x, v_y, and v_z be the velocity of water flowing in the x, y, and z directions, respectively; then, with θ being the volume of water per unit volume of bulk soil, from the equation of continuity,

$$-[(\partial v_x/\partial x) + (\partial v_y/\partial y) + (\partial v_z/\partial z) = \partial\theta/\partial t \qquad (7.3)$$

and from Darcy's law, one may, for incompressible steady-state flow in a porous medium where K is constant, derive the expression (Kirkham and Powers, 1972, p. 52)

$$(\partial^2 h/\partial x^2) + (\partial^2 h/\partial y^2) + (\partial^2 h/\partial z^2) = 0, \qquad (7.4)$$

as the expression governing groundwater flow. The equation is abbreviated $\nabla^2 h = 0$.

Charles S. Slichter (1899), a mathematician at the University of Wisconsin, was the first to show in 1899 that Laplace's equation applies to the motion of groundwater (Kirkham and Powers, 1972, p. 52). Many mathematical solutions for groundwater flow using Laplace's equation have been done by Don Kirkham of Iowa State University.

IV. ELLIPSE EQUATION

In addition to Darcy's equation and Laplace's equation, another important equation for saturated flow is called the Colding equation after the Danish engineer A. Colding, who published it in 1872 (van der Ploeg et al., 1997).

It is used to determine drain spacings. The equation is also called the ellipse equation, because it describes an ellipse. Therefore, before we look at the Colding equation, let us study an ellipse.

The locus of a point *P* that moves in a plane so that the sum of its distances from two fixed points in the plane is constant is called an *ellipse* (Ayers, 1958, p. 322). The fixed points *F* and *F'* are called the *foci* and their midpoint *C* is called the *center* of the ellipse (Fig. 7.2). The line *FF'* joining the foci intersects the ellipse in the points *V* and *V'*, called the *vertices*. The segment *V'V* intercepted on the line *FF'* by the ellipse is called its *major axis*; the segment *B'B* intercepted on the line through *C* perpendicular to *F'F* is called its *minor axis*.

A line segment in which the extremities are any two points on the ellipse is called a *chord*. A chord that passes through a focus is called a *focal chord*; a focal chord perpendicular to the major axis is called a *latus rectum*.

The equation of an ellipse assumes its simplest (*reduced*) form when its center is at the origin and its major axis lies along one of the coordinate axes. When the center is at the origin and the major axis lies along the *x*-axis, the equation of the ellipse is (Fig. 7.3)

$$(x^2/a^2) + (y^2/b^2) = 1. \tag{7.5}$$

Figure 7.3 is an oblate ellipse. *Oblate* comes from the Latin *oblatus*, which means "offered" or "thrust forward," and means "being thrust forward at the equator." In geometry, oblate means flattened at the poles.

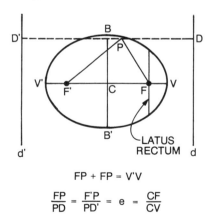

$$FP + FP = V'V$$

$$\frac{FP}{PD} = \frac{F'P}{PD'} = e = \frac{CF}{CV}$$

FIG. 7.2   The ellipse. It is the locus of a point P that moves in a plane so that the sum of its distances from two fixed points in the plane, F and F', is constant. (From Ayers, F., Jr., *Theory and Problems of First Year College Mathematics*, p. 322, ©1958, Schaum Publishing Co.: New York. This material is reproduced with permission of The McGraw-Hill Companies.)

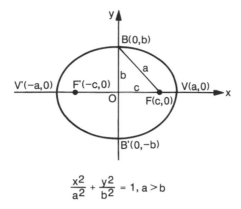

$$\frac{x^2}{a^2} + \frac{y^2}{b^2} = 1, a > b$$

FIG. 7.3   An oblate ellipse. (From Ayers, F., Jr., *Theory and Problems of First Year College Mathematics*, p. 323, ©1958, Schaum Publishing Co.: New York. This material reproduced with permission of The McGraw-Hill Companies.)

When the center is at the origin and the major axis lies along the *y*-axis, the equation of the ellipse is (Fig. 7.4)

$$(x^2/b^2) + (y^2/a^2) = 1. \qquad (7.6)$$

Figure 7.4 is a prolate ellipse. *Prolate* comes from the Latin *prolatus*, which is the past participle of *proferre*, "to bring forward." Prolate means "extended or elongated at the poles."

A circle is a special form of an ellipse in which the semimajor and semiminor axes are equal in length. The equation of a circle is

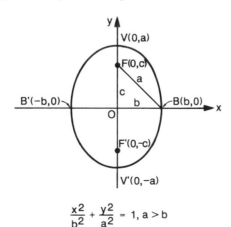

$$\frac{x^2}{b^2} + \frac{y^2}{a^2} = 1, a > b$$

FIG. 7.4   A prolate ellipse. (From Ayers, F., Jr., *Theory and Problems of First Year College Mathematics*, p. 323, ©1958, Schaum Publishing Co.: New York. This material reproduced with permission of The McGraw-Hill Companies.)

$$x^2 + y^2 = r^2, \tag{7.7}$$

where r is the radius of the circle and the circle has its center at the origin of the x, y coordinate system.

If we are dealing in three dimensions, we have an *ellipsoid*. The locus of the equation

$$x^2/a^2 + y^2/b^2 + z^2/c^2 = 1 \tag{7.8}$$

is called an ellipsoid (Fig. 7.5). If at least two of a, b, c are equal, the locus is called an ellipsoid of revolution, and if $a = b = c$, the locus is a sphere (Ayers, 1958, p. 387).

The need for soil drainage is widespread around the world, not only in the wet soils of northern Europe and in states of the United States that are wet in the spring (e.g., Iowa), but also in irrigated regions. It is generally accepted that the Danish engineer Colding was the first to derive a drain-spacing equation based on modern soil water flow concepts. For parallel, equally spaced tile (tube) drains resting on an impermeable barrier, and for steady-state flow conditions, Colding (1872) derived the following expression (van der Ploeg et al., 1997, 1999):

$$L^2 = [(4K)/R]b^2 \tag{7.9}$$

where L is the drain distance, K is the soil hydraulic conductivity, R is the constant rate of precipitation, and b is the maximum height of the water table above the drain level, midway between the drains (Fig. 7.6). Equation 7.9 describes an ellipse, with the drain distance L being the major axis and

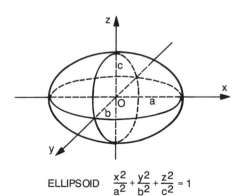

ELLIPSOID $\quad \dfrac{x^2}{a^2} + \dfrac{y^2}{b^2} + \dfrac{z^2}{c^2} = 1$

FIG. 7.5 The ellipsoid. (From Ayers, F., Jr., *Theory and Problems of First Year College Mathematics*, p. 387, ©1958, Schaum Publishing Co.: New York. This material is reproduced with permission of The McGraw-Hill Companies.)

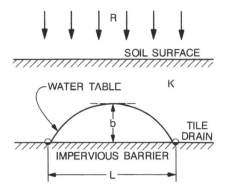

FIG. 7.6 Schematic representation of the subsurface drain problem, as considered by Colding. (From van der Ploeg et al., The Colding equation for soil drainage: Its origin, evolution, and use, ©1999, Soil Science Society of America: Madison, Wisconsin. Reprinted by permission of the Soil Science Society of America.)

the maximum water table height b above the drain level being the semiminor axis (Kirkham and Powers, 1972, pp. 90, 92).

It is not known if Colding was familiar with the work of Darcy (1856), but apparently he was not, because he does not mention him in his work. Darcy's work on hydraulic conductivity did not receive much attention until the second edition of the book by Dupuit (1863), and even then it took time for people to become familiar with it. Nevertheless, Colding was using Darcy-like theory to derive his equation.

The U.S. Bureau of Reclamation uses the Colding equation, as modified by Hooghoudt (1940) for design purposes (van der Ploeg et al., 1999). Instead of equally spaced tile drains, Hooghoudt (1937, 1940) considered equally spaced drainage ditches overlying an impervious layer (Fig. 7.7). In the Imperial Valley in California, the Colding equation, as modified by Aronovici and Donnan (1946) is used (van der Ploeg et al., 1999). Aronovici and Donnan, apparently unaware of the work of Hooghoudt, also developed a modified Colding equation almost identical to the Hooghoudt (1940) equation. It is important to recognize that some of today's most common drainage design practices are based on the Colding (ellipse) equation.

The ellipse is an important geometric form because of its widespread application in soil-water relations and other aspects of nature. Apollonian curves—that is, ellipses, parabolas, and hyperbolas—all have amazing relationships hidden in them (Anvar Kacimov, personal communication, December 9, 1999). (See the Appendix, Section VI, for a biography of Apollonius, a Greek geometer.) Johannes Kepler (1571–1639, German

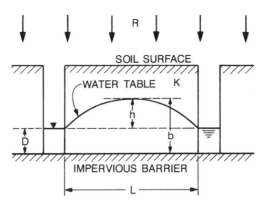

FIG. 7.7   Geometric representation of a homogeneous soil, underlain by an impervious barrier, that is drained by parallel, equally spaced ditches, where the ditches reach the impervious barrier, as considered by Hooghoudt. (From van der Ploeg et al., The Colding equation for soil drainage: Its origin, evolution, and use, ©1999, Soil Science Society of America: Madison, Wisconsin. Reprinted by permission of the Soil Science Society of America.)

astronomer and mathematician) also came to his celestial mechanics formula from the geometric side. He selected first an ellipse and then applied it to orbits (Goodstein and Goodstein, 1996). To understand the interception of solar radiation by plant leaves (Kirkham, 1986), we need to study Kepler's laws of planetary motion. He saw that the planets (earth included) orbit the sun in elliptical paths.

Other examples of Apollonian curves used in soil-water studies include the work by Kacimov (2000), who used the special case of an ellipsoid, the hemisphere, to study three-dimensional groundwater flow to a lake, and the work by Kirkham and Clothier (1994), who used the ellipsoidal equation (Eq. 7.8) to describe the shape of the wet front as it expands under a surface disc that is infiltrating water into the soil.

V. LINEAR FLOW LAWS

Darcy's law is a linear flow law. It is linear because the v, the Darcy velocity, of $v = -Ki$, varies linearly with the hydraulic gradient i (Kirkham and Powers, 1972, p. 74). Ohm's law is one of the most common linear flow laws and is used in problems concerning the flow of electricity. In Ohm's law, the current transported is linearly related to the difference in the driving potential across the system. We shall return to Ohm's law when we study electrical analogues (Chapter 20). Gauss's law, used in studying electrostatic fields, is another linear flow law (Kirkham, 1961a, p. 104).

Poiseuille's law for flow of a liquid through a capillary tube is not a linear flow law, because an exponent greater than one occurs in it. Poiseuille found that the volume of fluid moving in unit time along a cylinder is proportional to the fourth power of its radius. We will use Poiseuille's law to study flow of water in the vessel members in the xylem tissue (Chapter 14).

Table 7.1 shows that linear flow laws are similar. For example, Darcy's law is similar to Ohm's law, Fick's law, and Fourier's law. These laws are commonly used in soil physics; Darcy's law is used in studies of water flow, Fick's law in studies of gaseous flow (Kirkham, 1994), and Fourier's law in studies of heat flow. It is important to know the similarities, because, for water flow (Darcy's law) problems for which solutions are desired, there may exist analogous electrical flow (Ohm's law) or heat flow (Fourier's law) problems that have been already solved. These solutions can be used for writing down directly the solution of the desired water flow problem (Kirkham, 1961a, p. 104).

Linear flow phenomena are involved in other studies, as well, as listed by Moon and Spencer (1961):

1. **High-voltage engineering:** Design of high-tension transformer bushings, transmission-line insultators, electrostatic voltmeters, Van de Graaf generators
2. **Magnetostatics:** Calculation of generators, motors, lifting magnets, solenoids, synchrotrons
3. **Heat conduction:** Determination of temperature distributions in electric machinery, heating devices, cable ducts, refrigerators
4. **Fluid flow:** Calculation of flow about airfoils and other obstructions, seepage of fluids through sand
5. **Electrodynamics:** Determination of resistance of irregular-shaped conductors, electrical prospecting
6. **Electrostatics:** Design of vacuum tubes, electron microscopes, cathode-ray oscilloscopes, television tubes
7. **Elasticity:** Vibration engineering, structural engineering
8. **Diffusion:** Calculation of the heating and cooling of ingots, the annealing of glass, and the diffusion of fluids
9. **Acoustic waves:** Design of loud speakers and microphones
10. **Electromagnetic waves:** Calculation of wave guides and antennas

To the list of Moon and Spencer (1961) can be added an eleventh item:

11. **Mass flow of gas under a small pressure gradient**

TABLE 7.1 Linear flow laws encountered in soil-plant-water relationships. No exponents (other than one) appear in a linear flow law.

LINEAR FLOW LAWS

LAW	QUANTITY TRANSPORTED PER SEC ACROSS FLOW REGION	TRANSPORT COEFFICIENT OF FLOW REGION	DIFFERENCE IN DRIVING POTENTIAL ACROSS SYSTEM	FORM FACTOR F FOR SYSTEM (e.g., "rectangular box")[†]
OHM $I = \dfrac{\sigma \Delta V A}{L} = \dfrac{V}{R}$[‡]	I = AMPERES OR COULOMBS PER SEC	σ = SPECIFIC ELECTRICAL CONDUCTIVITY (S/cm)	ΔV = DIFFERENCE IN ELECTRICAL POTENTIAL BETWEEN INPUT AND OUTPUT (volts)	$F = \dfrac{A}{L}$
DARCY $Q = \dfrac{k \Delta h A}{L}$	$Q = cm^3/sec$	k = HYDRAULIC CONDUCTIVITY (cm/sec)	Δh = DIFFERENCE IN HEAD OF WATER COLUMN BETWEEN INPUT AND OUTPUT (cm)	$F = \dfrac{A}{L}$
FICK $Q = \dfrac{D \Delta C A}{L}$	Q = grams/sec	D = DIFFUSION COEFFICIENT (cm^2/sec)	ΔC = DIFFERENCE IN CONCENTRATION OF GAS AT INPUT AND OUTPUT $(grams/cm^3)$	$F = \dfrac{A}{L}$
FOURIER $Q = \dfrac{K \Delta T A}{L}$	Q = cal/sec	K = THERMAL CONDUCTIVITY (cal per sec per cm^2 for a thickness of 1cm and temp. diff. of 1°C)	ΔT = DIFFERENCE IN TEMPERATURE BETWEEN INPUT AND OUTPUT (°C)	$F = \dfrac{A}{L}$

[†]A = cross sectional area of the system (box) perpendicular to the direction of flow; L = length of system (box) through which the flow occurs. The form factor is for two-dimensional flow problems; it is the same for equal geometries of the flow region.

[‡]R = resistance in ohms, Ω. (1 S = $1/\Omega$ = mho)

VI. APPENDIX: BIOGRAPHY OF APOLLONIUS OF PERGA

Apollonius of Perga (Pergaeus), a Greek geometer of the Alexandrian school, was probably born about 25 years later than Archimedes (i.e., about 261 B.C.). He flourished in the reigns of Ptolemy Euergetes and Ptolemy Philopator (247–205 B.C.) (Heath and Neugebauer, 1971). His treatise on *Conics* gained him the title of the Great Geometer, and, through this work, his fame has been transmitted to modern times. Most of his other treatises were lost, although their titles and a general indication of their contents were passed on by later writers, especially Pappus. After Apollonius wrote the *Conics* in eight books in a first edition, he brought out a second edition, considerably revised with regard to books I and II.

The degree of originality of the *Conics* can best be judged from Apollonius's own prefaces. He made the fullest use of his predecessors' works, such as Euclid's four books on conics, which is clear from his allusions to Euclid, Conon, and Nicoteles. Books I through IV form an "elementary introduction" (i.e., contain the essential principles) and the rest of the books are specialized investigations. Apollonius introduced the names *parabola, ellipse*, and *hyperbola*. Books V through VII are highly original. Apollonius's genius takes its highest form in book V, where he treats normals as minimum and maximum straight lines drawn from given points to the curve (independently of tangent properties), discusses how many normals can be drawn from particular points, finds their feet by construction, and gives propositions determining the center of curvature at any point (Heath and Neugebauer, 1971).

Six other treatises by Apollonius (each in two books) were concerned with cutting off a ratio, cutting off an area, determinate sections, tangencies, inclinations, and plane loci. An Arabic version of the first treatise was found toward the end of the seventeenth century in the Bodleian library by Edward Bernard, who began a translation of it. (The Bodleian library is a famous library at Oxford University in England named after Sir Thomas Bodley, 1545–1613, an English diplomat and man of letters and founder of the library.) Edmund Halley (1656–1742), the English astronomer, finished the translation and published it with a restoration of the second treatise (1706).

Other works by Apollonius referred to by ancient writers include 1) *On the Burning-Mirror*, where the focal properties of the parabola probably were discussed; 2) *On the Cylindrical Helix*; 3) a comparison of the dodecahedron and the icosahedron inscribed in the same sphere; 4) a work which included Apollonius's criticisms and suggestions for the improvement of Euclid's *Elements*; 5) a work in which he showed how to find closer limits for the value of π than the 3 1/7 and 3 10/71 of Archimedes; 6) an arithmetic work on a system of expressing large numbers and showing how to multiply such large numbers; and 7) extensions of the theory of irrationals expounded in Euclid.

VII. APPENDIX: BIOGRAPHY OF HENRY DARCY

Henry Philibert Gaspard Darcy (1803–1858) is best known for his scientific work on pipe flow (Howland, 1971). He lived in Dijon for most of his life, where he was inspector general of bridges and highways. His father, a town functionary, died when he was 14 years old (Philip, 1995). His determined mother named him the English "Henry" instead of the French "Henri," and ensured that both Henry and his brother Hugues received the best education possible. Henry won a scholarship to the Dijon Polytechnic and in 1826 graduated brilliantly as a civil engineer.

Working as an engineer, he devoted his life to providing the town of Dijon with pure water. The waters of cities then, including Dijon, were often inadequate, always in short supply, and dirty. Dijon had at its disposition only wells plus the water of the Ouche, and well waters were not protected from contamination. The town of Dijon was crossed by the ancient bed of a small stream, the Suzon, which was uncovered over almost all its length, with no part paved, and served over a length of 1300 meters as the main sewer for wastes of every kind. It was never cleaned, and during hot weather, the town was poisoned by pestilential odors.

To clean up the water, Darcy substituted the method that became standard (Philip, 1995). In 1833, Darcy, on his own initiative, presented his plan to the municipal authorities. The Municipal Council adopted his recommendations, and the General Council of Bridges and Highways (Ponts et Chaussées) approved all parts of the proposed plan. He developed a network of underground conduits with underground reservoirs. On September 6, 1840, without any errors or mishaps, the beneficial waters reached the reservoir of the Porte Guillaume. By 1844, the whole network of underground conduits had been completed (Philip, 1995). Consequently, Dijon possessed from 1840 the benefits that Paris did not discover until 20 years later, and enjoyed an abundance of water. Other towns asked for Darcy's assistance, such as Brussels, which officially asked for his help in 1851 and 1852 and adopted the plan that he provided.

Darcy had given his native town the better part of his life. For this work of 12 to 15 years, he wished to receive no remuneration. He would not agree even to be reimbursed for his expenses. He accepted only a gold medal that commemorated his work.

In 1848, the revolution overthrew King Louis-Philippe and brought in the radical and short-lived Second Republic. Despite the facts that Darcy was apolitical and that he had given generously of his own money to set up workers' cooperatives, the Second Republic saw him as a dangerous and reactionary collaborator with the ancient regime. Darcy was stripped of his offices and banished from Dijon. In 1852, the Second Republic was succeeded by the Second Empire of Emperor Napoleon III, and Darcy was politically rehabilitated.

In 1854, Darcy was 51, but in poor health. Ever since his days as a young engineer, he had been prey to nervous troubles and to attacks producing symptoms of meningitis. Time and overwork gradually made these attacks more acute. He suffered a bad period in 1845, while he was directing the works at Blaisy. Later, in Paris, he lost consciousness during a conference, and in 1853 he fell down in the open street. He took leave for several months, but, as he continued to suffer intolerably, he despaired and asked to be released from his responsibilities. Nevertheless, unable to stay inactive, he pursued his hydraulic experiments, and it was during these last years that he was able, thanks to financial help from the Ministry of Public Works, to carry out the work that he wrote up and published in 1856.

In 1857, the Académie des Sciences wished to elect him to the vacancy left by the mathematician Cauchy who had just died. (Baron Augustin Louis Cauchy, 1789–1857, was one of the greatest of modern French mathematicians.) Darcy was elected without discussion, but on January 2, 1858, he succumbed to pleurisy aggravated by angina and died in Paris. Dijon gave him a public funeral appropriate to his great labors. The whole population went to receive his remains at the railway station, all the functionaries in uniform, armed soldiers lining the streets, the workers of cooperatives founded on his initiative carrying the coffin, and the bishop officiating. Sadly, 135 years after Darcy's death, when the whole town mourned it, nobody in Dijon knew who he was, even though his name appears in many places in Dijon (Philip, 1995). No biographical information that I found stated whether Darcy married or had children.

There is in existence a collection of letters from Darcy to Henri Émile Bazin (1829–1917). Bazin was 26 years younger than Darcy, a hydraulic engineer working in Dijon whose researches on channel and pipe flow are well known. Bazin, acting as Darcy's assistant, was trained to be a careful and assiduous experimenter. He carried on Darcy's original program of tests on open-channel resistance. His studies also extended to wave propagation, to flow over weirs, and to the contraction of the liquid vein coming from an orifice. Bazin was elected to the French Academy of Sciences in 1865. He died on February 7, 1917, at Dijon (Howland, 1971).

REFERENCES

Aronovici, V.S., and Donnan, W.W. (1946). Soil permeability as a criterion for drainage-design. *Trans Amer Geophys Union* 27; 95–101.

Ayers, F., Jr. (1958). *Schaum's Outline of Theory and Problems of First Year College Mathematics*. Schaum Publishing: New York.

Colding, A. (1872). On the laws of water movement in soil (in Danish, with a French summary). *Det kgl. Danske Videnskabernes Selskabs Skrifter*, 5te Raekke, naturvidenskabelig og mathematisk Afdeling, 9 B VIII, 563–621 (Copenhagen).

Darcy, H. (1856). *Les Fontaines Publiques de la Ville de Dijon*. (The public fountains of the city of Dijon). Victor Dalmont (Ed.), Libraire des Corps Impériaux des Ponts et Chaussées et des Mines: Quai des Augustins, 49, Paris. (In French.)

Dupuit, J. (1863). *Theoretical and Practical Studies about the Movement of Water in Open Channels and through Permeable Terrains*. (In French.) 2nd ed. Dunod: Paris.

Goodstein, D.L., and Goodstein, J.R. (1996). *Feynman's Lost Lecture: The Motion of Planets Around the Sun*. Jonathan Cape: London; Published in the US by Norton.

Heath, T.L., and Neugebauer, O.E. (1971). Apollonius. *Encyclopaedia Britannica* 2; 122–123.

Howland, W.E. (1971). Bazin, Henri Émile (1829–1917). *Encyclopaedia Britannica* 3; 314.

Hooghoudt, S.B. (1937). Contributions to the knowledge of some physical soil properties. no. 6. Determination of the conductivity of soils of the second kind. (In Dutch.) *Verslagen van Landbouwkundige Onderzoekingen*, no. 43(13) B, Dep. van Economische Zaken, Directie van den Landouw, Algemeene Landsdrukkerij: The Hague, The Netherlands.

Hooghoudt, S.B. (1940). Contributions to the knowledge of some physical soil properteis. no. 7. General discussion of the problem of drainage and infiltration by means of parallel drains, trenches, ditches and canals. (In Dutch). *Verslagen van Landbouwkundige Onderzoekingen*, no. 46 (14) B, Dep. van Landbouw en Visscherij, Directie van den Landbouw, Algemeene Landsdrukkerij: The Hague, The Netherlands.

Kacimov, A.R. (2000). Three-dimensional groundwater flow to a lake: An explicit analytical solution. *J Hydrol* 240; 80–89. (See his Fig. 1.)

Kirkham, D. (1961a). *Lectures on Agricultural Drainage*. Institute of Land Reclamation, College of Agriculture, Alexandria University: Alexandria, Egypt. (Copy in the Iowa State University Library, Ames, Iowa.)

Kirkham, D. (1961b). Soil physical properties. In *Agricultural Engineer's Handbook* (Richey, C.B., Jacobson, P., and Hall, C.W., Eds.), pp. 793–801. McGraw Hill: New York.

Kirkham, D., and Powers, W.L. (1972). *Advanced Soil Physics*. Wiley: New York.

Kirkham, M.B. (1986). Theoretical consideration of direct-beam solar radiation on plant leaves. *Int Agrophys* 2; 53–58.

Kirkham, M.B. (1994). Streamlines for diffusive flow in vertical and surface tillage: A model study. *Soil Sci Soc Amer J* 58; 85–93.

Kirkham, M.B., and Clothier, B.E. (1994). Ellipsoidal description of water flow into soil from a surface disc. *Trans Int Congr Soil Sci* 2b; 38–39.

Millar, C.E., Turk, L.M., and Foth, H.D. (1965). *Fundamentals of Soil Science*. 4th ed. Wiley: New York.

Moon, P., and Spencer, D.E. (1961). *Field Theory for Engineers*. D. Van Nostrand: Princeton, New Jersey. See pp. 2–3.

Muskat, M. (1946). *The Flow of Homogeneous Fluids through Porous Media*. J.W. Edwards: Ann Arbor, Michigan.

Philip, J.R. (1995). Desperately seeking Darcy in Dijon. *Soil Sci Soc Amer J* 59, 319–324.

Slichter, C.S. (1899). Theoretical investigations of the motion of groundwater. *US Geol Surv Annu Rep* 19; 295–384.

Soil Science Society of America. (1997). *Glossary of Soil Science Terms*. Soil Science Society of America: Madison, Wisconsin.

van der Ploeg, R.R., Marquardt, M., and Kirkham, D. (1997). On the history of the ellipse equation for soil drainage. *Soil Sci Soc Amer J* 61; 1604–1606.

van der Ploeg, R.R., Kirkham, M.B., and Marquardt, M. (1999). The Colding equation for soil drainage: Its origin, evolution, and use. *Soil Sci Soc Amer J* 63; 33–39.

Field Capacity, Wilting Point, Available Water, and the Non-Limiting Water Range

8

The amount of water available for plant uptake has been related to a soil's *water budget*. The three terms associated with the water budget are *field capacity* (FC), *wilting point* (WP), and *available water* (AW).

I. FIELD CAPACITY

To define field capacity we consider the following. In many soils, after a rain or irrigation, the soil immediately starts draining to the deeper depths. After one or two days the water content in the soil will reach, with time, for many soils, a nearly constant value for a particular depth in question. This somewhat arbitrary value of water content, expressed as a percent, is called the field capacity.

It is not known who first used the term *field capacity*. The term was not used by Briggs and Shantz, who developed the concept of the wilting point (see next section). Briggs (see the Appendix, Section E, for his biography) defined the "moisture equivalent," which was the amount of water held against centrifugation of soil at $3000 \times g$, where g is the acceleration due to gravity (Landa and Nimmo, 2003). The term is no longer accepted (Soil Science Society of America, 1997), but it was a precursor to the idea of field capacity.

Early researchers recognized that there was a point at which water moved slowly after a rain or irrigation (Taylor and Ashcroft, 1972, p. 299). They wanted to assign a value to this point, and therefore, the concept of field capacity developed. They recognized it as the amount of water that a well-drained soil holds against gravitational forces and when downward drainage is markedly decreased. They felt it was a true equilibrium and they felt it was the upper limit of available water for plants.

However, as time progressed, soil scientists realized that field capacity was an imprecise term. They saw that it was not a unique value, because equilibrium is never reached. Soil water is dynamic; removal of water occurs due to drainage, evaporation, and transpiration and addition of water occurs with dewdrops, rainfall, and irrigation (Taylor and Ashcroft, 1972, p. 300). The movement of water downward does not cease, but continues at a reduced rate for a long time. There is no real value for field capacity. Therefore, a range of values (soil water contents) are associated with field capacity (Fig. 8.1). Many factors influence field capacity, as follows (Hillel, 1971, p. 162–165).

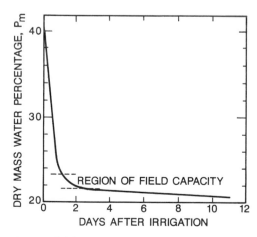

FIG. 8.1 Diagram showing field capacity as a range of values of soil water contents. (From Taylor, S.A., and Ashcroft, G.L., *Physical Edaphology: The Physics of Irrigated and Nonirrigated Soils*, p. 301, ©1972 by W.H. Freeman and Company. Used with permission.)

1. **Previous soil water history:** A wetting soil and a drying soil hold different amounts of water. A soil that is saturated and then dries has a higher field capacity than a soil that is being wetted. This is due to hysteresis (see Chapter 6).
2. **Soil texture and structure:** These change with soil horizon and influence water retention. Clayey soils retain more water, and longer, than sandy soils. The finer the texture is, the higher is the apparent field capacity, the slower is its attainment, and the less distinct is its value (Hillel, 1971, p. 164).
3. **Type of clay:** The higher the content of montmorillonite is, the greater is the content of water.
4. **Organic matter:** Soil organic matter helps retain water.
5. **Temperature:** The temperature influences the amount of water held, particularly if the soil has been previously wetted. The amount of water retained at field capacity decreases as the soil temperature increases (Kramer, 1983, p. 71). This results in increased runoff from a watershed as soil warms.
6. **Water table:** The term "field capacity" is of doubtful value in soils with a water table near the surface. The term applies to free-draining soils.
7. **Depth of wetting:** Usually, the wetter the profile is at the outset, the greater is the depth of wetting during infiltration, the slower is the rate of redistribution, and the greater is the apparent field capacity.
8. **Presence of impeding layers (e.g., clay, sand, gravel):** The layers inhibit redistribution and increase the apparent field capacity. Again, the term "field capacity" is of questionable value for soils having layers of widely differing hydraulic conductivities.
9. **Evapotranspiration:** The rate and pattern of extraction of water by plant roots from soil can affect the gradients and flow directions in the profile and modify redistribution (Hillel, 1971, p. 165).

People have suggested abandoning the concept of field capacity, because it has caused misleading conclusions. For example, if it is assumed that no drainage occurs, when in fact it is, drainage is included in consumptive use by plants. This leads to consumptive use values that are too large.

Until the 1984 edition of the *Glossary of Soil Science Terms* (Soil Science Society of America, various years), the term "field capacity" was labeled "obsolete." Current glossaries no longer call it obsolete, and the term is widely used in the literature. One is often asked to provide the field capacity for a soil when publishing a paper. The term is useful for qualitative, not quantitative, understanding of water in the soil.

Field capacity is not the upper limit of available water to plants because all water that is not held tightly by soil can be used by plants while it is in contact with roots, even if water is rushing by during rapid drainage. What limits uptake is soil aeration, and, as we shall see (Chapter 10), the air-filled pore space must be at least 10% by volume for most roots to survive (Wesseling and van Wijk, 1957).

Note that field capacity does not apply to pots in a greenhouse. Field capacity refers only to field conditions. Greenhouse pots do not have underlying soil that pulls water down deep into the soil profile by capillarity. However, one can talk of "pot capacity," which is the amount of water remaining in a pot after an irrigation and visible drainage has ceased.

One should always try to measure field capacity in the field for each soil. The matric potential associated with field capacity can be as high as −0.0005 MPa in a highly stratified soil or as low as −0.06 MPa in a deep, dryland soil (Baver et al., 1972, p. 382). If one cannot measure field capacity in the field, it is often estimated to be the soil water content at a soil matric potential of −0.03 MPa (one-third bar).

II. WILTING POINT

The wilting point, also called the *permanent wilting point*, may be defined as the amount of water per unit weight or per unit soil bulk volume in the soil, expressed in percent, that is held so tightly by the soil matrix that roots cannot absorb this water and a plant will wilt.

Unlike field capacity, the term *wilting point* is associated with known scientists, Briggs and Shantz (1912). They defined the "wilting coefficient" (wilting point) as "the moisture content of the soil (expressed as a percentage of the dry weight) at the time when the leaves of the plant growing in that soil first undergo a permanent reduction in their moisture content as the result of a deficiency in the soil-moisture supply" (Briggs and Shantz, 1912, p. 9). As with field capacity, early workers felt that wilting point was a precise value.

The method of determining permanent wilting point is as follows (Taylor and Ashcroft, 1972, p. 303). An indicator plant, usually sunflower (*Helianthus annuus*), is put in 500 grams of soil in a metal can. The plant grows and is given adequate moisture until the third pair of true leaves is formed. Then the top of the can is sealed with wax. The sunflower grows in a greenhouse or outdoors until it wilts. Then it is transferred to a dark, humid chamber for recovery. If the plant recovers, it is put out again. The procedure is repeated until the plant remains wilted overnight (24 hours) in the humid chamber. The soil water content then is at the permanent wilting point.

For plants that have leaves that do not wilt, like cacti, Briggs and Shantz (1912) developed special procedures to determine the wilting point. For example, they put a plant with water-storage tissue in a glass container with soil. They glued a knitting needle to one side of the glass. They put the glass with knitting needle in a horizontal position by propping it between two other containers sitting on a table. The needle was free to move up and down a scale. As the cactus used water in the soil, the needle moved in one direction. Then the motion along the scale was gradually reversed, as the cactus shoot itself started to lose water. The wilting point was the point of reversal of needle movement (Briggs and Shantz, 1912, pp. 47–53).

As with field capacity, later researchers realized that the wilting point is not a unique value. It is dynamic, like field capacity. There are a range of values at which the rate of water supply to a plant is not sufficient to prevent wilting, depending on the soil profile (soil texture, compaction, stratification); the amounts of water in the soil at different depths, which affect root distribution; the transpiration rate of a plant; and the temperature (Table 8.1). One should use a water bath to determine the wilting point, to control the temperature. Also, leaves wilt differently. Usually the basal leaves wilt first (Taylor and Ashcroft, 1972, p. 303), so one can refer to the "first permanent wilting point," at which the basal leaves do not recover, and the "ultimate permanent wilting point," at which the apical leaves do not recover. The permanent wilting point depends upon plant osmotic adjustment. Therefore, we recognize that there is a range of values for permanent wilting point, and it is not a unique value (Fig. 8.2).

TABLE 8.1   Influence of temperature on the soil water percentage at which sunflowers will wilt permanently. (From Taylor, S.A., and Ashcroft, G.L., *Physical Edaphology: The Physics of Irrigated and Nonirrigated Soils*, p. 303, ©1972 by W.H. Freeman and Company. Used with permission.)

	Permanent wilting percentage for three soils		
Temperature	Millville silt loam	Benjamin silty clay loam	Yolo fine sandy loam
°C			
5	. . .	. . .	9.0 ± 0.11
12.8	. . .	. . .	8.5 ± 0.03
15	8.38 ± 0.13	11.63 ± 0.23	. . .
25	7.34 ± 0.17	10.46 ± 0.26	. . .
35	6.66 ± 0.16	. . .	. . .

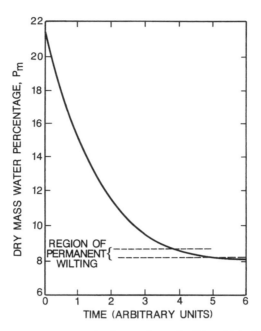

FIG. 8.2 Average water percentage in the top foot of soil in which alfalfa is rooted to a depth of 3 m. The permanent wilting percentage is a range of values of soil water contents over which the removal rate is slow. (From Taylor, S.A., and Ashcroft, G.L., *Physical Edaphology: The Physics of Irrigated and Nonirrigated Soils*, p. 302, ©1972 by W.H. Freeman and Company. Used with permission.)

If one cannot measure the permanent wilting point, it is usually esti- mated to be the water content at a soil matric potential of −1.5 MPa (−15 bars). However, plants can absorb water from soil at potentials much lower than this; creosote bush (*Larrea divaricata*) can absorb water to −6.0 MPa (Salisbury and Ross, 1978, p. 389). But the amount of water actually held by the soil between −1.5 MPa and −6.0 MPa is small.

The point at which the water content at the soil-root interface reaches the wilting point is of interest mathematically for root models (Philip, 1957; Gardner, 1960). In the models, the wilting point is dependent not only on the soil water content at wilting, but also the diffusivity of the soil, the radius of the root, and the transpiration rate. In his 1957 model of water uptake by plant roots, Philip pointed out that uncritical use of the wilting point as an invariant index of the lower limit of the availability of soil mois- ture to plants can be misleading (Philip, 1957; Raats et al., 2002, p. 18).

However, permanent wilting point still needs to be determined to cal- culate available water, which we shall discuss in the next section. Soil water content can be measured directly in the field using a hydraulically inserted heavy-duty time domain reflectometry probe (Long et al., 2002). This allows fine-scale resolution of water content for site-specific management. Equipment is under development to monitor soil water content using time domain reflectometry probes mounted on planters (Karlheinz Köller, Professor, Institute for Agricultural Engineering in the Tropics and Subtropics, University of Hohenheim, Stuttgart, Germany; personal communication, July 17, 2003). The problem is that tractor wheels go at least 1 m/s, and a probe turning through the field is not in contact long enough with the soil to measure water content; for the hydraulically inserted probe, 5 minutes are required for a reading (Long et al., 2002). However, technical difficulties will be overcome. When water content can be measured continuously as a piece of farm equipment moves through the field, available water can be calculated and then irrigations can be applied on the patches of land that need it—just like yield monitors allow application of fertilizers to sites in a field with low productivity.

III. AVAILABLE WATER

Plant available water, AW, may be defined as the difference between field capacity, FC, and wilting point, WP. The formula is:

$$AW = FC - WP. \qquad (8.1)$$

The field capacity might be measured as 5% of water per unit volume of bulk soil for a sand, which we shall label A, and might be measured as 50% per unit volume of bulk soil for a heavy clay, which we shall call B. The wilting point might be 2% water per unit volume for the sand A, and it might be 20% per unit volume for the heavy clay B. Using the numerical values of FC and WP for the sand A and heavy clay B, we find available water as:

$$\text{(Sand } A\text{) } AW = 5\% - 2\% = 3\%$$
$$\text{(Heavy clay } B\text{) } AW = 50\% - 20\% = 30\%.$$

The above two AWs are in percentages referred to a volume of bulk soil. These AWs may be considered to mean that, in 100 cm of the sand A profile, there are 3 cm of equivalent surface water in the plant available form; and in 100 cm of heavy clay B, there are 30 cm of equivalent surface water in plant available form. The clay soil B stores $(30 - 3) = 27$ cm more of equivalent surface water per meter depth of soil profile than does the sand A. From this example, we see that soil texture can have a large effect on soil water availability.

TABLE 8.2   Yield (metric ton/ha) of alfalfa, potatoes, and sugar beets at different soil moisture contents. (Data obtained from Taylor, 1952.)

Crop	Moisture content in cm of water per 100 cm of soil depth at time of irrigation			
	30	18	15	5
Alfalfa	14.3	14.3	13.4	10.3
Potatoes	33.8	35.7	32.2	7.8
Sugar beets	43.2	42.3	40.5	28.9

As noted in the preceding section, the terms field capacity and wilting point should be used with caution. Field capacity should be based on moisture measurements made in the field to a depth of interest, say 100 to 150 cm, and not on laboratory measurements. Equation (8.1) implies to some agronomists that water can be taken up by plant roots with equal ease, from field capacity to the wilting point. This view was promulgated by F.J. Veihmeyer and A.H. Hendrickson at the University of California in Davis, who collaborated for many years starting in the 1920s. For some plants this may be true, because for them the energy of getting water from the soil into the plant will be small compared to the energy required to get the water through the plant and through the stomata on leaves, and then into an evaporated form into the atmosphere. For such plants, one would not worry if the soil were to approach fairly close to the wilting point before rainfall or irrigation water was supplied. For most crops, however, yields are reduced if the water in the soil approaches the wilting point before water is supplied. This is illustrated in Table 8.2, where yields of alfalfa, potatoes, and sugar beets are shown when irrigation water was applied at four different moisture levels: 30, 18, 15, and 5% (30, 18, 15, and 5 cm of equivalent surface water per 100 cm of soil profile). The wilting point of this soil was 3 percent and field capacity was about 30 percent. Yields were reduced before the permanent wilting point was reached, showing that water is not equally available between field capacity and the wilting point (Taylor, 1952).

IV. NON-LIMITING WATER RANGE

In 1985, John Letey, a soil physicist at the University of California in Riverside, developed a concept called the *non-limiting water range* (NLWR), which acknowledges that water may not be equally available to plants between field capacity and the permanent wilting point. The interaction between water and other physical factors that affect plant growth must be considered. Bulk density and pore size distribution affect the relationship between water and both aeration and mechanical resistance. The relationship between water and aeration is opposite to that between water and mechanical resistance. Increasing water content decreases aeration, which is undesirable, but decreases mechanical resistance, which is desirable. The non-limiting water range may be affected by aeration and/or mechanical resistance (Fig. 8.3). The NLWR becomes narrower as bulk density and aeration limit plant growth. On one end of the scale, oxygen limits root growth and on the other end of the scale, mechanical resistance restricts root growth. The restriction may occur at a water content higher than the

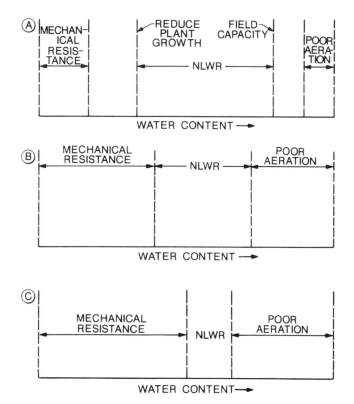

FIG. 8.3 Generalized relationships between soil water content and restricting factors for plant growth in soils with increasing bulk density and decreasing structure in going from case A to C. The non-limiting water range is abbreviated NLWR. (From Letey, J., Relationship between soil physical properties and crop production. *Adv Soil Sci* 1; 277–294, Fig. 4, 1985 ©Springer-Verlag, Heidelberg, Germany. This figure is used by permission of Springer-Verlag and John Letey.)

value that would be considered limiting to plants on the basis of plant available water.

To determine the NLWR, one must determine the matric potential at which the oxygen diffusion rate (ODR) limits root growth. The oxygen diffusion ratemeter is used to determine this value (see Chapter 10). Then one needs to determine the matric potential at which root growth is inhibited due to too high a resistance. This is done with a penetrometer (see Chapter 9). For example, in a coastal plain soil in South Carolina, researchers found that corn roots stopped growing at a matric potential of –0.08 bar, as determined using the ODR method, and stopped growing at a matric potential of –0.4 bar due to too high a resistance (Letey, 1985). One needs to use a soil moisture release curve to find the soil water contents associated with these matric potentials. The NLWR is the difference between the two water contents: the larger soil water content minus the lower soil water content.

The NLWR, also called the least limiting water range, is now often cited in the literature. Scientists especially in Canada and Oceania, are using it to predict crop production and indicate soil quality (da Silva et al., 1994; da Silva and Kay, 1996, 1997a, 1997b; Zou et al., 2000; Groenevelt et al., 2001).

V. BIOGRAPHIES OF BRIGGS AND SHANTZ

Dr. Lyman James Briggs, a physicist, was born May 7, 1874, in Assyria, Michigan, the son of Chauncey L. and Isabella (McKelvey) Briggs. He got his B.S. degree at Michigan State College in 1893, his M.S. degree at the University of Michigan in 1895, and his Ph.D. at Johns Hopkins in 1901 (Cattell, 1944). He received a Doctor of Science (Sc.D.) degree from Michigan State in 1932; a Doctor of Engineering degree from the South Dakota School of Mines in 1935; a Doctor of Laws (LL.D.) degree from the University of Michigan in 1936; a Sc.D. from George Washington University in 1937; a Sc.D. from Georgetown University in 1939; and a Sc.D. from Columbia University in 1944 (Debus, 1968).

He was in charge of the Physics Laboratory Division (now Bureau of Soils) for the U.S. Department of Agriculture (USDA), 1896–1906. He was physicist in charge of the Biophysical Laboratory, Bureau of Plant Industry, 1906–1912, and from 1912 to 1920, he was in charge of biophysical investigations. He was detailed to the Bureau of Standards by executive order in 1917–1919. He was chief of Division of Mechanics and Sound, Bureau of Standards from 1920–1933 and its assistant director of research and testing from 1926–1933. He was director of the Bureau of Standards from 1933 to 1945, and was director emeritus from 1945 until his death in 1963. He was a member of the National Advisory Committee for Aeronautics (1933–1945), and was its vice chairman from 1942–1945. He was chairman of the subcommittee on aircraft structures, 1937–1945; a member of the aerodynamics subcommittee, 1922–1930; chairman of the Federal Specifications Board, 1932–1940, and of the Federal Fire Council, 1933–1939; president of the National Conference on Weights and Measures, 1935–1945; member of the International Ice Patrol Board, 1933–1945; chairman of the Washington Biophysical Institute Council, 1933–1939; on the board of directors of the American Standards Association, 1933–1945; member of the U.S. National Committee for the International Geophysical Year; on the executive committee of the engineering division of the National Research Council, 1945–1950, and on its Committee of Fundamental Physical Constants; and director of the scientific program for stratospheric balloon flights. He was a trustee of George Washington University from 1945 until his death (Debus, 1968).

He shared the Magellan medal with Paul R. Heyl in 1922, received the Medal of Merit in 1948, and the Gold Medal of the U.S. Department of Commerce for exceptional service. He was an honorary Fellow of the American College of Dentists; a Fellow of the American Association for the Advancement of Science; a Fellow of the American Physical Society (and its vice president in 1937 and president in 1938). He was a member of the National Academy of Sciences; American Society of Mechanical Engineers; Washington Academy of Science (its president in 1917); Philosophical Society of Washington (its president in 1916); American Philosophical Society; American Academy of Arts and Sciences; Institute of Aeronautical Science; Newcomen Society (engineering society); Washington Academy of Medicine (its president, 1945–1946); and an honorary member of the Physical Society of Engineering. He was a member of Tau Beta Pi, Sigma Xi, and Sigma Pi Sigma (Debus, 1968).

His areas of research interest were aerodynamic characteristics of projectiles, bombs, and aerofoils in a high-speed windstream; acceleration of gravity at sea; gyroscopic stabilization; soil analysis; properties of liquids under negative pressures; and defense projects. He collaborated with Paul R. Heyl on the development of an earth inductor compass (Debus, 1968).

Briggs married Katherine E. Cook on December 23, 1896, and they had two children: Mrs. Isabel Myers and Albert Cook (deceased) (Debus, 1968). Lyman Briggs died on March 25, 1963. His scientific contributions have been described in detail by Landa and Nimmo (2003).

Dr. Homer LeRoy Shantz, a botanist, was born in Kent County, Michigan, on January 24, 1876, the son of Abraham K. and Mary E. (Ankney) Shantz. He got his B.S. degree at Colorado College, Colorado Springs, in 1901, and his Ph.D. at the University of Nebraska in 1905. In 1926, he received a Sc.D. from Colorado College (Debus, 1968).

He was an instructor of botany and zoology at Colorado College, 1901–1902; of botany at the School of Agriculture in Nebraska, 1903–1904; and in Missouri, 1905–1906. He was professor of botany and bacteriology at the University of Louisiana in 1907. He worked for the Bureau of Plant Industries, USDA, first as an expert in alkali and drought-resistant plant breeding investigations (1908–1909); then as a plant physiologist (1910–1920); and then was in charge of plant geography and plant physiology (1920–1926). He was special lecturer on plant geography in the Graduate School of Geography, Clark University, 1922–1926. Between 1926 and 1928, he was professor of botany and head of the department at Illinois. He was president of the University of Arizona, 1928–1936. He was Chief of the Division of Wildlife Management, U.S. Forest Service, 1936–1944 (Cattell, 1944), and was annuitant collaborator with the USDA from 1945 until his death in 1958. In 1956, he was a professor of botany at the University of Arizona, and in 1956–1957, he was principal investigator for the Arizona African Expedition (Debus, 1968).

He was a Fellow of the American Society of Agronomy and of the Royal Society of Arts. He was a member of the Phytographic Society of Sweden and honorary president of the 7th International Botanical Congress in Stockholm in 1950, and in Paris in 1954. He was a member of the Botanical Society; Washington Association of American Geographers; the American Society of Plant Physiologists (he received its Charles Reid Barnes life membership); Ecological Society; Wildlife Society; the Society pro Fauna et Flora Fennica; International Society for Protection of Nature; the International Institution of African Languages and Cultures; Sigma Xi; and Phi Beta Kappa (Cattell, 1944; Debus, 1968).

He was involved with many special projects. In 1918, he was part of the plant resources "Inquiry" in Africa and Latin America, formed to determine natural plant resources and crop producing possibilites of large portions of Africa and Latin America for use by the American Commission to Negotiate Peace, 1918–1919. In 1924, he was on the Education Committee of East Africa. In 1931–1934, he was a USDA member of the National Land Use Planning Committee of the U.S. Geological Survey, and was an explorer in the Smithsonian Institution expedition to Africa in 1919–1920. He was a member of the Educational Commission to East Africa under the auspices of the Phelps Stokes Fund and the International Education Board in 1924 (Debus, 1968).

His research interests included the vegetation of the Great Plains and the Great Basin; the indicator value of natural vegetation; the physiology of drought resistance; biological study of the lakes in the Pike's Peak region; North American Branchinecta and their habitats; plant geography of Africa and Latin America; plant geography and plant industry; agriculture of the African natives; wildlife management; and agricultural geography of Africa (Cattell, 1944; Debus, 1968).

He married Lucia Moore Soper on December 25, 1901, and they had two children: Homer LeRoy and Benjamin Soper. He died June 23, 1958 (Debus, 1968).

Importance of Briggs and Shantz

Both Briggs and Shantz are cited in a book listing the most important scientists from antiquity to the present (Debus, 1968). In the seventh edition of *American Men of Science* (Cattell, 1944), they had stars by their names. (A star was prefixed to 1000 biographical entries out of about 34,000 names listed.) The areas of science were broken down into 12 disciplines, and the number of people ranked in each discipline, of the 1000 men ranked, were as follows:

Chemistry, 175
Physics, 150
Zoology, 150
Botany, 100
Geology, 100
Mathematics, 80
Pathology, 60
Astronomy, 50
Psychology, 50
Physiology, 40
Anatomy, 25
Anthropology, 20

In each of the 12 principal sciences, the names were arranged in the order of merit by ten leading scientists of the discipline, and the position of each scientist then was ranked in his specialty. Briggs was ranked first in physics, and Shantz was ranked third in Botany. (Briggs was ranked even above I.I. Rabi, who was ranked sixth in physics. Rabi won a Nobel Prize in Physics in 1944 for his resonance method, using molecular beams, for recording the magnetic properties of atomic nuclei. Rabi's work laid the basis for NMR, now routinely used in medical diagnosis.) The biographies make clear the importance of Briggs and Shantz, who were two of the most important scientists in the United States.

REFERENCES

Baver, L.D., Gardner, W.H., and Gardner, W.R. (1972). *Soil Physics*. 4th ed. Wiley: New York.
Briggs, L.J., and Shantz, H.L. (1912). The wilting coefficient for different plants and its indirect determination. *USDA Bur Plant Industry Bull No.* 230. U.S. Dept. Agr: Washington, DC.
Cattell, J. (Ed.) (1944). *American Men of Science. A Biographical Directory*. 7th ed. Science Press: Lancaster, Pennsylvania.
da Silva, A.P., Kay, B.D., and Perfect, E. (1994). Characterization of the least limiting water range of soils. *Soil Sci Soc Amer J* 58; 1775–1781.
da Silva, A.P., and Kay, B.D. (1996). The sensitivity of shoot growth of corn to the least limiting water range of soils. *Plant Soil* 184; 323–330.
da Silva, A.P., and Kay, B.D. (1997a). Estimating the least limiting water range of soils from properties and management. *Soil Sci Soc Amer J* 61; 877–883.
da Silva, A.P., and Kay, B.D. (1997b). Effect of soil water content variation on the least limiting water range. *Soil Sci Soc Amer J* 61; 884–888.
Debus, A.G. (Ed.). (1968). *World Who's Who in Science. From Antiquity to the Present*. Marquis-Who's Who: Chicago, Illinois.
Gardner, W.R. (1960). Dynamic aspects of water availability to plants. *Soil Sci* 89; 63–73.
Groenevelt, P.H., Grant, C.D., and Semetsa, S. (2001). A new procedure to determine soil water availability. *Aust J Soil Res* 39; 577–598.
Hillel, D. (1971). *Soil and Water. Physical Principles and Processes*. Academic Press: New York.
Kramer, P.J. (1983). "Water Relations of Plants." Academic Press, New York. 489 pp.
Landa, E.R., and Nimmo, J.R. (2003). The life and scientific contributions of Lyman J. Briggs. *Soil Sci Soc Amer J* 67; 681–693.
Letey, J. (1985). Relationship between soil physical properties and crop production. *Adv Soil Sci* 1; 277–294.
Long, D.S., Wraith, J.M., and Kegel, G. (2002). A heavy-duty time domain reflectometry soil moisture probe for use in intensive field sampling. *Soil Sci Soc Amer J* 66; 396–401.
Philip, J.R. (1957). The physical principles of soil water movement during the irrigation cycle. *Proc Congr Int Comm Irrigation Drainage*, 3rd, San Francisco, 8; 125–153.
Raats, P.A.C., Smiles, D.E., and Warrick, A.W. (2002). Contributions to environmental mechanics: Introduction. In *Environmental Mechanics: Water, Mass and Energy Transfer in the Biosphere* (Raats, P.A.C., Smiles, D.E., and Warrick, A.W., Eds.), pp. 1–28. Geophys. Monogr 129. American Geophysical Union: Washington, DC.
Salisbury, F.B., and Ross, C.W. (1978). *Plant Physiology*. 2nd ed. Wadsworth Pub: Belmont, California.
Soil Science Society of America. Editions dated (1975), (1978), (1979), (1984), (1987), and (1997). *Glossary of Soil Science Terms*. Soil Science Society of America: Madison, Wisconsin.
Taylor, S.A. (1952). Use of mean soil moisture tension to evaluate the effect of soil moisture on crop yields. *Soil Sci* 74; 217–226.

Taylor, S.A., and Ashcroft, G.L. (1972). *Physical Edaphology: The Physics of Irrigated and Nonirrigated Soils*. W.H. Freeman: San Francisco.

Wesseling, J., and van Wijk, W.R. (1957). Soil physical conditions in relation to drain depth. In *Drainage of Agricultural Lands* (Luthin, J.N., Ed.), pp. 461–504. American Society of Agronomy: Madison, Wisconsin. (See especially Fig. 4, p. 468).

Zou, C., Sands, R., Buchan, G., and Hudson, I. (2000). Least limiting water range: A potential indicator of physical quality of forest soils. *Aust J Soil Res* 38; 947–958.

9

Penetrometer Measurements

As we pointed out in Chapter 1, the four soil physical factors that affect plant growth are mechanical impedance, water, aeration, and temperature (Kirkham, 1973). In Chapter 4, we learned how to measure matric potential of water in the soil using tensiometers. In later chapters, we shall study other techniques to measure water in soils and plants, and in the next chapter we shall see how to measure soil aeration. In this chapter, we learn how to measure mechanical impedance using penetrometer measurements.

We first will define a penetrometer and then look at different kinds of instruments and their uses. We will consider the type of tests that are done with penetrometers and what factors affect the measurements, and then look specifically at the cone penetrometer.

I. DEFINITION, TYPES OF PENETROMETERS, AND USES

A penetrometer is any device forced into the soil to measure resistance to vertical penetration (Davidson, 1965). The earliest soil penetrometers were fists, thumbs, fingernails, pointed sticks, and metal rods. They are still used for qualitative measurements.

Results of such tests are expressed in terms such as "loose," "soft," "stiff," and "hard." However, penetrometers are designed to give quantitative measurements of soil penetration resistance for a more precise correlation with properties such as bearing value, safe soil pressure, rolling resistance, trafficability of wheels or crawler tracks on soil, relative density, crop yield, and tilth (Davidson, 1965). *Tilth* is from the Anglo-Saxon word *tilthe* and means a tilling or cultivation of land. Dr. Jerry L. Hatfield, a Kansas native and the director of the United States National Soil Tilth

Laboratory, located on the campus of Iowa State University in Ames, Iowa, has a nonscientific definition of tilth: "The wellness of the seedbed" (Muhm, 1990). The $11.9 million Tilth Laboratory opened in April, 1989, and has the goal of quantifying the effects of tillage on the soil. Using penetrometers is one method to do this quantification.

II. TYPES OF TESTS

Two types of tests are done, when making penetration-resistance measurements: a static test or a dynamic test. In a static penetration test, the penetrometer is pushed steadily into the soil. A static penetration test is exemplified by the use of the cone penetrometer, which we discuss in detail in Section IV. In a dynamic penetration test, the penetrometer is driven into the soil by a hammer or falling weight. A dynamic penetration test is done with a spray-tainer or spra-tainer. The apparatus was designed in the 1950s by Professor Champ B. Tanner of the University of Wisconsin in Madison, Wisconsin. For a biography of Tanner, see the Appendix, Section V.

    The spra-tainer is shown in Fig. 9.1 (Kirkham et al., 1959b). It is a thin-walled can of 12-ounce size (341 grams) manufactured to dispense products such as shaving cream and bug spray under pressure. The bottom of the can is removed and the top is left open. The can, which is 8 cm long and 6.9 cm in diameter, is driven into the soil with a special hammer weighing 2.35 kilograms and dropped from a height of 42.5 cm (Kirkham et al.,

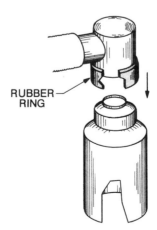

RUBBER RING

FIG. 9.1   The spra-tainer can. (From Kirkham et al., 1959b; Reprinted by permission of Marcel de Boodt for the Landbouwhogeschool en de Opzoekingsstations van de Staat te Gent, Ghent, Belgium.)

1959a). This driving of the cans into the soil is done when the soil is at or near field capacity. The cans are driven entirely into the soil (to a depth of 8 cm). The cans are steel, and, unless they encounter rocks in the soil, they may be used repeatedly. A thin coating of petroleum jelly is wiped on the cans before each use. The special hammer is used, together with a special driving head and driving tube, in order that the driving be done the same way by all operators. The spra-tainer can and the driving head fit in a driving tube, the latter having triangular-shaped legs with spikes in their ends to hold the driving tube vertically on the soil surface. The top of the driving head has, extending upward on its axis, a guide rod. A photograph of the setup is shown by Kirkham et al. (1959b). The number of blows to drive in the cans (acting as penetrometers) are counted, and this number is the quantified measurement. The seamless tube cans are an important feature of the equipment. Because of the thin and sharp walls of the spra-tainer cans, the soil is relatively undisturbed, and, if soil samples are taken after getting the penetration resistance, the samples (8 cm long and 6.9 cm in diameter) may be called undisturbed.

III. WHAT PENETROMETER MEASUREMENTS DEPEND UPON

All penetrometer measurements depend upon two factors: the water content of the soil and time. Above freezing, differences in measurements due to temperature are not detectable (Loyd Stone, personal communication, February 4, 1983). Therefore, measurements depend on temperature only if the soil is frozen.

Figure 9.2 (Davidson, 1965) shows that as the water content increases, the penetration resistance decreases. As noted above, measurements with the spra-tainers are made when the soil is near field capacity. A measurement made with the cone penetrometer in a cohesive, fine-grained soil is an inverse function of water content (Davidson, 1965). In humid climates, trafficability measurements are made during the wet season. In dry or hard soils, or in soils containing pebbles and stones, any operator will find it difficult to obtain consistent and reliable penetrometer measurements, especially as penetration depth increases.

Measurements depend upon time because of impulse. In physics, impulse is defined as follows (Schaum, 1961, p. 62):

$$\text{Impulse} = \text{force} \times \text{length of time the force acts} = Ft. \qquad (9.1)$$

Units of impulse are the nt-s in the mks system, the dyne-s in the cgs system, and the lb-s in the English system.

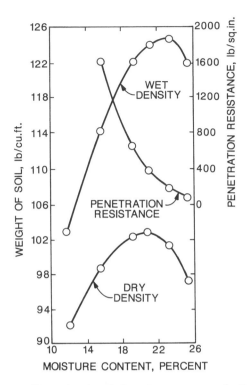

FIG. 9.2 Typical curves illustrating the relation of water content of soil to density and penetration resistance. (From Davidson, D.T., ©1965, American Society of Agronomy, Madison, Wisconsin. Reprinted by permission of the American Society of Agronomy.)

Impulse and momentum are related. The change of momentum produced by an impulse is equal to the impulse. Thus if an unbalanced force F acting for a time t on a body of mass m changes its velocity from an initial value v_o to a final value v_t, then

$$\text{Impulse} = \text{change in momentum,}$$
$$Ft = m(v_t - v_o). \tag{9.2}$$

This equation indicates that the unit of impulse in any system is equal to the corresponding unit of momentum. Therefore, 1 nt-s = 1 (kg-m)/s and l lb-s = 1 (slug-ft)/s (Schaum, 1961, p. 62).

Because of impulse (dependent upon time), a penetrometer, like a cone penetrometer, must be pushed at a steady rate into the soil. It should take about 15 s to go 24 inches (4 cm/s) (Davidson, 1965, p. 480). According to Don Kirkham (personal communication, February 20, 1982), penetrometer measurements are a "pain in the neck" for two reasons: their dependence upon time and water content.

IV. CONE PENETROMETER

Now let us look at the cone penetrometer (SoilTest, 1978a), which is a penetrometer that has gained wide acceptance. It was developed by the U.S. Army Corps of Engineers for predicting the carrying capacity of cohesive, fine-grained soils for army vehicles in off-road military operations (Davidson, 1965).

The applied force required to press the cone penetrometer into a soil is an index of the resistance or impedance of the soil and is called the *cone index* (CI). Cone index readings are taken to depths of 24 inches (61 cm) to permit plotting of a cone index curve, which, in addition to its signficance in trafficability studies, gives quantitative information on soil compactness or density that can be correlated with other soil physical properties or with crop yields (Davidson, 1965).

The parts of the cone penetrometer made in the United States consist of the handle, proving ring, dial gauge, rod graduated in 6-inch (15-cm) or 12-inch (30-cm) intervals, and a stainless steel cone (Fig. 9.3) (SoilTest, 1978b). The operator's handle is mounted at the top of the proving ring. The staff is 19″ long (48.3 cm), making it possible to take readings to that depth. The cone is 1.5 inches (3.8 cm) in height and has a 30-degree apex angle and a base area of 0.5 inch squared (3.14 cm^2). The diameter of the base of the cone is 0.79 inches (2.0 cm). The cone index or force per unit area required to move the cone to a given plane of soil to show the shearing resistance of that soil is indicated on the proving ring dial. The proving ring has 150-pound capacity and the dial indicator reads the cone index in the range of 0 to 300 pounds per square inch (psi). (See next paragraph for SI units.) In Europe, Eijkelkamp Agrisearch Equipment (P.O. Box 4, 6987 ZB Giesbeek, The Netherlands) sells penetrometers. Hartge et al. (1985) in Germany report results using the Eijkelkamp penetrometer. Gauges in Europe read in kg/cm^2 (Don Kirkham, personal communication, February 18, 1982). In Australia, Rimik Agricultural Electronics (14 Molloy Street, Toowoomba, Queensland 4352) makes a cone penetrometer that has a cone index read-out in kPa.

Because the Corps of Engineers' cone penetrometer is made in the United States, its dial gauge reads out in the English units of pounds per square inch (psi). Therefore, we need to know how to convert the dial readings, in lb/in^2, into SI units. Remember $F = ma$ (force = mass times acceleration) and in

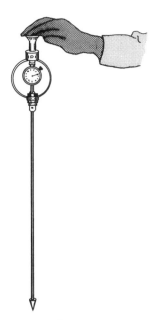

FIG. 9.3   The cone penetrometer. (From SoilTest, 1978b; Reprinted by permission of ELE International, Loveland, Colorado.)

a gravitational field, $w = mg$ (weight = mass times acceleration due to gravity). So each gram has an earth-pull on it of 980 dynes and each kilogram has an earth-pull on it of 9.8 newtons. In the cgs system of units, we make the following calculations (remember 1×10^6 dynes/cm^2 = 1 bar).

To convert from the English system to the cgs system:

$$1 \text{ psi} = 1 \text{ lb/in}^2 = [(454 \times 980) \text{ dynes}]/(2.54 \text{ cm})^2 = 68962.7 \text{ dynes/cm}^2 =$$
$$68962.7/10^6 \text{ bars} = 0.0689627 \text{ bars or, to 4 significant figures,}$$
$$0.06896 \text{ bars.}$$

This agrees with the value that Taylor and Ashcroft (1972, p. 511) give in their extensive list of conversion factors: 1 psi = 0.06895 bar, the slight difference (0.06895 bar vs. 0.06896 being due to rounding of values).

We know that 10 bars = 1 MPa. Thus,

$$0.0689627 \text{ bars} = 0.00689627 \text{ MPa} = 6896.27 \text{ Pascals.}$$

To convert from the English system to the mks system:

$$1 \text{ psi} = [(0.454 \times 9.8) \text{ newtons}]/[(2.54/100) \text{ m}]^2 = 0.68962 \times 10^4 \text{ newtons/m}^2 = 6.896 \times 10^3 \text{ newtons/m}^2 = 6896 \text{ nt/m}^2.$$

Note the conversion units:

$$14.7 \text{ psi} = 1 \text{ atm}; 0.987 \text{ atm} = 1 \text{ bar}.$$

1 Pascal = 1 nt/m^2 and 14.5 lb/in^2 = 1 bar; 0.06896 bar per psi × 14.5 psi per bar = 1.0. The value 0.06896 bar per psi checks out.

The cone penetrometer that Loyd R. Stone in the Department of Agronomy at Kansas State University uses was made by the Physics Shop in at Kansas State University (personal communication, March 6, 1990). He has penetrometers with different cone tips and base areas. Dr. Stone's penetrometers are calibrated by pressing the cone on a balance with known masses in kilograms. The probe scale has no units, just numbers. The read-out (number) on the probe is calibrated against kilograms. The value in kilograms is divided by area for that cone tip, and he gets probe readings in units of kg/cm^2 (mass per unit area) (Intrawech et al., 1982). Others also use a cone index in units of kg/cm^2 (Cruse et al., 1981; Bradford, 1986).

Note that acceleration due to gravity is not included when one gets a reading of mass per unit area (kg/cm^2). As noted in Chapter 2, we must express values in SI units, and journals require them for publication. But either kg/cm^2 or the units converted to SI units from the English units on the U.S. Army Corps of Engineers' cone penetrometer (force per unit area or MPa) are all right. Engineers around the world (like those in the U.S. Army Corps of Engineers) think in terms of living on earth and talk of weight on earth. For example, they say, "This man weighs 185 lb." They do not say, "This man has a mass of 5.8 slugs." The 185 pounds is the man's force on the surface of a floor. It is a valid reading, because the springs on a calibrated bathroom scale will give the man's weight in pounds. If the man were standing on scales on the moon, where surface gravity is 0.17 of the earth's, he would weigh 31 pounds. So when the astronauts were on the moon, they needed lead weights in their boots to hold them down.

In sum concerning units, we need to recognize that the U.S. Army Corps of Engineers' gauge is reporting a force per unit area or a weight per unit area (remember $w = mg$), and units of kg/cm^2 report a mass per unit area. A gravity constant is associated with the Corps of Engineers' gauge, and it is not with a reading given in kg/cm^2.

Loyd Stone often uses a penetrometer with a cone angle of 45 degrees. He prefers a wider angle than that on the penetrometer of the Corps of Engineers (45 vs. 30 degrees). With the wider angle, the soil does not get so compressed as the cone moves in, especially at lower depths (personal communication, March 6, 1990). To get more accurate readings, he uses a smaller area on the cone and a more sensitive proving ring. If a proving ring needs a 500 pound (227 kg) force to move it, it is no good, because a man cannot push 500 pounds. So Dr. Stone uses a 0 to 50 lb proving ring and has a small cone area. The proving ring and the cone area must be matched. Some people like to go to larger cone areas, which are harder to push into the ground, to get better representation of the soil, because a larger area is sampled (Loyd Stone, personal communication, March 6, 1990). Dr. Stone's meter has a brake and holds the reading until it is released. The Corp of Engineers' penetrometer does not hold the reading. It is not necessary to core soil first when using a cone penetrometer (Loyd Stone, personal communication, February 17, 1982), but to use a blunt-end penetrometer it is necessary to core the soil to the depth of interest because the soil becomes compacted.

When reviewing a paper describing a study in which a cone penetrometer has been used, make sure that the authors give 1) the cone angle; 2) the rate of penetration; and 3) the physical meaning of their units (i.e., whether or not gravity is taken into account in the units).

The correlation between readings made with cone penetrometers is good, if the same model of penetrometer is used at the same location, and even if two different people make the measurements (Loyd Stone, personal communication, February 7, 1983). Differences in readings occur due to fractures in the soil (e.g., holes), which result in much variability between readings. However, there is still some variability due to operators. An electronically driven cone penetrometer (Fig. 9.4) has been developed to overcome differences due to human operators (American Society of Agronomy, 1987; Christensen et al., 1998).

In this chapter, only commonly used penetrometers have been noted. Specifically designed ones for laboratory experimentation have been developed. For example, see Whiteley et al. (1981). Perumpral (1987) reviews applications of cone penetrometers in engineering.

V. APPENDIX: BIOGRAPHY OF CHAMP TANNER

Champ Bean Tanner, the inventor of the spra-tainer, was born in Idaho Falls, Idaho, on November 16, 1920, the son of Bertrand Myron Tanner and Orea Bean Tanner. After the death of his father in 1924, he was raised by his widowed mother. The family moved to Teton City and then to Rexburg, Idaho, where his mother taught high school until 1930. In 1930 the family (Champ, two brothers, and his mother) moved to Provo, Utah, to continue Orea's education at Brigham Young University. After earning

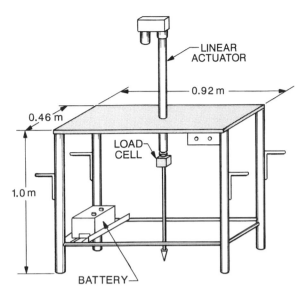

FIG. 9.4 A portable penetrometer, designed by Kansas State University researchers. It bores into ground at a constant speed of 30 cm (12 inches) every 30 seconds. (From American Society of Agronomy, ©1987. Reprinted by permission of the American Society of Agronomy.)

her B.S., Mrs. Tanner taught at Provo High until 1938, when she joined the English Department at Brigham Young University.

Tanner graduated from Provo High School in 1938. He received his undergraduate degree from Brigham Young University in 1942 with high honors in chemistry and soil science. After four years of service in the U.S. Army (1942–1946), he entered graduate school at the University of Wisconsin in Madison. He earned his Ph.D. in soils in 1950 under the joint direction of Professors E.E. Miller and M.L. Jackson (American Society of Agronomy, 1988). He joined the Department of Soil Science as the first agricultural physicist employed since F.H. King's retirement in 1901. (For a biography of King, see Tanner and Simonson, 1993). He remained at the University of Wisconsin for 40 years, and served as chair of the department of soil science from 1984 until his retirement in 1988.

In soil physics, he studied water flux in unsaturated soils, the thermal regime in soils, and soil aeration and redox potentials. His ability to develop instrumentation such as the spra-tainer for the dynamic penetration test was recognized by his colleagues (Don Kirkham, personal communication, undated). Tanner was the first to make in situ measurements of oxygen tension in the field. As a pioneer in micrometeorology, he dedicated much of his research to near-ground measurements of heat and water vapor transport from soil, water, and plant surfaces. He was the first to apply approaches of energy balance and the Bowen ratio to agronomic crops, and he devised the instruments for the necessary measurements. He developed the measurement of net radiation absorbance in crop foliar canopies and estimated soil evaporation and plant evaporation as functions of plant density and row spacing (Walsh et al., 1991).

In the area of plant-water relations, Tanner provided fundamental information on the relationship between water availability and plant growth. He created original instruments and techniques for estimating plant physiological responses, including the use of pressure chambers to measure water potential in plant storage organs and in situ water potential measurements of potato tubers and other root crops (Walsh et al., 1991). The paper describing the stomatal meter that he made with graduate students Edward T. Kanemasu and George W. Thurtell (Kanemasu et al., 1969) became a citation classic (Institute for Scientific Information, 1979). His *Soils Bulletin No. 6* (Tanner, 1963) is still regularly referred to.

Tanner directed the research for 25 Ph.D. and 15 M.S. students (American Society of Agronomy, 1990) and worked with several postdoctoral scientists. His students became leaders in agricultural meteorology and soil physics. He took pleasure in their achievements, but little credit, because he believed that the qualities ensuring success, such as integrity, imagination, deep curiosity, and hard work, are native and not taught (American Society of Agronomy, 1988). I worked in Tanner's laboratory when I was a graduate student studying under Wilford R. Gardner at the University of Wisconsin. There Tanner taught me how to weld thermocouples and make thermocouple psychrometers. His attention to detail was well known, and both field and laboratory measurements had to be done exactly right. He started work early in the morning. The going bet was that some day Tanner would arrive so early that he would meet Marvin L. Wesely (Gaffney, 2003), one of his students who worked late into the nights.

Tanner was the first soil scientist to be elected to the National Academy of Sciences (1981). He received the Award for Outstanding Achievement in Biometeorology from the American Meteorological Society in 1980 and the Soil Science Society of America's Soil Science Research award in 1978. He was a Fellow of the American Meteorological Society, the American Society of Agronomy, the Soil Science Society of America, the Crop Science Society of America, and the American Association for the Advancement of Science (American Society of Agronomy, 1988). He was awarded the

Emil-Truog named professorship at the University of Wisconsin in 1979. He was a Fulbright lecturer in Australia and Papua New Guinea. He served as editor for the American Meteorological Society, the Soil Science Society of America, the American Society of Agronomy, and the American Society of Plant Physiologists.

A symposium on the subject of biophysical measurements was held at the annual meetings of the American Society of Agronomy in November, 1988, to honor Tanner. Papers from the symposium were published in an issue of *Theoretical and Applied Climatology* (Campbell, 1990).

Tanner married Kay (Catherine May Cox) on September 24, 1941. They had five children: three sons, Bertrand D. Myron S., and Clark B.; and two daughters, Catherine and Terry Lee. Clark, born in 1960, died in 1977 of acute leukemia. Bertrand, like his father, is skilled in instrumentation, and is an executive at Campbell Scientific, Inc., the company best known for its data loggers. Champ Tanner's accomplishments were all the more remarkable because he got polio in the early 1950s, and, although he recovered, he walked with difficulty. He died of cancer on September 22, 1990, at the age of 69.

REFERENCES

American Society of Agronomy. (1987). Kansas State researchers design new penetrometer. *Crops Soils Mag* 39(6); 21.

American Society of Agronomy. (1988). Retirements. C.B. Tanner. *Agronomy News*, August, 1988: 13.

American Society of Agronomy. (1990). Deaths. Champ B. Tanner. *Agronomy News*, November, 1990; 27.

Bradford, J.M. (1986). Penetrability. In *Methods of Soil Analysis. Part 1. Physical and Mineralogical Methods* (Klute, A., Ed.) 2nd ed., pp. 463–478. American Society of Agronomy and Soil Science Society of America: Madison, Wisconsin.

Campbell, G.S. (1990). Introduction. Tanner Symposium on Biophysical Measurements and Instrumentation. *Theoretical Appl Climatol* 42; 201–202.

Christensen, N.B., Sisson, J.B., Sweeney, D.W., and Swallow, C.W. (1998). Electronically-controlled, portable, cone penetrometer. *Commun Soil Sci Plant Anal* 29; 1177–1182.

Cruse, R.M., Cassel, D.K., Stitt, R.E., and Averette, F.G. (1981). Effect of particle surface roughness on mechanical impedance of coarse-textured soil materials. *Soil Sci Soc Amer J* 45; 1210–1214.

Davidson, D.T. (1965). Penetrometer measurements. In *Methods of Soil Analysis. Part 1. Physical and Mineralogical Properties, Including Statistics of Measurement and Sampling* (Black, C.A., Evans, D.D., Ensminger, L.E., White, J.L., and Clark, F.E., Eds.), pp. 472–484. American Society of Agronomy: Madison, Wisconsin.

Gaffney, J.S. (2003). Marvin Wesely. 1944–2003. *Bull Amer Meteorol Soc* 84; 812–814.

Hartge, K.H., Bohne, H., Schrey, H.P., and Extra, H. (1985). Penetrometer measurements for screening soil physical variability. *Soil Tillage Res* 5; 343–350.

Institute for Scientific Information. (1979). This Week's Citation Classic: Kanemasu, E.T., Thurtell, G.W., and Tanner, C.B. Design, calibration, and field use of a stomatal diffusion porometer. Plant Physiol 44:881–885, 1969. *Current Contents, Agr, Biol, Environ Sci* 10(41); 10.

Intrawech, A., Stone, L.R., Ellis, Jr., R., and Whitney, D.A. (1982). Influence of fertilizer nitrogen source on soil physical and chemical properties. *Soil Sci Soc Amer J* 46, 832–836.

Kanemasu, E.T., Thurtell, G.W., and Tanner, C.B. (1969). Design, calibration and field use of a stomatal diffusion porometer. *Plant Physiol* 44; 881–885.

Kirkham, D. 1973. Soil physics and soil fertility. *Bulletin des Recherches Agronomiques de Gembloux Faculté des Sciences Agronomiques de l'État* (new series) 8(2); 60–88.

Kirkham, D., De Boodt, M., and De Leenheer, L. (1959a). Modulus of rupture determination on cylindrical soil core samples. *Overdruk uit Mededelingen van de Landbouwhogeschool en de Opzoekingsstations van de Staat te Gent* 24(1); 369–376.

Kirkham, D., De Boodt, M., and De Leenheer, L. (1959b). Air permeability at the field capacity as related to soil structure and yields. *Overdruk uit Mededelingen van de Landbouwhogeschool en de Opzoekingsstations van de Staat te Gent* 24(1); 377–391.

Muhm, D. (1990). Ames tilth lab delves into mysteries of soil. *The Des Moines (Iowa) Register*, page 1J and 2J. February 11, 1990.

Perumpral, J.V. (1987). Cone pentrometer applications—A review. *Trans Amer Soc Agr Eng* 30; 939–944.

Schaum, D. (1961). *College Physics.* Schaum Publishing: New York.

SoilTest. (1978a). *Corps of Engineers—Cone Penetrometer. Model CN-973. Operating Instructions.* SoilTest, Inc., 2205 Lee Street, Evanston, Illinois. (SoilTest is now owned by ELE International, P.O. Box 389, Loveland, Colorado 80539.)

SoilTest. (1978b). *Classification and Rapid Testing Equipment.* (Product bulletin). SoilTest, Inc., 2205 Lee Street, Evanston, Illinois. (SoilTest is now owned by ELE International, P.O. Box 389, Loveland, Colorado 80539.)

Tanner, C.B. (1963). Basic instrumentation and measurements for plant environment and micrometeorology. *Soils Bull.* 6, Univ of Wisconsin Soils Dept: Madison. (Univ. Microfilms #OP 62 102).

Tanner, C.B., and Simonson, R.W. (1993). Franklin Hiram King—pioneer scientist. *Soil Sci Soc Amer J* 57; 286–292.

Taylor, S.A., and Ashcroft, G.L. (1972). *Physical Edaphology: The Physics of Irrigated and Nonirrigated Soils.* W.H. Freeman: San Francisco.

Walsh, L.M., Kutzbach, J.E., Miller, E.E., Murdock, J.T., and Peterson, A.E. (1991). Champ Bean Tanner (1920–1990). *Bull Amer Meteorol Soc* 72; 1562–1565.

Whiteley, G.M., Utomo, W.H., and Dexter, A.R. (1981). A comparison of penetrometer pressures and the pressures exerted by roots. *Plant Soil* 61; 351–364.

10

Measurement of Oxygen Diffusion Rate

Air and water comprise a large part of the soil. For an average soil, air and water take up 50% of the space (Fig. 10.1). Organic matter and mineral matter take up the other 50%. At optimum moisture content for plant growth, the air and water space are about equal, each about 25 percent of the soil volume (Kirkham and Powers, 1972, p. 1). With so much of the soil volume taken up by air and water, it is obvious that air and water must play a major part in soil and plant-water relations.

We have looked at methods to measure the soil-water matric potential (tensiometers and pressure plates; see Chapter 4) and to measure mechanical impedance (penetrometers; see Chapter 9). In this chapter, we look at the most widely used method to monitor aeration status of the soil, the oxygen diffusion rate (ODR) method.

I. THE OXYGEN DIFFUSION RATE METHOD

Respiration by plant roots depends on soil oxygen. Roots, like animals, do not photosynthesize, and they give off carbon dioxide and take in oxygen during respiration. Diffusion of gases in the soil practically stops when the fraction of air-filled pores is less than 10% (Wesseling and van Wijk, 1957, p. 468) (Fig. 10.2). Therefore, roots need at least 10% by volume air space in the soil to survive (Kirkham, 1994).

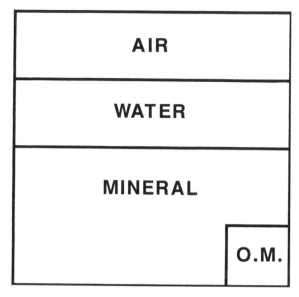

FIG. 10.1   Space in a soil. (From *Advanced Soil Physics* by Kirkham, D., and Powers, W.L., p. 1, ©1972, John Wiley & Sons, Inc.: New York. This material is used by permission of John Wiley & Sons, Inc. and William L. Powers.)

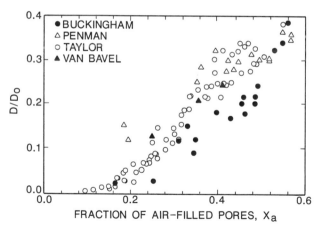

FIG. 10.2   Scatter diagram of the relation between the ratio D/D_o and the fraction of air-filled pores x_a calculated from data of Buckingham (1904) (closed circles), Penman (1940) (open triangles), Taylor (1949) (open circles), and van Bavel (1952) (closed triangles). D_o is the coefficient of diffusion of CO_2 in still air. D is the coefficient of diffusion of CO_2 in the soil. If x_a decreases to 0.1 to 0.2, D appears to become zero. (From Wesseling, J., and van Wijk, W.R., ©1957, American Society of Agronomy: Madison, Wisconsin. Reprinted by permission of the American Society of Agronomy.)

Evaluation of the aeration conditions at the interface between the root and the soil system presents the greatest possibility of defining the influence of aeration on plant growth (Phene, 1986). In the process of respiration, plants quickly take up the oxygen surrounding the roots in the rhizosphere, and an increasing oxygen concentration gradient develops between the soil atmosphere and the atmosphere next to the root surrounded by the water film. Movement of oxygen from the atmosphere to a respiring root involves diffusion through the following three phases (Lemon and Erickson, 1952):

1. The gaseous phase of the soil;
2. The gas-liquid phase boundary; and
3. The liquid phase of the water film around the root.

Because the diffusion coefficient of oxygen in water is about 2.4×10^{-5} cm^2 s^{-1} and the diffusion coefficient of oxygen in air is about 1.8×10^{-1} cm^2 s^{-1}, the limiting factor for this transport of oxygen is usually the diffusion rate through the water film rather than through the gas-filled pore space (Phene, 1986). (We can check out how fast one gas diffuses through another by spilling ammonia, for example, in a back corner of a large lecture hall. The speaker would smell the ammonia within a matter of seconds.)

When we are considering oxygen diffusion through air, we need to know the components of air, which are given in Table 10.1 (Weast, 1964, p. F-88). Air contains 20.946% oxygen.

In view of these concepts, a method to measure the oxygen diffusion rate through the liquid phase to a reducing surface approximating that of the plant root should be useful for assessment of soil aeration (Phene, 1986). And, indeed, the ODR method is the best index of oxygen availability for plant roots in soil (Gliński and Stęphiewski, 1985, p. 189). In the method, developed by Lemon and Erickson (1952), we use a platinum microelectrode to simulate the root in an electrolytic solution (the soil water with its dissolved solutes).

II. ELECTROLYSIS

Because we are dealing with an electrolytic solution, we need to understand electrolysis (Fig. 10.3). *Electrolysis* is defined as the decomposition into ions of a chemical compound in solution by the action of an electric current passing through the solution. In electrolysis, we have the electrolytic solution, an anode, a cathode, and a battery. Electrons flow from the positive terminal, the anode, to the negative terminal, the cathode. The anode is defined as a positive electrode or positive terminal of an electric source.

TABLE 10.1 Components of atmospheric air (exclusive of water vapor) (numbers come from Weast, 1964, p. F-88)

Constituents	Percent	Parts per million[*]
N_2 (nitrogen)	78.084	
O_2 (oxygen)	20.946	
CO_2 (carbon dioxide)[+]	0.033	
Ar (argon)[#]	0.934	
Ne (neon)		18.18
He (helium)		5.24
Kr (krypton)		1.14
Xe (xenon)		0.087
H_2 (hydrogen)		0.5
CH_4 (methane)		2
N_2O (nitrous oxide)		0.5

[*] one percent = 10,000 ppm.

[+] The concentration of carbon dioxide is increasing in the atmosphere at a rate of 1.5–2 ppm per year and has been doing this since the 1980s. This increase is not noted in a 1994 Handbook of Chemistry and Physics (Lide, 1994), which lists the components of air as above, except the concentration of CO_2 is given as 314 ppm or 0.0314 percent, a value even lower than the 1964 handbook (see above; 330 ppm or 0.033 percent). The concentration of CO_2 in the air in Manhattan, Kansas, in year 2003 averages 370 ppm.

[#]Argon, neon, helium, krypton, and xenon are inert gases in Group VIIIA of the periodic table. The sixth element in this group is radon, which is of environmental concern because it is toxic. It is a radioactive gaseous chemical element formed, together with alpha rays, as a first product in the atomic disintegration of radium and used in the treatment of cancer. It is produced naturally in the ground in some regions of the U.S. and can build up in basements. Homes need to be checked to make sure that they do not have radon.

The cathode is defined as the negative pole or electrode of an electrolytic cell. (*Anode* comes from the Greek *anodos*, meaning "a way up," which comes from *ana*, which means "up" and *hodos*, which means "way." *Cathode* comes from the Greek *kathodos*, going down, which comes from *kata-*, which means "down" and *hodos*, which means "way.")

Anions, which are negatively charged, are attracted to the anode, and cations, which are positively charged, are attracted to the cathode. The chemical in a solution, therefore, can be separated into its components (anions going to the anode; cations going to the cathode), if an electrical current passes through the solution. In the ODR method, the root is going to be simulated by a platinum microelectrode, which is the cathode. The root (Pt microelectrode) is a reducing surface. At the cathode, reduction takes place. At the anode, oxidation takes place. Reduction is the gain of electrons, and "to reduce" means to remove O_2 from. In oxidation, there is a loss of electrons.

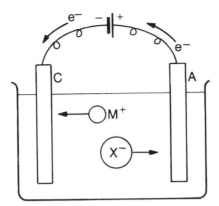

FIG. 10.3 Electrolysis. A = anode; C = cathode; battery at top; X^- = anion; M^+ = cation. Reduction occurs at cathode. (From Sienko, M.J., and Plane, R.A., *Chemistry*, p. 280, ©1957, McGraw Hill Book Co., Inc.: New York. This material is reproduced with permission of The McGraw-Hill Companies.)

We now need to review the laws of electrolysis. Michael Faraday (1791–1867) (see the Appendix, Section V, for his biography), an English chemist and physicist, who is best known for discovering electromagnetic induction, formulated the laws of electrolysis, which are as follows (Schaum, 1961, p. 169).

A. Faraday's Laws of Electrolysis

1. The mass of a substance liberated or deposited at an electrode is proportional to the quantity of electricity (i.e., to the number of coulombs) that has passed through the electrolyte. The coulomb is named after Charles Augustin de Coulomb (1736–1806), a French physicist. A coulomb is a unit for measuring the quantity of an electric current. It is the amount of electricity provided by a current of one ampere flowing for one second; one ampere = one coulomb per second; in symbols, I (ampere) = q (coulomb)/t (s).

2. The masses of different substances liberated or deposited by the same quantity of electricity are proportional to their equivalent weights. The *equivalent weight* of an element is its atomic weight divided by its valence. Thus, the equivalent weight of copper is 1/2 of its atomic weight for the electrolysis of solutions containing Cu^{++}, because the reaction at the cathode is

$$Cu^{++} + 2e^- \rightarrow Cu.$$

If a solution of Cu^+ were electrolyzed, the equivalent weight of copper would be the same as the atomic weight, because only 1 electron would be involved in the electrode reaction:

$$Cu^+ + 1e^- \rightarrow Cu.$$

When the equivalent weight of a substance is expressed in grams, it is called the *gram-equivalent weight*.

One *faraday*, or 96,500 coulombs, is the quantity of electricity that will deposit 1 gram-equivalent weight of any substance. Thus, the mass m in grams of any substance liberated in electrolysis is

m = gram-equivalent weight × number of faradays transferred.

III. MODEL AND PRINCIPLES OF THE ODR METHOD

Let us now consider the model, which is assumed in the ODR method (Fig. 10.4) (Phene, 1986, p. 1139). As noted, a cylindrical platinum microelectrode simulates the root. Around it are soil particles and the water film, which is right up against the root. The radius of the root (Pt microelectrode) is a and the radius of the root plus the water film is b, so the thickness of the water film is the length *(b − a)*.

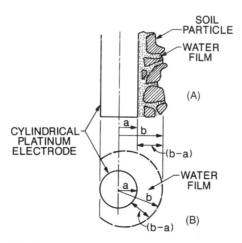

FIG. 10.4 The model that is assumed in order to explain microelectrode behavior. (A) Particles and solution separating the electrode from gas-filled pores. (B) Coaxial cylindrical model with water film of mean thickness *(b − a)*. (From Phene, C.J., ©1986, American Society of Agronomy and Soil Science Society of America: Madison, Wisconsin. Reprinted by permission of the American Society of Agronomy.)

The governing equations are as follows (Phene, 1986, his Equation 5):

$$Q/(At) = (D_e c_2)/[a(\ln b - \ln a)] \tag{10.1}$$

where

Q = amount of oxygen flowing to root [grams]

A = surface area of electrode [cm^2]

t = time [s]

D_e = effective diffusion coefficient of oxygen through the medium surrounding the electrode [cm^2 s^{-1}]

a = radius of root (or of the microelectrode) [cm]

b = radius of root (or of the microelectrode) + water film [cm]

c_2 = concentration of oxygen at the liquid-gas interface (radius = b) [grams cm^{-3}])

We note that there is no c_1 in the equation, and this is because c_1, the concentration of oxygen at the root surface or at the Pt microelectrode, is considered to be zero, which we shall see later.

D_e is dependent on the properties of the soil medium surrounding the microelectrode.

$$D_e = D_o \theta(L/L_e)^2 \tag{10.2}$$

where

D_o = the diffusion coefficient of oxygen through pure water

θ = fraction of the surface area of the microelectrode covered with water as opposed to solid

L/L_e = tortuosity factor of the diffusion path

Troeh et al. (1982) illustrate the effect of tortuosity on diffusion (Fig. 10.5). An oxygen molecule must move through the tortuous (twisted) paths of

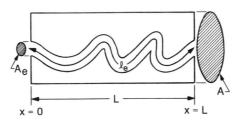

FIG. 10.5 Diffusion model relating the actual path length (l_e) and area (A_e) to the block length (L) and cross section (A). (From Gaseous diffusion equation for porous materials by Troeh, F.R., Jabro, J.D., and Kirkham, D., *Geoderma* 27; 239–253, ©1982, Elsevier Scientific Publishing Co.: Amsterdam. Reprinted by permission of Elsevier, Amsterdam.)

pores in the soil, which is the length L_e (or l_e, as shown by Troeh et al., 1982). The length of the way that the crow flies is L. The tortuosity factor, therefore, is L/L_e (a ratio and dimensionless).

If the electrical potential of the platinum microelectrode is lowered with respect to a reference electrode (the potential at the reference electrode is known), the oxygen at the microelectrode surface is reduced electrolytically until the oxygen concentration at the surface is zero. The reduction rate and the diffusion rate of oxygen are equal. They must be independent of the voltage. Experimentation shows that the current is independent of the voltage when the applied electrical potential (volts) is between 0.3 and 0.7 V (Fig. 10.6). In this range, the current is independent of the voltage and a function only of the diffusion rate of oxygen to the microelectrode surface.

The resulting electrical current is expressed as follows:

$$i = nFAf_{a,t} \tag{10.3}$$

where

i = current (microamperes)

n = the number of electrons required to reduce one molecule of oxygen ($n = 4$)

F = Faraday constant (F = 96,500 coulumbs/mol of oxygen)

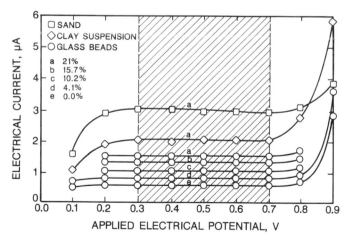

FIG. 10.6 Electrical current-voltage relations for water-saturated media: open squares, sand; open diamonds, clay suspension; open circles, glass beads 18 μm median diameter. Letters a–e refer to concentration of O_2 in equilibrium with saturating solution. (From Phene, C.J., ©1986, American Society of Agronomy and Soil Science Society of America: Madison, Wisconsin. Reprinted by permission of the American Society of Agronomy.)

A = surface area of the microelectrode

$f_{a,t}$ = flux of oxygen at the surface of the microelectrode of radius a at time t

The oxygen flux ($f_{a,t}$) is calculated by measuring the steady-state current (i) after 4 or 5 minutes, assuming that the rate of oxygen reduction is limited by the rate of oxygen diffusion and equal to it. The transport is strictly a diffusion process.

$$ODR = f_{a,t,} = (iM)/(nFA) = [D_o\theta(L/L_e)^2 c_2]/[a(\ln b - \ln a)] \quad (10.4)$$

where M = molecular weight of oxygen (M = 32 grams/mole) and is inserted to convert the units from mole to grams.

Substitution of the values for M, n, and F into Equation 10.4 gives

$$ODR = C(i/A) \; \mu g \; cm^{-2} \; min^{-1} \quad (10.5)$$

where ODR = oxygen diffusion rate ($\mu g \; cm^{-2} \; min^{-1}$)

$$C = (M60)/(nF) = 0.00497 \; \mu g \; \mu A^{-1} \; min^{-1}, \quad (10.6)$$

where A is the SI symbol for ampere (A does not stand for area, as we defined it above).

The 60 is used to convert from seconds to minutes. Remember one coulomb per second is an ampere:

$(q/t) = I$ where q = coulombs; t = seconds; and I = amperes.

Or we can write

$$ODR \; (\mu g \; cm^{-2} \; min^{-1}) = (60Mi)/(nFA) =$$
$$(60 \times 32 \times i)/(4 \times 96,500) \times A = 0.00497 \; (i/A) \quad (10.7)$$

where i is the current and A is the surface area of the microelectrode.

Equation 10.5 or Equation 10.7 includes all physical factors that affect the ODR to a single root of constant dimensions similar to that of the platinum microelectrode.

IV. METHOD

To determine ODR, we use the Soil Oxygen Diffusion Ratemeter, Model D, supplied by Jensen Instruments (Tacoma, Washington). The Jensen Instrument platinum microelectrodes have a length of 4.0 mm and a diameter of 0.65 mm. The formula given in Equation 10.7 then reduces to

$$ODR = 0.059i \; (\mu g \; cm^{-2} \; min^{-1}). \quad (10.8)$$

The ODR measurement system consists of the following components (Fig. 10.7):

1. The platinum microelectrode (the cathode)
2. The Ag-AgCl half-cell (reference electrode)
3. The anode (Jensen Instruments has a brass anode)
4. The electrical circuit
5. A milliammeter to measure the output current.

Oxidation potentials are available from the literature (for example, see Sienko and Plane, 1957, pp. 600–601), so we know the potential of the reference Ag-AgCl half cell.

Any spatial arrangement or maximum distance among the three electrodes (cathode, anode, reference electrode) is permissible, as long as they all contact the same body of soil water. That is, there must be an electrically conductive pathway among electrodes in the soil (Jensen Instruments, undated). For example, one could not do a split-root experiment, with a root system split between two boxes of soil, and put, say, two electrodes in one soil box and one electrode in the other box. The electrolytic solution (the soil water with its dissolved ions) must be in common contact with all three electrodes.

Measurements need to be made at the same voltage. A voltage of 0.65 V relative to a Ag/AgCl reference electrode has become a fairly standard voltage for ODR measurements (Jensen Instruments, undated). The initial current following application of the selected voltage (0.65 V) will be high and will decrease rapidly as oxygen in the immediate vicinity of the

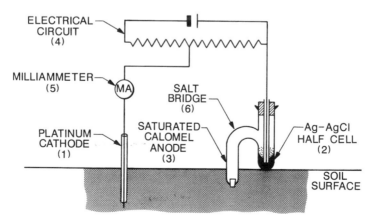

FIG. 10.7 Diagram of apparatus used to make in situ soil measurement of O_2 diffusion. (From Phene, C.J., ©1986, American Society of Agronomy and Soil Science Society of America: Madison, Wisconsin. Reprinted by permission of the American Society of Agronomy.)

microelectrode is depleted. As the current decreases, the rate of decrease will lessen, and the current will tend toward an equilibrium with the rate of oxygen diffusion to the electrodes. After 4 or 5 minutes, the rate of decrease of the current will be small enough that the current can be observed with sufficient accuracy. The current then is read out on the display. The measurements are simple, and a technician easily can be trained to take the readings. However, they are tedious to take because one needs to wait 4–5 minutes between each reading.

Let us assume that we measure a current of 7.94 µA with the Jensen Instrument. What is the ODR? Using Equation 10,8, we find

$$0.059(7.94) = 0.468 \ \mu g \ cm^{-2} \ min^{-1}.$$

We wish to compare this value with others in the literature. We look at values given by Gliński and Stępniewski (1985) in their Fig. 20 on page 80 (Fig. 10.8). On their ordinate, they give units in $\mu g \ O_2 \ m^{-2} \ s^{-1}$; 50 $\mu g \ O_2$

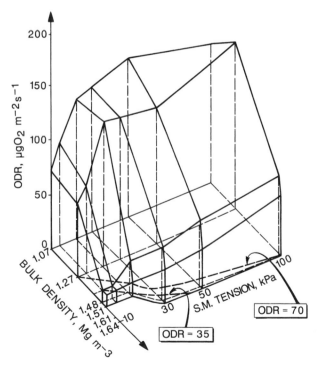

FIG. 10.8 Oxygen diffusion rate versus bulk density and soil moisture tension in a loamy textured black earth. (From Gliński, J., and Stępniewski, W., *Soil Aeration and Its Role for Plants,* p. 80, ©1985, CRC Press: Boca Raton, Florida. Reprinted by permission of CRC Press.)

m^{-2} s^{-1} = 0.3 μg cm^{-2} min^{-1}; 100 μg O_2 m^{-2} s^{-1} = 0.6 μg cm^{-2} min^{-1}; and 150 μg O_2 m^{-2} s^{-1} = 0.9 μg cm^{-2} min^{-1}. So our value of 0.468 μg cm^{-2} min^{-1} falls within values they give, and, consequently, seems reasonable.

We can also compare our value with those presented by Huang et al. (1998), who studied the effect of temperature and aeration status on creeping bentgrass (*Agrostis palustris* Huds.) grown in sand and fritted clay in polyvinyl tubes in a growth chamber. Plants were grown under two temperature regimes (22/15°C day/night and 35/25°C day/night) and two soil aeration treatments, a well-aerated one in which the soil oxygen status was maintained at a sufficient level (ODR = 1.5 μg cm^{-2} min^{-1}) and an over-watered one in which oxygen diffusion rate was maintained at a hypoxic level of about 0.2 μg cm^{-2} min^{-1} (Fig. 10.9). Our ODR value of 0.468 μg cm^{-2} min^{-1} again falls within the range of values reported by Huang et al., so we can be reassured that our value is reasonable.

For the factors affecting ODR measurements, see Phene (1986, p. 1145–1150). Only one factor will be mentioned here: electrode poisoning. This expression is used to indicate anything that can happen to the electrodes other than breakage. Most often, poisoning results from a chemical deposit that changes the characteristics of the Pt surface. Rickman et al.

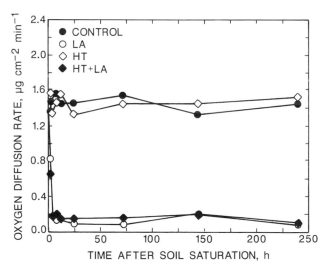

FIG. 10.9 Soil aeration status for creeping bentgrass expressed as oxygen diffusion rates during the experimental period in the well-aerated and optimum temperature treatment (control), low aeration (LA) treatment, high temperature (HT) treatment, and combination of high temperature and low aeration treatments (HT + LA). (From Huang et al., ©1998, Crop Science Society of America: Madison, Wisconsin. Reprinted by permission of the Crop Science Society of America.)

(1968) studied two of the phenomena that cause electrode poisoning: 1) oxide plating at the Pt tip; and 2) the magnitude and nature of the poisoning affecting electrodes installed for periods of 4 weeks or more in different soil. They found that salts (principally calcium bicarbonate) and clay particles (principally biotite) were deposited on Pt microelectrodes left in place in a loamy sand for 2 months, and reduced the ODR by an average of 50% when compared to ODRs from periodically reinserted electrodes in the same soil. Devitt et al. (1989) show excellent pictures of electrodes that have been poisoned.

V. APPENDIX: BIOGRAPHY OF MICHAEL FARADAY

Michael Faraday (1791–1867), who according to Cajori (1929, p. 240–245) was the greatest experimentalist of the nineteenth century in the field of electricity and magnetism, was born on September 22, 1791, at Newington, Surrey, which later became part of the borough of Southwark in south London, but was then in the country. He, himself, would have said that he was a "natural philosopher" instead of a physicist and chemist (da Costa Andrade, 1971). He was the son of a blacksmith. "My education," he said, "was of the most ordinary description, consisting of little more than the rudiments of reading, writing, and arithmetic at a common day-school. My hours out of school were passed at home and in the streets" (Cajori, 1929).

In 1804, he served as errand boy at a bookstore and bookbindery near his home. The following year, when he was 14 years old, he became an apprentice to the bookbinder. At that time he liked to read scientific books, which happened to pass through his hands. "I made such simple experiments in chemistry," he said, "as could be defrayed in their expense by a few pence per week, and also constructed an electrical machine" (Cajori, 1929). At the age of 19 he sometimes in the evening attended lectures given by a Mr. Tatum on natural philosophy, his brother paying the admission fee for him. In 1812, he had the good fortune to hear four lectures delivered at the Royal Institution by Sir Humphry Davy, the great English chemist (1778–1829).

About this time, Faraday started to work as a journeyman bookbinder for a Frenchman in London. His new work was uncongenial. "My desire," he said, "to escape from trade, which I thought vicious and selfish, and to enter into the service of science, which I imagined made its pursuers amiable and liberal, induced me at last to make the bold and simple step of writing to Sir H. Davy, expressing my wishes, and a hope that if an opportunity came in his way he would favor my views; at the same time, I sent the notes I had taken of his lectures" (Cajori, 1929). Davy replied, "I am far from displeased with the proof you have given me of your confidence. . . ." Faraday became Davy's assistant at the Royal Institution in 1813. In the autumn of that year, Davy and his wife started on a tour abroad, Faraday going with them as amanuensis. Even though Faraday had many menial duties to perform on this trip, he saw much of the active scientific research in Europe, and the trip expanded his view. "His University was Europe; his professors the master whom he served, and those illustrious men to whom the renown of Davy introduced the travellers" (da Costa Andrade, 1971).

After being with Davy in France, Italy, and Switzerland, he returned to the Royal Institution in 1815. Soon after his return he began original researches, and published his first paper in 1816. He also began to lecture before the City Philosophical Society. In a letter he wrote about "the glorious opportunity I enjoy of improving in the knowledge of chemistry and the sciences with Sir H. Davy." In 1821, in his thirtieth year, Faraday married and brought his wife to his rooms at the Royal Institution, where they lived together for 46 years. They had no children. In 1824, he was elected member of the Royal Society at a time when Davy was its president, in spite of Davy's jealous opposition to Faraday's election (Cajori, 1929). In 1823, Faraday had liquified chlorine, which aroused the jealousy of Davy, who considered that he had initiated the work and was entitled to the credit (da Costa Andrade, 1971). Nevertheless, Faraday always spoke with respect and admiration for the talents of the man who had done so much to start him in his early scientific career (Cajori, 1929).

In 1825, Faraday made a chemical discovery of the first importance by isolating benzene from a liquid obtained in the production of oil gas (da Costa Andrade, 1971). In that same year Faraday's position at the Royal Institution was improved by his promotion to the post of director of the laboratory. The next year he began to give formal lectures for the members of the Institution on Friday evenings, and those Friday evening discourses have continued ever since (da Costa Andrade, 1971). He also initiated the Christmas lectures for young people, known formally as Christmas Courses of Lectures Adapted to a Juvenile Auditory, of which he himself gave 19 courses. As an inspiring lecturer and deviser of effective lecture experiments Faraday was supreme, and there are many contemporary accounts of the interest and enthusiasm that his discourses aroused.

Faraday's conceptions of electric and magnetic force and their interrelations, expressed in terms of his lines of force, were fundamental. It was from them that James Clerk Maxwell (Scottish physicist, 1831–1879) developed his equations and the concept of electromagnetic waves, which lie at the base of all modern theories of electromagnetic phenomena (da Costa Andrade, 1971).

Cajori (1929, pp. 244-245) says:

> Faraday's first magnetoelectric apparatus, the forerunner of the dynamo, produced such insignificant results that Faraday after lecturing upon it, was asked what on earth was the use of it. A church dignitary had a conception of its dangerous possibilities in the hands of incendiaries, and deplored the discovery. Knowledge antedates understanding. . . . We live forward, but we understand backwards. . . . There is an aspect here of our physical research that is often lost sight of, namely, the small proportion of successful discoveries compared with the number of investigators. Certainly the number of unsuccessful attempts, even in the case of those fortunate individuals who make the great discoveries, is very much greater than the number of their successful attempts. Faraday's reputed satisfaction with 1/10 percent return comes to mind.

Faraday was a Sandemanian, a Christian group prevalent in the Pennines (hills in northern England on the Scottish border) that has now died out. Tanford (1991) says, "There is no question about Faraday's faith and that it guided him when addressing moral issues, but Faraday himself never claimed a connection with his scientific work and indeed emphasized the need to separate religious and ordinary beliefs in an 1854 lecture."

It was shortly after his work on electromagnetic induction that Faraday, always in search of unity, showed that the five kinds of electricity then distinguished—frictional, galvanic, voltaic, magnetic (induced current), and thermal—were fundamentally the same. In this same period of his researches (c. 1831–1844), he arrived at the basic laws of electrolysis that bear his name, and introduced the terms that are universally used: anode, cathode, anion, cation, and electrode.

In 1858, he retired to live hear Hampton Court, Surrey, but retained a lively interest in science. His health gradually waned, and there he died on August 25, 1867. Faraday was possibly the greatest experimental genius the world has known (da Costa Andrade, 1971). He was incessantly prompted by the belief that certain fundamental relations were waiting to be found. He was not dismayed by dozens of fruitless experiments, and he persisted until basic discoveries were established. To all his other gifts he added the ability to describe his ideas in clear and simple language. Five books about Faraday were published in 1991 and 1992, and have been reviewed by Hunt (1992).

REFERENCES

Buckingham, E. (1904). Contributions to our knowledge of the aeration of soils. *USDA Bur Soils Bull* 25; 52 pp.

Cajori, F. (1929). *A History of Physics*. Macmillan: New York.

Cantor, G. (1991). *Michael Faraday: Sandemanian and Scientist. A Study of Science and Religion in the Nineteenth Century*. St. Martin's: New York.

da Costa Andrade, E.N. (1971). Faraday, Michael. *Encyclopaedia Britannica* 9; 66 67.

Devitt, D.A., Stolzy, L.H., Miller, W.W., Campana, J.E., and Sternberg, P. (1989). Influence of salinity, leaching fraction, and soil type on oxygen diffusion rate measurements and electrode "poisoning." *Soil Sci* 148; 327–335.

Gliński, J., and Stępniewski, W. (1985). *Soil Aeration and Its Role for Plants*. CRC Press: Boca Raton, Florida.

Huang, B., X. Liu, and J.D. Fry. 1998. Shoot physiological responses to two bentgrass cultivars to high temperature and poor soil aeration. *Crop Sci* 38; 1219–1224.

Hunt, B.J. (1992). Faraday at home and abroad. *Nature* 256, 1059–1060.

Jensen Instruments. (Undated). *Operating Instructions. Soil Oxygen Diffusion Ratemeter, Model D*. 10 pp. Jensen Instruments, 2021 South Seventh Street, Tacoma, Washington 98405-3014.

Kirkham, D., and Powers, W.L. (1972). *Advanced Soil Physics*. Wiley: New York.

Kirkham, M.B. 1994. Streamlines for diffusive flow in vertical and surface tillage: A model study. *Soil Sci Soc Amer J* 58; 85–93.

Lemon, E.R., and Erickson, A.E. (1952). The measurement of oxygen diffusion in the soil with a platinum microelectode. *Soil Sci Soc Amer Proc* 16; 160–163.

Lide, D.R. (1994). *CRC Handbook of Chemistry and Physics*. CRC Press: Boca Raton, Florida.

Penman, H.L. (1940). Gas and vapor movement in the soil. I. *J Agr Sci* 30; 437–462.

Phene, C.J. (1986). Oxygen electrode measurement. In *Methods of Soil Analysis. Part 1. Physical and Mineralogical Methods* (Klute, A., Ed.) 2nd ed., pp. 1137–1159. American Society of Agronomy, Soil Science Society of America: Madison, Wisconsin.

Rickman, R.W., Letey, J., Aubertin, G.M., and Stolzy, L.H. (1968). Platinum microelectrode poisoning factors. *Soil Sci Soc Amer Proc* 32; 204–208.

Schaum, D. (1961). *Theory and Problems of College Physics*. 6th ed. Schaum Publishing: New York.

Sienko, M.J., and Plane, R.A. (1957). *Chemistry*. McGraw-Hill: New York.

Tanford, C. (1991). Religious resonances. Book review of *Michael Faraday: Sandemanian and Scientist*. By Geoffrey Cantor. Macmillan, New York, 359 pp. *Nature* 351; 705.

Taylor, S.A. (1949). Oxygen diffusion in porous media as a measure of soil aeration. *Soil Sci Soc Amer Proc* 14; 55–61.

Troeh, F.R., Jabro, J.D., and Kirkham, D. (1982). Gaseous diffusion equations for porous materials. *Geoderma* 27; 239–253.

van Bavel, C.H.M. (1952). Gaseous diffusion and porosity in porous media. *Soil Sci* 73; 91–96.

Weast, R.C. (1964). *Handbook of Chemistry and Physics*. Chemical Rubber Co.: Cleveland, Ohio.

Wesseling, J., and van Wijk, W.R. (1957). Soil physical conditions in relation to drain depth. In *Drainage of Agricultural Lands* (Luthin, J.N., Ed.), pp. 461–504. American Society of Agronomy: Madison, Wisconsin.

11

Infiltration

In previous chapters we have introduced fundamental concepts about water in soil, including static water in soil and water movement in saturated soil. In this chapter we consider infiltration, the tension infiltrometer, and four soil characters (unsaturated hydraulic conductivity, sorptivity, repellency, and mobility) that we can measure with the tension infiltrometer.

I. DEFINITION OF INFILTRATION

Infiltration rate may be defined as the meters per unit time of water entering into the soil regardless of the types or values of forces or gradients. The term *hydraulic conductivity*, which has been defined as the meters per day of water seeping into the soil under the pull of gravity or under a unit hydraulic gradient, should not be confused with infiltration rate. Infiltration rate need not refer to saturated conditions. If two rain drops of total volume 2 mm^3 = 0.000002 m^3 fall per day on a m^2 of soil and are absorbed into the soil, the infiltration rate is 0.000002 m/day.

Water entry into soil is caused by matric and gravitational forces. Therefore, this entry may occur in the lateral and upward directions as well as the downward one (Baver et al., 1972, p. 365). Infiltration normally refers to the downward movement. The matric force usually predominates over the gravitational force during the early stages of water entry into soil, so that observations made during the early stages of infiltration are valid when considering the absence of gravity.

If water infiltrates into a dry soil, a definite *wetting front*, also called a *wet front*, can be observed. This is the boundary between the wetted upper part of the soil and the dry lower part of the soil. If water is infiltrating into soil contained in a clear plastic column, one can observe the progress of the wet front and mark wet fronts as they change with time (Fig. 11.1). At present, it is impossible to measure the matric potential exactly at the wet front, because it progresses too rapidly into the soil. However, one can measure the amount of water infiltrated and the depth and shape of the wet front, and come to important conclusions about the entry of water into the soil. Infiltration is extremely important, because it determines not only the amount of water that will enter a soil, but also the entrainment of the "passenger" chemicals (nutrients, pollutants) dissolved in it.

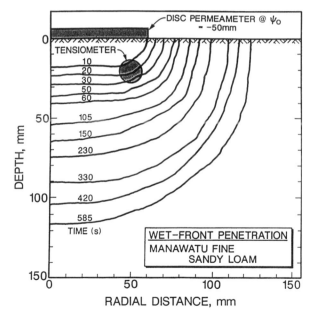

FIG. 11.1   Wet fronts for a sandy loam soil. (From Kirkham, M.B., and Clothier, B.E. ©1994a. Ellipsoidal description of water flow into soil from a surface disc. *Trans Int Congr Soil Sci* 2b; 38–39. Reprinted by permission of The International Society of Soil Science.)

II. FOUR MODELS OF ONE-DIMENSIONAL INFILTRATION

Four models for infiltration into the soil have been developed. They all deal with one-dimensional, downward infiltration into the soil (Baver et al., 1972, pp. 366–371).

A. Lewis Equation

From work initiated in 1926, Mortimer Reed Lewis, an irrigation engineer at Oregon State College, used the following equation for infiltration:

$$I = \gamma t^a \tag{11.1}$$

where I is the cumulative infiltration between time zero and t, and γ and a are constants. Equation 11.1 has been erroneously attributed to A.N. Kostiakov, and often appears in the literature as the "Kostiakov" equation (Swarzendruber, 1993). The parameters in Eq. (11.1) are evaluated by fitting the model to experimental data. By definition, the infiltration rate $i = dI/dt$. Thus, the infiltration rate for the Lewis equation is given by

$$i = a\gamma t^{a-1}. \tag{11.2}$$

B. Horton Equation

In the 1930s, Robert E. Horton, a pioneer in the study of infiltration in the field, developed the following equation:

$$i = i_f + (i_o - i_f) \exp(-\beta t) \tag{11.3}$$

where i_o is the initial infiltration rate at $t = 0$, i_f is the final constant infiltration rate that is achieved at large times, and β is a soil parameter that describes the rate of decrease of infiltration.

Horton felt that the reduction in infiltration rate with time was largely controlled by factors operating at the soil surface. These included swelling of soil colloids and the closing of small cracks, which progressively sealed the soil surface. He also recognized that a bare soil surface was compacted by raindrops, but crop cover mitigated their effect. Horton's field data showed that the infiltration rate eventually approached a constant value, which was often somewhat smaller than the saturated permeability of the soil. The latter observation was thought to be due to air entrapment.

C. Green and Ampt Equation

The preceding models are empirical. W. Heber Green and G.A. Ampt in Australia published in 1911 an infiltration equation that was based on a simple physical model of the soil. It has the advantage that the parameters in the equation can be related to physical properties of the soil. Physically, Green and Ampt assumed that the soil was saturated behind the wetting front and that one could define some "effective" matric potential at the wetting front. During infiltration, if the soil surface is held at a constant matric potential or head h_o with associated water content θ_o (e.g., by ponding water over it), water enters the soil behind a sharply defined wet front that moves downward with time (Fig. 11.2A) (Jury et al., 1991, pp. 131–134). Green and Ampt replaced this process with one that has a discontinuous change in water content at the wetting front (Fig. 11-2B). In addition, they made the following assumptions: 1) The soil in the wetted region has constant properties (K_o, θ_o, D_o, h_o, where K_o and D_o are the

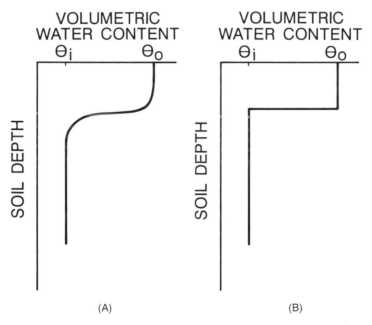

(A)   (B)

FIG. 11.2 Water content profiles during infiltration. (A) A profile that actually occurs during infiltration. (B) A profile corresponding to the Green-Ampt infiltration model. (From Jury, W.A., Gardner, W.R., and Gardner, W.H. ©1991. *Soil Physics*, 5th ed., p. 132. John Wiley & Sons: New York. This material is used by permission of John Wiley & Sons, Inc.)

hydraulic conductivity and water diffusivity in the Green-Ampt model, respectively) and 2) the matric potential (head) at the moving front is constant and equal to h_F.

The Green-Ampt model can be used to calculate the infiltration rate into a horizontal soil column initially at a uniform water content θ_i such that $\theta_o > \theta_i$ and an associated matric potential or head h_o maintained at the entry surface for all times > 0. Using the assumptions of the model and Darcy's law, the following equation can be derived:

$$i = (dI/dt) = \Delta\theta \, (D_o/2t)^{1/2} \tag{11.4}$$

where i = infiltration rate; I = infiltration; t = time; $\Delta\theta = \theta_o - \theta_i > 0$, $D_o = K_o \Delta h / \Delta\theta$ is the soil water diffusivity of the wet soil region $0 < x < L$, the depth of the wetting front, and $\Delta h = h_o - h_F > 0$, and K_o is the constant hydraulic conductivity of the wet region, and h_F is the matric potential or head of the moving front. Note in this model that the infiltration rate into the soil is proportional to $t^{-1/2}$. A similar expression is obtained for infiltration into a vertical soil column at short times after infiltration begins.

The model has been used as a conceptual aid in visualizing a complex process. Indirect evaluation of h_F has permitted the model to be used in practical applications.

D. Philip Infiltration Model

J.R. Philip in 1957 suggested an approximate algebraic equation (based on sound physical reasoning) for vertical infiltration under ponded conditions. (See the Appendix, Section XI, for a biography of Philip.) The equation, which is simple yet physically well founded, is as follows:

$$I = St^{1/2} + At \tag{11.5}$$

where I is the cumulative infiltration (mm), S is the sorptivity (mm hr$^{-1/2}$), and A is an empirical constant (mm/hr). The first term on the right-hand side of Equation 11.5 gives the gravity-free absorption into a ponded soil due to capillarity and adsorption. The second term represents the infiltration due to the downward force of gravity. S and A may be found empirically by fitting Equation 11.5 to infiltration data. Alternatively, these parameters may be derived from the hydraulic properties of the soil. This is not possible for other empirical infiltration equations. For horizontal (gravity-free) infiltration, cumulative infiltration I is given by

$$I = St^{1/2} \tag{11.6}$$

III. TWO- AND THREE-DIMENSIONAL INFILTRATION

The previous discussion dealt with one-dimensional infiltration in which water is assumed to flow vertically (or more rarely horizontally) into the soil. Multidimensional infiltration theory is an area of soil physics research dominated by the works of J.R. Philip, who published on the topic in a major paper in 1969. Sequels to his work have been carried out by Peter A.C. Raats (1971) in the Netherlands and Robin A. Wooding (1968) in New Zealand.

According to Jury et al. (1991, p. 143), Wooding (1968) derived an approximate expression for the steady rate of infiltration from a circular pond of radius r_o, overlying a soil in which the hydraulic conductivity-matric potential function was assumed to be

$$K(h) = K_o \exp(\alpha h) \tag{11.7}$$

where K_o and α are constants representing the soil properties and h is the matric potential (head). Wooding (1968) says that the parameter α is defined as the logarithmic derivative of the hydraulic conductivity with respect to capillary potential (what we now call matric potential). Since $K = K_o$ when $h = 0$, K_o represents the saturated hydraulic conductivity of a soil. Using this expression and a simplified form of the three-dimensional water flow equation, Wooding derived the following equation for the steady infiltration flux rate:

$$i_f = K_o [1 + (4/\pi\alpha r_o)]. \tag{11.8}$$

[Note that in the equation on p. 143 of Jury et al., there is a mistake. The last symbol, r_o is squared in Jury et al. This is incorrect. Wooding's equation is written correctly in Equation 11.8 above.] It is notable that, contrary to one-dimensional flow, the final infiltration rate in Wooding's equation (Eq. 11.8) exceeds K_o. This occurs because water may enter and move laterally as well as vertically.

Multidimensional infiltration models have utilized difficult mathematics. However, practical advances in infiltration can be made with simple models. For example, a simple, ellipsoidal description of the pattern of wetting to approximate the depth to the wetting front underneath a disc permeameter, set at Ψ_0 and supplying water to soil initially at water content θ_n, will be described in Section X of this chapter.

IV. REDISTRIBUTION

The term *redistribution* refers to the continued movement of water through a soil profile after irrigation or rainfall has stopped at the soil surface.

Redistribution occurs after infiltration and is complex, because the lower part of the profile ahead of the wet front will increase its water content and the upper part of the profile near the surface will decrease its water content, after infiltration ceases. Thus, hysteresis can have an effect on the overall shape of the water content profile. See Jury et al. (1991, p. 144 and following) for a discussion of redistribution and figures illustrating it.

V. TENSION INFILTROMETER OR DISC PERMEAMETER

Recognition of the importance of macropores and preferential flow has led to the development of instruments that can be used in the field to control preferential water flow through macropores and soil cracks. Let us first define macropores and see their size in relation to other soil pores. Pores in the soil can be classified into five categores (Clothier, 2004): macropores, with diameters ranging from 75 to >5000 µm; mesopores with diameters ranging from 30 to 75 µm; micropores with diameters ranging from 5 to 30 µm; ultramicropores with diameters ranging from 0.5 to 5 µm; and cryptopores with diameters <0.1 µm. Xylem vessels, by comparison, range in diameter from 8 to 500 µm, with 40 µm a reasonable value to use in calculations (Nobel, 1983, p. 493).

The first practical instrument to control macropore flow was developed in 1981 by Brent E. Clothier of New Zealand and Ian White of Australia (Clothier and White, 1981). This simple instrument was known as the *sorptivity tube* (Fig. 11.3), then as the *tension infiltrometer*, and later still it evolved into the *disc permeameter* (Fig. 11.4), as described by Perroux and White (1988). Originally, the term "disc permeameter" was used when three-dimensional infiltration was being considered, and the term "tension infiltrometer" was used when one-dimensional infiltration was being considered, but today the terms are used interchangeably. One must state if one is considering one-dimensional or three-dimensional flow when using the instruments. With these instruments, the amount of macropore flow measured is controlled by applying water to soil at water potentials Ψ_o, less than 0. The maximum diameter of vertical pores, connected to the soil surface, through which water can enter is given by the capillary rise equation [also given in Chapter 6 as Equation 6.9]:

$$h_c = (2\sigma \cos \alpha)/(r\rho g), \tag{11.9}$$

where σ = surface tension (surface tension coefficient) of the liquid (units of g/s^2 or dyne/cm); α = contact angle between the liquid and tube;

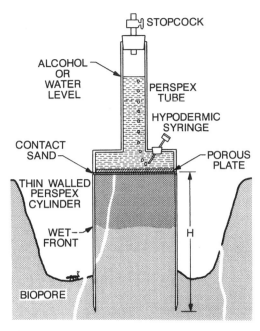

FIG. 11.3   The sorptivity tube. An ant hole of greater than 0.75 mm in diameter is shown, which has no effect on the infiltration process. $H = 9.5$ cm. The original sorptivity tube was made out of glass. Alcohol can be put in the glass supply tower to determine if a soil is repellent. (From Clothier, B.E., and White, I., ©1988, Soil Science Society of America. Reprinted by permission of the Soil Science Society of America.)

r = radius of tube (cm); ρ = density of liquid (g/cm^3); and g = acceleration due to gravity (cm/s^2).

This maximum diameter is proportional to the matric potential, $(-\Psi_o)^{-1}$. The more negative the Ψ_o, the smaller is the maximum diameter of a pore that can participate in flow from the soil surface. These two instruments (tension infiltrometer and disc permeameter) are being used to supply water to soil in situ at readily selectable zero or negative pressures. A "ready reckoner" of the relationship between the negative pressure Ψ, where Ψ is in terms of energy per unit weight (in cm of H$_2$O head), and the capillary diameter d in mm is $-3/d$ (Clothier, 2004). For example, a 4-cm head will fill pores up to 0.75 mm.

In Figure 11.4 we see that a disc permeameter consists of two towers: one that is open to the air and one that is sealed off from the air. The one that is sealed off from the air supplies water to the soil under tension. The amount of tension is determined by the depth, d, that the air-inlet tube is below the water surface. The two towers are connected by an air-supply

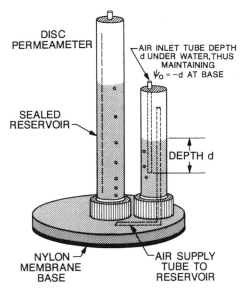

FIG. 11.4 The disc permeameter. (From a talk given by Brent E. Clothier at the International Soil Science Congress, Acapulco, Mexico, July, 1994. The paper from the Congress is published as Clothier et al., 1994, but the figure is not published in the paper. Reprinted by permission of Brent E. Clothier.)

reservoir at the bottom. But note that the bubble tower (with air-inlet tube) is not connected to the reservoir. Hence, one needs only to replenish the reservoir with a tracer solution such as KBr, when determining mobility (see Section IX). The nylon membrane base determines the tension that can be held, and is similar to the porous cup on a tensiometer. The amount of tension that a tensiometer can hold is determined by the size of the pores in the ceramic cup. The smaller the pores, the more tension can be held. The nylon membrane used in the type of disc permeameter shown in Fig. 11.4 is Nybolt Nylon Monofil Mesh, Reference PA40/23, 40 micrometer opening and 23% free surface, purchased from Ure Pacific, Auckland, New Zealand (Kirkham and Clothier, 2000).

The maximum tension that a tension infiltrometer can hold is about −150 mm, but on a good day it can hold −200 mm (B.E. Clothier, personal communication, October 18, 1996). Although this might not seem like much compared to one bar (−1020 cm water), when we consider the height of water being held (15 to 20 cm), this is a significant amount of water.

Remember that we are infiltrating water into soil under tension, but the pores are filled with water. The water in the pores is under tension. Whether the water under the tension infiltrometer is part of the vadose zone depends on the definition of the term. The *Glossary of Soil Science Terms* (Soil Science Society of America, 1997) defines *vadose zone* as "The aerated region of soil above the permanent water table." Sometimes it is defined as the "unsaturated" zone. The *Glossary*'s definition would not include the water under a tension infiltrometer. The water under a tension infiltrometer is similar to water in the falling-water-table situation, when the water table has fallen below the soil surface, and the capillary tubes (soil pores) hold water under tension (see Fig. 6.9). When the water table is at the soil surface, the water is not under tension. Under these conditions, or saturated conditions with no tension, Darcy's law applies.

When using the tension infiltrometer, vegetation is scraped away from the soil and a thin layer of contact sand is put between the tension infiltrometer and the soil surface, as shown in Fig. 11.3. The tension infiltrometer is used to determine the unsaturated hydraulic conductivity and sorptivity; the repellency of soils; and the mobility of chemicals through the soil, which we shall discuss in Sections VII, VIII, and IX, respectively.

The tension infiltrometer is sold in the United States by Soil Measurements Systems (SMS), Tucson, Arizona, with a system patented by Iowa State University (Ankeny et al., 1989). SMS pays a royalty to Iowa State University for sale of the equipment. SMS makes two models, one with an 8-cm diameter base plate (Model No. SW-080) and one with a 20-cm base plate (Model No. SW-808B). The reason the equipment could be patented was because Ankeny et al. (1989) added an adjustment to the bubble tower to provide variable tension. The Clothier and White (1981) and Perroux and White (1988) models (Figs. 11.3 and 11.4; not patented) have only one air-inlet tube and provide only one tension, but a tension that can be adjusted by moving the air-inlet tube up and down in the water reservoir. The Ankeny et al. (1989) model has three air-inlet tubes. Two are clamped off while the third tension is used.

VI. MINIDISK INFILTROMETER

The minidisk infiltrometer is made by Decagon Devices (Pullman, Washington). It consists of a plastic tube, 22.5 cm long and 3.1 cm in outside diameter, marked with milliliter gradation (0 to 100 mL), a rubber stopper placed in the top, and a styrofoam-looking base that holds the tension. One-half centimeter above the base is an air-inlet tube. The original minidisk infiltrometer, first sold in 1997, infiltrates water at a set suction (tension) of 2.0 cm and has a radius of 1.59 cm. Decagon Devices developed two more minidisk infiltrometers at set suctions of 0.5 cm and 6.0 cm (each with a radius of 1.59 cm). The minidisk infiltrometer is especially suited for greenhouse work with pots, even though it can be used in the field, too. The hydraulic conductivity of soil can be measured with it using the method of Zhang (1997).

What the base is made out of is proprietary information, but it corresponds to the nylon membrane at the bottom of the tension infiltrometer. The base of the minidisk infiltrometer can get clogged when working on organic soils, such as occurred in the experiments of Kirkham and Clothier (2000), but the disk can be modified by puncturing the bottom with small needle holes and replacing the tension it held with the nylon used in the tension infiltrometer designed by Clothier and White (1981). Kirkham and Clothier (2000) did this modification to maintain functionality of the minidisk infiltrometer. The original base was not removed because it was necessary to have a base to place on the soil during infiltration.

The tension of 2 cm is established by the length of the plastic tube inserted near the bottom of the infiltrometer. There are two forces that balance out to establish the 2 cm. First, there is a gravitational head that must be offset. This is the height of the air entry point from the base of the minidisk. This head is offset by the capillarity of the thin plastic tube, plus the hydraulic resistance of the length of the tube. Personnel at Decagon Devices have worked out this length, so that the tension is 2 cm (B.E. Clothier, personal communication, March 28, 2001).

Some people have misunderstood the minidisk infiltrometer and think that water in it is under pressure. Water is not under pressure, but has to be under tension; if it were under pressure, the water would jump out of the tube. To get water into the infiltrometer before a run, one places the minidisk infiltrometer in a large bucket of water and plugs the top with the rubber stopper. The water stays in the minidisk infiltrometer until it is placed on soil, when it is then sucked out of the tube.

The minidisk infiltrometer has the advantage of being portable. It can be carried to any soil on the globe in a purse or carry-on suitcase, where it can be used to determine the hydraulic conductivity of the soil. No checking of baggage is needed, as for the larger versions of the tension infiltrometer.

VII. MEASUREMENT OF UNSATURATED HYDRAULIC CONDUCTIVITY AND SORPTIVITY WITH THE TENSION INFILTROMETER

Even though the minidisk infiltrometer can be used to get hydraulic conductivity, the tension infiltrometer (Fig. 11.4) is more widely used.

Two methods can be used to get unsaturated hydraulic conductivity with the tension infiltrometer: the method of Smettem and Clothier (1989) and the method of Ankeny et al. (1991). In the Smettem and Clothier method, two tension infiltrometers with different radii are used to get both unsaturated hydraulic conductivity and sorptivity. In their method, two equations with two unknowns (unsaturated hydraulic conductivity and sorptivity) are solved simultaneously. The Ankeny et al. (1991) method requires the use of the equation developed by Gardner (1958) to get the relationship between the hydraulic conductivity and water content. In both methods, three-dimensional infiltration is assumed and the Wooding equation is used. For the Wooding equation to be used, infiltration must be a steady state. Thus, before either method is used, one must make sure that the water is infiltrating into the soil at steady state (i.e., the infiltration rate, mm/s, is a constant value). Let us first look at sorptivity, because we will be determining it along with hydraulic conductivity when we consider the method of Smettem and Clothier (1989).

The concept of sorptivity comes from the theoretical work that Philip (1969) did on infiltration. The sorptivity of a soil is a measure of the ability of the soil to attract water by capillary action. When considering sorptivity, think of the soil as a sponge. Each soil (sponge) has its ability to absorb water. The units of sorptivity are length/square root of time (e.g., mm/s^2). Infiltration into a soil is affected by gravity and capillarity. Equation 11.5 gives Philip's equation for vertical infiltration and Equation 11.6 gives Philip's equation for gravity-free infiltration. In both equations, infiltration goes as the square root of time and from Equation 11.6 we can see that the units of sorptivity are length divided by the square root of time.

Here is the theoretical development of Smettem and Clothier (1989) based on theory developed by Philip (1969):

$$S_o^2 = [(\theta_o - \theta_n)/b] \int_{\theta_n}^{\theta_o} D \, d\theta, \tag{11.10}$$

where

   S_o = sorptivity
   θ_n = initial soil water content
   θ_o = water content to which the soil surface is wetted
   D = diffusivity
   $D = k \, (\partial \varphi / \partial \theta)$
   k = hydraulic conductivity under unsaturated conditions

b is a parameter and is a function of the slope of the diffusivity function; $\frac{1}{2} < b < \pi/4$; b can be approximated as 0.55.

Combining Wooding's (1968) equation (Equation 11.8) and Philip's theory (1969), Smettem and Clothier (1989) obtained the following equation:

$$q_\infty/(\pi r^2) = K_o + (2.2S_o^2) / (\pi r \Delta\theta), \qquad (11.11)$$

where q_∞ = flow at long times (steady conditions); this is the same as i_f in Equation (11.8).

$$\Delta\theta = \theta(\psi_o) - \theta(\psi_n). \qquad (11.12)$$

ψ_o is the supply potential (i.e., the tension with which the water is applied to the soil). Smettem and Clothier (1989) used a supply potential of −35 mm when they did their experiment. ψ_n is the potential of the soil before the tension infiltrometer is put on top of it. For practical purposes, $\Delta\theta = \theta_o - \theta_n$.

The theoretical development is difficult, but the resulting equation is simple to use for the experimentor. Two tension infiltrometers with different radii are taken to the field where one wants to determine unsaturated hydraulic conductivity and sorptivity. The antecedent volumetric water content is determined (θ_n) and the tension infiltrometers are set close by. After steady infiltration has been reached for each, one records the value. One then takes the tension infiltrometers off the soil and gets the volumetric water content of soil right under the tension infiltrometers. This is θ_o. One then determines $\Delta\theta$ and solves Equation 11.11 for K_o and S_o by simultaneous solution using data from the discs with two different radii. We have the known parameters, $\Delta\theta$, r, and q, and we have the two unknowns, S_o and K_o.

The method of Smettem and Clothier (1989) is simple and results in two values (K_o and S_o). However, many people cannot obtain or afford two tension infiltrometers with different radii, hence the method of Ankeny et al. (1991) provides a means of obtaining unsaturated hydraulic conductivity with just one tension infiltrometer.

The determination of unsaturated hydraulic conductivity using the method of Ankeny et al. (1991) is not straightforward or obvious from reading the paper. On November 7, 1995, Dr. Brent E. Clothier visited Kansas State University and demonstrated to students how to determine unsaturated hydraulic conductivity with this method. We now go through the procedure, step by step, as he explained it.

We need to know the hydraulic conductivity as a function of head, h. We use the following equation [rewritten from Equation 11.7 where K_s is used instead of K_o used by Jury et al. (1991)]:

$$K(h) = K_s \exp (\alpha h), \tag{11.13}$$

where K_s and α are constants. In Equation 11.13, K_s represents gravity and α represents capillarity; α also is a slope (the slope of the hydraulic conductivity versus head). We can relate K in Equation 11.13 to the K in a form of Darcy's law, as follows:

$$J_w = K (\partial \psi)/(\psi z), \tag{11.14}$$

where J_w is the soil water flux.

The soil that Clothier used in his demonstration was a Haynie sandy loam. It has 65% sand, 24% silt, and 11% clay.

In the Ankeny et al. (1991) method, we use two heads. In this experiment, the two heads will be as follows:

$h_1 = -10\ cm$
$h_2 = -2\ cm$

We will have two unknowns and two equations and will solve the equations to get the unsaturated hydraulic conductivity. We need to run the tension infiltrometer two times; therefore, we shall put it two times on the soil. Water flows into the soil by gravity and capillarity.

In his analysis, Clothier used the following form of Wooding's (1968) equation:

$$Q = \pi r^2 K_o + (4rK_o)/\alpha, \tag{11.15}$$

where Q (m^3/s) is the flow through a disc, which is proportional to the surface area of the disc times the hydraulic conductivity plus capillarity (movement of water that goes off the perimeter). The first term on the right-hand side of Equation 11.15 represents the gravity component, and the second term on the right-hand side of Equation 11.15 represents the capillary component. We can measure Q, how fast the water in the sealed reservoir drops in the tension infiltrometer.

From Equation 11.15, we write

$$q = K_o[1 + 4/\pi\alpha r]. \tag{11.16}$$

We shall apply this equation at one head and then another head.

$$K = K_1 \exp (\alpha h_1) = K_2 \exp (\alpha h_2) \tag{11.17}$$

$$K_1/K_2 = \exp [\alpha (h_1 - h_2)] \tag{11.18}$$

$$\alpha = [\ln(K_1/K_2)]/(h_1 - h_2) = [\ln (q_1/q_2)/(h_1 - h_2). \tag{11.19}$$

We are taking the functional form, $K = K_s \exp(\alpha h)$. We have a fixed radius and α is constant.

$$(K_1/K_2) = q_1/q_2. \tag{11.20}$$

So we shall measure two q's (q_1 and q_2).

If we have a clay soil, capillarity dominates and α (slope of the hydraulic conductivity versus head) is small. If we have a sandy soil, α is large. For our first head, $h_1 = -10$ cm (the air tube is 10 cm under water). We apply the tension infiltrometer to the soil. We wait until steady state flow.

We need to know the ratio, R, of the areas of the reservoir to the disc (the area that touches the soil) for the tension infiltrometer. The nylon base of the tension infiltrometer had a diameter of 6.57 cm. The reservoir had a diameter of 3.37 cm.

$$(3.37/6.57)^2 = R = 0.263. \tag{11.21}$$

We time the drop in the sealed reservoir and get the steady state rate. We multiply this rate by 0.263 to get q_1. We find the following:

$$q_1 = 0.115 \text{ cm/s}. \tag{11.22}$$

We have done the −10 cm head.

Now we set the tension infiltrometer at the −2 cm head. We find the following:

$$q_2 = 0.175 \text{ cm/s}. \tag{11.23}$$

From Equation 11.19, we have

$$\alpha = [(\ln(q_1/q_2))] / (h_1 - h_2)$$

$$\alpha = [\ln(0.115 / 0.175)]/[-10 \text{ cm} - (-2 \text{ cm})] = 0.05 \text{ cm}^{-1} \tag{11.24}$$

$$K = \exp(0.05h). \tag{11.25}$$

We use Equation 11.16, Wooding's equation, to calculate K_o:

$$q = 0.175 \text{ cm/s}$$
$$\alpha = 0.05 \text{ cm}^{-1}$$

$$r = 6.57 \text{ cm}/2 = 3.285 \text{ cm}.$$

We solve for K_o:

$$K_o = 0.02 \text{ cm/s} \tag{11.26}$$

at 0.175 cm/s (q_2 for −2 cm head).

$$K = 0.02 \exp(0.05h).$$ (11.27)

Equation 11.27 is for $K(h)$ for $h > 2$ cm.

VIII. MEASUREMENT OF REPELLENCY WITH THE TENSION INFILTROMETER

Although some sands and peats are observed to become water repellent when dry, most soils show no obvious reluctance to wet, so it is assumed that they absorb water and ethanol freely (Scotter et al., 1989). Given the same effective contact angle, sorptivity of a liquid into a porous material should be proportional to $(\sigma/\mu)^{1/2}$, where σ and μ are the surface tension and viscosity of the liquid, respectively (Philip, 1969; see his Equation 45 on p. 238). Thus, dry soil should imbibe water about twice as fast as alcohol. Organic coatings on peds or particles in some soils can induce some water repellence, but do not affect their ability to absorb alcohol. So in water-repellent dry soil the sorptivity of alcohol will be greater than water. Scotter et al. (1989) checked this observation out in the field, where they took a sorptivity tube (Fig. 11.3) to measure the sorptivity of ethanol and water by a fine sandy loam, a common agricultural soil in the region of Palmerston North, New Zealand. They did as much as they could to reduce the likelihood of water repellence. They removed the top 50 mm of soil because organic coatings are most likely to be at the surface. They did the experiments in November in the first drying cycle following an unusually wet winter, when the initial water content of the soil was 0.24 m³/m³. The soil appeared to absorb water normally, proportional to the square root of time and with a uniform wet front. But to their surprise the ethanol was absorbed an order of magnitude faster than the water, indicating significant water repellence.

On the same day of the first experiment, Scotter et al. (1989) took some of the soil back to the laboratory, sieved it, packed it into a tube and again measured the sorptivity of the two liquids. This time the water went in faster than the ethanol, presumably due to the abrasion during sieving. Other experiments showed that after a few days sieved soil again became water repellent, apparently due to a reorientation of the organic coatings on the soil particles or aggregates.

Further work on repellent soils of New Zealand (Clothier et al., 2000) confirmed the earlier work (Scotter et al., 1989). Clothier et al. (2000) showed that a Ramiha silt loam, another agricultural soil in the region of Palmerston North, was ephemerally hydrophobic. Temporal changes in the measured infiltration rate changed as the repellency broke down. Clothier et al. (2000) cautioned that repellency might not be evident in a soil if infiltration is not observed over a long period. It may take time for the repellency to break down before the infiltration rate climbs to a rate characteristic of the nonrepellent soil.

Peat soils are notoriously repellent. They are common in The Netherlands and much research has been done in that country to study repellency (Ritsema, 1998; Dekker, 1998). When the polders in The Netherlands were drained, the peat soils were allowed to get too dry. They never could be wetted up because of their extreme repellency. So the engineers learned that the peat soils had to be kept wet, if farmers were going to be able to plant seeds and make the soils arable (Don Kirkham, personal communication, undated). In Turkey highly organic soils have been seen to catch on fire (Don Kirkham, personal communication, undated). Forest soils are also repellent because they are organic (Kirkham and Clothier, 2000). John Letey at the University of California at Riverside, in the heart of orange-grove country before population expansion, is well known for his studies of the repellency of soils (Letey et al., 1975). Organic drippings from citrus groves make orchard soils repellent.

One cannot use alcohol in the commercially available tension infiltrometers from Soil Measurement Systems (Tucson, Arizona) and Decagon Devices (Pullman, Washington), because the towers are made of plastic and the alcohol would corrode the plastic. One needs a tension infiltrometer made out of glass, as was the original sorptivity tube (Fig. 11.3) to carry out experiments in which one infiltrates alcohol into the soil to determine repellency.

IX. MEASUREMENT OF MOBILITY WITH THE TENSION INFILTROMETER

Many water flow processes of interest such as groundwater recharge are concerned only with area-averaged water input. Therefore, preferential flow of water through structural voids does not necessarily invalidate equations that assume homogeneous flow, like Darcy's law. However, preferential flow is of critical importance in solute transport, because it enhances chemical mobility and can increase pollution hazards. Many times we need to monitor chemical mobility along with hydraulic properties. We now shall see how we can determine mobility of chemicals.

Water and nutrients not taken up by roots move to depth and eventually to groundwater with deleterious consequences. With the tension infiltrometer (disc permeameter), we can measure hydraulic properties of soil that control infiltration and retention. In particular, we can distinguish mobile and immobile water with it.

The soil water content, θ, is made up of the mobile water content, θ_m, and the immobile water content, θ_{im}. Figure 11.5 shows a mobile-water soil, and Fig. 11.6 shows a mobile-immobile water soil. Water is immobilized due to several factors: it can be in an occluded pore, it can be bound water, it can be in a dead-end pore, or it can be in the soil's microporosity and unable to move. θ_m is active in chemical transport, while θ_{im} is not. If we know θ_m, we can develop management strategies to minimize leaching losses of chemicals.

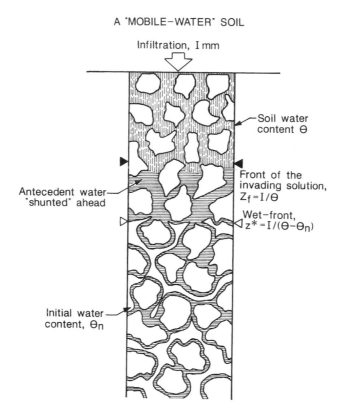

FIG. 11.5   A "mobile-water" soil. (Redrawn from a slide by Brent E. Clothier. Reprinted by permission of Brent E. Clothier.)

The method of using the disc permeameter to get θ_m and θ_{im} relies on using it to supply a tracer solution to soil (Clothier et al., 1992). The tracer is not originally in the soil. If a tracer is added to a disc permeameter at a concentration c_m, then from the observed solution concentration c^* in soil samples extracted from underneath the disc, θ_m can be calculated from the dilution by the water of the immobile phase that must have remained in place during the passage past it of the invading solution of tracer. So we have:

$$c^* \theta = c_m \theta_m + c_{im} \theta_{im}. \tag{11.28}$$

But if $c_{im} = 0$ (there was no tracer in the soil to begin with), then the last term on the right-hand side of Equation 11.28 drops out and we have:

$$\theta_m = \theta(c^*/c_m). \tag{11.29}$$

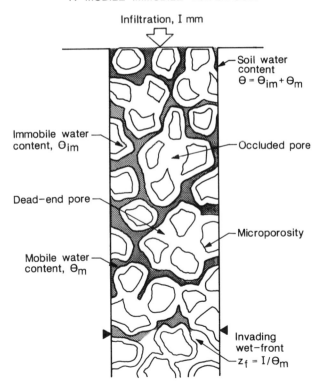

FIG. 11.6 A "mobile-immobile" water soil. (Redrawn from a slide by Brent E. Clothier. Reprinted by permission of Brent E. Clothier.)

In other words, the fraction, θ_m/θ, is directly proportional to the relative concentration of solute found in the soil, c^*, to that applied, c_m:

$$c^*/c_m = \theta_m/\theta. \tag{11.30}$$

In the experiments of Clothier et al. (1992), the tracer was bromide. (Chloride could not be used, because of the proximity of the Pacific Ocean and salt in the air that lands on the soil.) They applied water to the soil with a tension infiltrometer with $h_o = \psi_o = -20$ mm. (See Fig. 11.4.) The air-inlet tube was 20 mm under the surface of the water in the tower on the right-hand side of the figure. The soil was near saturation.

In the first experiment, pure water was used to first wet the soil. Then a tracer at concentration 0.1 mol/L KBr or 0.1 M KBr was drawn into this already wet soil predominantly by gravity. The original water content of the soil, a Manawatu fine sandy loam, was 0.414 m^3/m^3. Three runs were done. Bromide in the soil was determined after each run. The method is shown in Fig. 11.7. The ratio c^*/c_m was determined for the 3 runs, and they were 0.46, 0.48, and 0.50, with an average of 0.49. Therefore,

$$\theta_m/0.414 = 0.49 \ m^3/m^3 \tag{11.31}$$

$$\theta_m = 0.203 \ m^3/m^3 \tag{11.32}$$

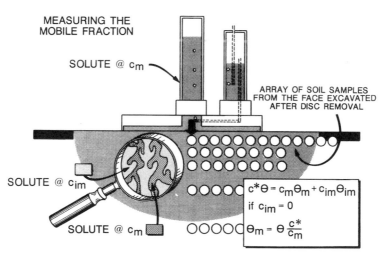

FIG. 11.7 Measuring the mobile fraction. (From Clothier, B.E., Green, S.R., and Magesan, G.N., ©1994. Soil and plant factors that determine efficient use of irrigation water and act to minimise leaching losses. *Trans Int Congr Soil Sci* 2a; 41–47. Reprinted by permission of The International Society of Soil Science.)

so only half of the soil water was mobile (0.414 m³/m³ vs. 0.203 m³/m³).

We can determine the depth of penetration of the solute using the following equation:

$$Z_s^* = I/\theta_m \qquad (11.33)$$

where

Z_s^* = depth of the solute front (the asterisk, *, after the s for solute indicates that we are dealing with a tracer)

I = cumulative infiltration

θ_m = mobile water content.

For the three runs, average I was 15.7 mm. So Z_s^* = 15.7 mm/(0.203 m³/m³) = 77.3 mm.

If all the water were mobile, then $\theta_m = \theta$.

$$15.7 \text{ mm}/(0.414 \text{ m}^3/\text{m}^3) = 37.9 \text{ mm}.$$

The observed front of the bromide, determined experimentally at the end of the experiment, showed that it was at about 77 mm, not at about 38 mm. So the experimental data backed up the calculation. Because not all the water was mobile, the tracer penetrated to deeper depths than it would have, had $\theta_m = \theta$.

The longitudinal mobility (the depth of penetration) is inversely related to θ_m. This is sometimes hard for people to understand when they are using the method initially, because they think that the more mobile the water the deeper will be the penetration of a solute. This is not so. The larger the mobile volume fraction, (i.e., θ_m) then the less longitudinally mobile the solute (the smaller Z_s^*). In other words, the solution carrying the dissolved solute travels through a larger volume fraction of the soil's wetted pore space, such that for a given amount of water infiltrated, I, the less the penetration into the soil. As θ_m becomes a smaller fraction of the wetted θ (total water content), then the smaller is the volume fraction of the soil's wetted pore space that transports the invading solution. Hence, for an equivalent amount of I, the greater the depth, due to the fact that Z_s^* is inversely related to θ_m.

In another experiment (Clothier et al. 1992), the tracer was put on a dry soil. The tension infiltrometer was still set at −20 mm, but there was no pre-wetting. The tracer solution was now drawn into the soil more by capillarity than by gravity. The mobile fraction measured right under the disc rose to $\theta_m = 0.291$ m³/m³. (Remember that the previous value, when a wet soil was used, $\theta_m = 0.203$ m³/m³.) This rise is attributed to the direct, capillary-induced movement of the invading solute into more of the soils microporosity. The solute invasion depth was less (shallower) than half of the former case (pre-wetted, $\varphi_o = -20$ mm). The pre-wetting had destroyed the capillary attractiveness of the micropores. The "dry-soil" effect keeps fertilizers near the soil surface. One can think of bromide as a tracer for nitrate. Thus effluent and other nutrient-laden fertilizer waters should be applied to dry soil. This applied chemical can be drawn by capillarity into more of the soil's microporosity, close to the soil surface (where roots are) and not escape to depth and groundwater. Once in the soil's microporosity, the applied nutrients are rendered less likely to be leached by subsequent rainfall.

X. ELLIPSOIDAL DESCRIPTION OF WATER FLOW INTO SOIL FROM A SURFACE DISC

The disc permeameter is being widely used to characterize the hydraulic properties of the surface of the soil. It is important to describe the multidimensional flow of water away from the circular source. Little work has been done to analyze precisely the flow of water away from a circular source of water applied at a constant negative potential, ψ_o. The Wooding equation (1968; his Equation 64) does describe the steady rate of three-dimensional infiltration from a circular pond. His equation applies to profiles at infinite time and does not give information about the shape of wet fronts of transient wetting. Kirkham and Clothier (1994a, 1994b) used a unique approach to analyze the flow pattern, because they assumed that the three-dimensional wetting fronts under the circular source were ellipsoidal.

They described mathematically the three-dimensional flow of water away from a disc source placed upon the soil's surface at constant negative potential, ψ_o less than 0, when the water is being applied to the soil at a steady rate, q in mm^3/s. The wet fronts analyzed are shown in Fig. 11.1. The wet fronts resulted from water being infiltrated from a quarter-disc permeameter set in the corner of a plastic box with soil. The setup is shown as Fig. 1 in Kirkham and Clothier (1994b). It allowed markings of the wet fronts as they penetrated the soil under the quarter-disc permeameter set at -50 mm supply potential (ψ_o). The reason for using this suction was to allow the soil to be unsaturated but not too far away from saturation, such that the flow would be unrealistically slow.

The mathematical development is given by Kirkham and Clothier (1994a, 1994b). Assuming the volume wetted in each wetted area (delineated in Fig. 11.1) is an ellipsoid, the following equation can be used for the volume of water, $V(t)$, that infiltrated from the disc:

$$V(t) = (2/3\pi) \, (\theta_m - \theta_n) \, R^2(t) \, Z(t), \tag{11.34}$$

where the extent of radial wetting at the surface is R at any time t, and vertically under the disc is Z. The initial volumetric water content of the soil is θ_n and the disc wets the soil surface to water content θ_o, a function of ψ_o. A weighted-average water content can be ascribed to the wetted field, designated θ_m.

It can be seen from the wet fronts (Fig. 11.1) that the ellipsoid describing the spatial pattern of wetting goes from being an oblate (egg on its side; see Fig. 7.3) at early times, through to being spheroidal by about the end of the experiment. It would eventually become prolate (egg on its end; see Fig. 7.4), with further extension being limited to the vertical.

As noted above, profiles predicted from Wooding's (1968) equation hold only at infinite time, and thus, they cannot give any information about transient wet fronts. Where the ellipsoidal idea has merit over Wooding's equation is in the practical operation of the disc permeameter. The ellipsoidal equation can give answers to the following questions: How deep is the wet front at any time? By what means can it be reckoned simply? Also, we can answer other questions relating to the modus operandi of the disc permeameter. How much of the soil's volume has been wetted during a disc experiment? What volume of soil has been sampled, and, hence, to what depth of soil do the Wooding K and S values apply?

The disc permeameter might be practically applied to different tillage situations. If the soil will not suck in water, as determined by the disc permeameter, then we need to till the soil. If there is a crust on the soil surface or the soil surface is repellent, water cannot infiltrate. Water runs off and moves preferentially through the macropores instead of through the soil matrix. Tillage of the soil is critical so that rainwater will not sit on the surface and evaporate (Fig. 11.8). We want water to go into the soil, where it will be used by plants. The disc permeameter also can be practically applied to different irrigation systems, including drip and furrow irrigation. Perhaps one could put the disc permeameter, set at a constant negative potential, on a dry soil and see how far water is sucked out to the side and calculate the volume wetted, based on the ellipsoidal equation. The distance could be measured with a ruler. Then one could place irrigation sources at distances from each other based on the measurements.

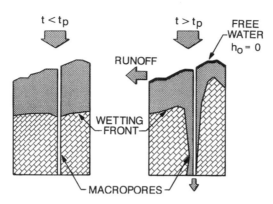

FIG. 11.8   Infiltration of an applied flux of water into soil. Left: Non-ponding infiltration when the flux through the surface is less than the saturated hydraulic conductivity, K_s, or ponding infiltration when the flux through the surface is greater than K_s prior to the time of ponding t_p. Right: Pattern of infiltration after incipient ponding, $t > t_p$, when the possibility of runoff exists, as does the entry of free water into macropores. (From Clothier, B.E., Infiltration. In *Soil and Environmental Analysis. Physical Methods*, 2nd ed., Marcel Dekker, Inc.: New York, ©2001. This material is reprinted with permission of Marcel Dekker, Inc.)

The ellipsoidal equation has another advantage. It is simple to use and people with essentially no mathematical training can apply it. A small hand-held calculator could be programmed easily to make the calculations.

XI. APPENDIX: BIOGRAPHY OF JOHN PHILIP

John Robert Philip, soil physicist, was born January 13, 1927, in Ballarat in rural Victoria, Australia (Burges et al., 1999). He acquired a love of learning from his schoolteacher mother, and won a scholarship to the prestigious Scotch College for boys in Melbourne, where his mathematical ability was recognized and developed. He also was encouraged to write poetry, and this remained a lifelong passion, his poems appearing in many literary publications. He graduated from Scotch College at 13, and he spent another two years at school before being deemed old enough (at 16) to study civil engineering at the University of Melbourne. Bored by the undemanding engineering courses, which he described as merely "learning which handbook to look up," he spent much of his time reading and writing poetry. He earned his Bachelors of Civil Engineering degree at age 19, the youngest-ever engineering graduate.

He was offered a research assistantship by the University of Melbourne and was sent to the Council for Industrial and Scientific Research (renamed CSIRO in 1949) station at Griffith to work on problems of furrow irrigation.

His fellowship ended after a year and he left to work as an engineer in Queensland. In 1951, he was asked to join the CSIRO Division of Plant Industry at Deniliquin, in New South Wales. Otto Frankel became Philip's boss. Frankel, a distinguished plant geneticist, was charged with revitalizing the division. He consulted with John Jaeger and Pat Moran, two famous mathematicians at the Australian National University, who reported positively on Philip's proposed research plans in agricultural physics. Frankel gave Philip freedom to proceed (Burges et al., 1999), and Philip praised Frankel for allowing the environment in which he could thrive (personal communication, December 9, 1997). Philip's papers at Deniliquin dealt with the analysis of environmental water and heat flow, which earned him a D.Sc. in physics from the University of Melbourne. Modern theory is based on his analysis (American Society of Agronomy, 1999).

In 1964 he moved from Deniliquin to Canberra to head the Agricultural Physics section of the Division of Plant Industry (headed by Frankel) (Burges et al., 1999). A bequest to CSIRO provided the funds to build a laboratory to house the team of researchers Philip assembled to work on fluid mechanics of porous media, micrometeorology, plant physical ecology, and soil physics. Philip helped to design the building, called the F.C. Pye Laboratory, built in 1966. The productivity of the team was aided by the architecture of the Pye Lab, which encouraged collaboration; it was open and such that one easily encountered colleagues in the sunny walkways. In 1970, the Division of Enviromental Mechanics was created, and Philip was the chief until his retirement in 1992 (American Society of Agronomy, 1999).

Philip was a strong defender of scientific autonomy. In 1975, he chaired the Science Task Force of the Royal Commission on Australian Government Administration. The report he drafted for this task force argued for governmental science characterized by freedom of action and outlined the environment necessary for effective and creative scientific research (American Society of Agronomy, 1999; Burges et al., 1999). In the article he prepared for the seventy-fifth anniversary issue of *Soil Science* (Philip, 1991), he criticized the lack of freedom that scientists now have.

Philip was the preeminent mathematician who solved difficult unsaturated flow problems. The solutions led to practical benefits, including how to handle infiltrating irrigation water. His work was the basis for numerous theoretical advances in infiltration and soil-plant-water relationships and his vertical infiltration model is used worldwide. He also dominated multidimensional infiltration theory. He published more than 300 papers. While he is best known for his work in soil and porous media physics, fluid mechanics, and hydrology, his papers in the plant-physiology literature are classic. He published on osmotic and turgor properties of plant cells. He pioneered the concept of the *soil*, *plant*, *atmosphere* as a thermodynamic *continuum* (SPAC) for water transfer. Everyone now talks about the SPAC, most of the time without acknowledging that Philip was the person who first used the term.

Philip received numerous honors, including honorary doctor's degrees from the University of Melbourne, the Agricultural University of Athens, and the University of Guelph in Canada. He was a Fellow of the Australian Academy of Sciences and, in 1974, became a Fellow of the Royal Society of London, the highest scientific honor in the Commonwealth. In 1991, he was elected corresponding member of the All Union (now Russian) Academy of Sciences. In 1995, he received the International Hydrology prize awarded jointly by UNESCO, the World Meteorological Association, and the International Association for Hydrological Science. He also received much recognition in the United States. He was named Fellow of the Soil Science Society and honorary member of the American Water Resources Association and was the first non-American to receive the Robert E. Horton Medal, the highest award for hydrology from the American Geophysical Union (American Society of Agronomy, 1999). In 1995, he was elected as a foreign associate by the National Academy of Engineering in Washington, D.C., for his pioneering research contributions to soil-water hydrology.

Philip and his wife, Francis, had two sons and a daughter. Philip was struck by a car as he stepped off a bus and was killed on June 26, 1999, in Amsterdam, The Netherlands, where he was visiting the Centre for Mathematics and Information Science (American Society of Agronomy, 1999). The driver was sentenced to one month in jail and suspension of his driver's license for a year. A volume honoring Philip was published by the American Geophysical Union (Raats et al., 2002), and a tribute plus complete bibliography appeared in the *Australian Journal of Soil Research* (Smiles, 2001).

REFERENCES

American Society of Agronomy. (1999). Deaths. John Robert Philip. *Crop Science-Soil Science-Agronomy News*, December; 27.

Ankeny, M.D., Kaspar, T.C., and Horton, Jr., R. (1989). Automated Tension Infiltrometer. United States Patent No. 4,884,436. Date of Patent: Dec. 5, 1989. Filed: Oct. 20, 1988. Assignee: Iowa State University Research Foundation, Inc., Ames, Iowa.

Ankeny, M.D., Ahmed, M., Kaspar, T.C., and Horton, R. (1991). Simple field method for determining unsaturated hydraulic conductivity. *Soil Sci Soc Amer J* 55; 467–470.

Baver, L.D., Gardner, W.H., and Gardner, W.R. (1972). *Soil Physics*. 4th ed. Wiley: New York.

Burges, S.J., Ford, P.W., and White, I. (1999). John Robert Philip (1927–1999). *EOS, Trans, Amer Geophys Union* 80; 574, 577.

Clothier, B.E. (2001). Infiltration. In *Soil and Environmental Analysis. Physical Methods* (Smith, K.A., and Mullins, C.E., Eds.), 2nd ed., pp. 239–280. Marcel Dekker: New York.

Clothier, B.E. (2004). Soil pores. In *The Encyclopedia of Soil Science* (Chesworth, W., Ed.) Kluwer: Dordrecht, The Netherlands (In press).

Clothier, B.E., and White, I. (1981). Measurement of sorptivity and soil water diffusivity in the field. *Soil Sci Soc Amer J* 45; 241–245.

Clothier, B.E., Green, S.R., and Magesan, G.N. (1994). Soil and plant factors that determine efficient use of irrigation water and act to minimise leaching losses. *Trans Int Congr Soil Sci* 2a; 41–47.

Clothier, B.E., Kirkham, M.B., and McLean, J.E. (1992). In situ measurement of the effective transport volume for solute moving through soil. *Soil Sci Soc Amer J* 56; 733–736.

Clothier, B.E., Vogeler, I., and Magesan, G.N. (2000). The breakdown of water repellency and solute transport through a hydrophobic soil. *J Hydrol* 231–232 (double volume issue); 255–264.

Dekker, L.W. (1998). Moisture Variability Resulting from Water Repellency in Dutch Soils. Doctoral thesis. Wageningen Agricultural University: The Netherlands.

Gardner, W.R. (1958). Some steady-state solutions of the unsaturated moisture flow equation with applications to evaporation from a water table. *Soil Sci* 85; 228–232.

Jury, W.A., Gardner, W.R., and Gardner, W.H. (1991). *Soil Physics*. 5th ed. Wiley: New York.

Kirkham, M.B., and Clothier, B.E. (1994a). Ellipsoidal description of water flow into soil from a surface disc. *Trans Int Congr Soil Sci* 2b; 38–39.

Kirkham, M.B., and Clothier, B.E. (1994b). Wetted soil volume under a circular source. In *Proceedings of the 13th International Conference, International Soil Tillage Research Organization* (Jensen, H.E., Schjønning, P., Mikkelsen, S.A., and Madsen, K.B., Eds.), pp. 573–578. The Royal Veterinary and Agricultural University and the Danish Institute of Plant and Soil Science, Lyngby and Copenhagen, Denmark.

Kirkham, M.B., and Clothier, B.E. (2000). Infiltration into a New Zealand native forest soil. In *A Spectrum of Achievements in Agronomy* (Rosenzweig, C. Ed.), pp. 13–26. ASA Special Pub. No. 62. American Society of Agronomy, Crop Science Society of America, and Soil Science Society of America: Madison, Wisconsin.

Letey, J., Osborn, J.F., and Valoras, N. (1975). *Soil Water Repellency and the Use of Nonionic Surfactants*. Technical Completion Report. Office of Water Research and Technology, U.S. Department of the Interior; Office of Water Research and Technology; and Water Resources Center, University of California, Davis. Contrib. No. 154. Department of Soil Science and Agricultural Engineering: University of California, Riverside.

Nobel, P.S. (1983). *Biophysical Plant Physiology and Ecology*. W.H. Freeman: San Francisco.

Perroux, K.M., and White, I. (1988). Designs for disc permeameters. *Soil Sci Soc Amer J* 52; 1205–1215.

Philip, J.R. (1969). Theory of infiltration. *Advance Hydrosci* 5; 215–296.

Philip, J.R. (1991). Soils, natural science, and models. *Soil Sci* 151; 91–98.

Raats, P.A.C. (1971). Steady infiltration from point sources, cavities, and basins. *Soil Sci Soc Amer Proc* 35; 689–694.

Raats, P.A.C., Smiles, D., and Warrick, A.W. (Eds.). (2002). *Environmental Mechanics. Water, Mass and Energy Transfer in the Biosphere. The Philip Volume*. Geophysical Monograph 129. American Geophysical Union: Washington, DC.

Ritsema, C.J. (1998). Flow and Transport in Water Repellent Sandy Soils. Doctoral thesis. Wageningen Agricultural University, The Netherlands.

Scotter, D., Tillman, R., Wallis, M., and Clothier, B. (1989). Alcophilic soils. *WISPAS. A Newsletter about Water in the Soil-Plant Atmosphere System*. No. 43, p. 3. Plant

Physiology Division, Department of Scientific and Industrial Research: Palmerston North, New Zealand.

Smettem, K.R.J., and Clothier, B.E. (1989). Measuring unsaturated sorptivity and hydraulic conductivity using multiple disc permeameters. *J Soil Sci* 40; 563–568.

Smiles, D. (Coordinating Ed.). (2001). The Environmental Mechanic. *Australian J Soil Res* 39; 649–681.

Soil Science Society of America. (1997). *Glossary of Soil Science Terms.* Soil Science Society of America: Madison, Wisconsin.

Swartzendruber, D. (1993). Revised attribution of the power form infiltration equation. *Water Resources Res* 29; 2455–2456.

Wooding, R.A. (1968). Steady infiltration from a shallow circular pond. *Water Resources Res* 4; 1259–1273. (See Eq. 64 for Wooding's equation.)

Zhang, R. (1997). Determination of soil sorptivity and hydraulic conductivity from the disk infiltrometer. *Soil Sci Soc Amer J* 61; 1024–1030.

12

Pore Volume

Soil not only permits entry of water and nutrients and their storage there, but it also allows wasteful passage past roots. We have seen how with the tension infiltrometer we can measure mobility of solutes (see Section IX, Chapter 11). A older method to analyze movement of solutes through the soil is based on measuring pore volumes. This method does not divide the soil water into mobile, θ_m, and immobile, θ_{im}, regions, but considers the water in the soil to be the sum of both, the total volumetric water content, θ. By measuring pore volume, we consider the physical process of the movement of solutes in fluids (Kirkham and Powers, 1972, p. 380). It gives us information concerning what takes place in a porous medium as one fluid displaces another.

I. DEFINITIONS

The *Glossary of Soil Science Terms* (Soil Science Society of America, 1997) defines *pore volume* as "pore space." And then the definition of *pore space* is given as follows: "The portion of soil bulk volume occupied by soil pores." This is a valid definition of pore volume. We see the pore volume defined this way in Fig. 10.1. Pores are filled with air or water. However, pore volume has another, special meaning in soil physical work. It is important to understand this meaning, because pore volumes are widely determined when one is studying the movement of solutes (e.g., dissolved fertilizers or pollutants) through a soil. The term has been misapplied, so it is important to know how to calculate pore volumes.

First let us review definitions. From Darcy's law, we remember that the quantity of flow Q (cm^3/hr) is the volume of fluid passing through a porous medium in an hour. We also recall that the *Darcy velocity*,

which, on dividing the numerator and denominator by A, yields

$$p = [(Qt)/A] / [(\alpha V/A)] \qquad (12.4)$$

In Equation 12.4 we can substitute LA for V, where L is the length of the soil column, and we have $v_d = Q/A$. We can thus write

$$p = v_d t/\alpha L \qquad (12.5)$$

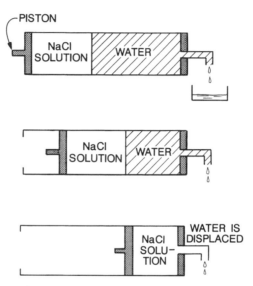

FIG. 12.1   Schematic drawing of piston flow in a tube. (From *Advanced Soil Physics* by Kirkham, D., and Powers, W.L., p. 382, ©1972, John Wiley & Sons, Inc.: New York. This material is used by permission of John Wiley & Sons, Inc. and William L. Powers.)

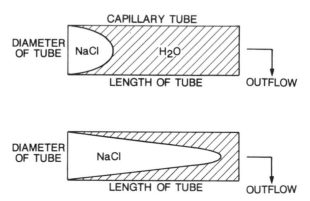

FIG. 12.2   Schematic drawing of actual flow in a tube. (From *Advanced Soil Physics* by Kirkham, D., and Powers, W.L., p. 383, ©1972, John Wiley & Sons, Inc.: New York. This material is used by permission of John Wiley & Sons, Inc. and William L. Powers.)

Knowing the equation for average pore velocity, we can substitute v for v_d/α, and we can write Equation 12.5 as

$$p = vt/L \qquad (12.6)$$

Kirkham and Powers (1972, p. 401 and following) then use Equation 12.6 to express C/C_o in terms of pore volumes p. However, the derivation involves calculus (the use of an error function), so we will not derive the

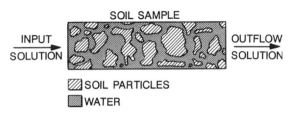

FIG. 12.3 Schematic drawing of flow in a saturated soil. (From *Advanced Soil Physics* by Kirkham, D., and Powers, W.L., p. 384, ©1972, John Wiley & Sons, Inc.: New York. This material is used by permission of John Wiley & Sons, Inc. and William L. Powers.)

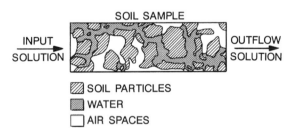

FIG. 12.4 Schematic drawing of flow in an unsaturated soil. (From *Advanced Soil Physics* by Kirkham, D., and Powers, W.L., p. 384, ©1972, John Wiley & Sons, Inc.: New York. This material is used by permission of John Wiley & Sons, Inc. and William L. Powers.)

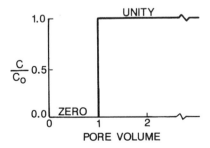

FIG. 12.5 Breakthrough curve for piston flow. (From *Advanced Soil Physics* by Kirkham, D., and Powers, W.L., p. 385, ©1972, John Wiley & Sons, Inc.: New York. This material is used by permission of John Wiley & Sons, Inc. and William L. Powers.)

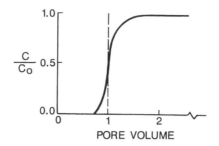

FIG. 12.6 Breakthrough curve for flow in a tube. (From *Advanced Soil Physics* by Kirkham, D., and Powers, W.L., p. 385, ©1972, John Wiley & Sons, Inc.: New York. This material is used by permission of John Wiley & Sons, Inc. and William L. Powers.)

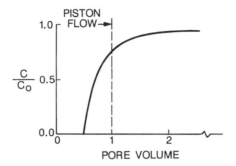

FIG. 12.7 Breakthrough curve for wide range in pore velocities as in a saturated soil. (From *Advanced Soil Physics* by Kirkham, D., and Powers, W.L., p. 386, ©1972, John Wiley & Sons, Inc.: New York. This material is used by permission of John Wiley & Sons, Inc. and William L. Powers.)

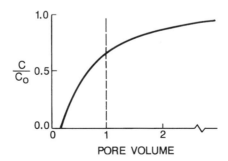

FIG. 12.8 Breakthrough curve characteristic of an unsaturated soil. (From *Advanced Soil Physics* by Kirkham, D., and Powers, W.L., p. 386, ©1972, John Wiley & Sons, Inc.: New York. This material is used by permission of John Wiley & Sons, Inc. and William L. Powers.)

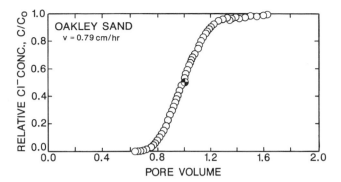

FIG. 12.9 Chloride breakthrough curve for Oakley sand. The bicolored circle shows where 0.5 pore volume would occur. (From *Advanced Soil Physics* by Kirkham, D., and Powers, W.L., p. 387, ©1972, John Wiley & Sons, Inc.: New York. This material is used by permission of John Wiley & Sons, Inc. and William L. Powers.)

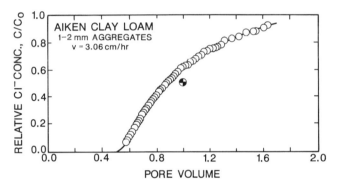

FIG. 12.10 Chloride breakthrough curve for Aiken clay loam. The bicolored circle shows where 0.5 pore volume would occur. (From *Advanced Soil Physics* by Kirkham, D., and Powers, W.L., p. 387, ©1972, John Wiley & Sons, Inc.: New York. This material is used by permission of John Wiley & Sons, Inc. and William L. Powers.)

equation. The advanced mathematical reader is referred to this derivation. For those without knowledge of calculus, a method is needed to calculate pore volumes.

IV. CALCULATION OF A PORE VOLUME

To learn how to calculate pore volumes, let us use an example from the experiment by Vogeler et al. (2001). They studied phytoremediation of soil contaminated with copper using poplar. The lysimeter with the poplar is shown on the cover of the book edited by Iskandar and Kirkham (2001).

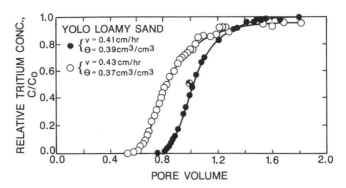

FIG. 12.11 Tritium breakthrough curve for Yolo loamy sand at two different water contents. The bicolored circle shows where 0.5 pore volume would occur. (From *Advanced Soil Physics* by Kirkham, D., and Powers, W.L., p. 388, ©1972, John Wiley & Sons, Inc.: New York. This material is used by permission of John Wiley & Sons, Inc. and William L. Powers.)

A solution of copper in the form of $Cu(NO_3)_2$ was added to the surface of the soil (Manawatu fine sandy loam) in the lysimeter. Then, everyday, the lysimeter was irrigated and drainage water was collected. After running the experiment for one week, Vogeler et al. (2001) wanted to know how many pore volumes had passed through the lysimeter. One pore volume was calculated as follows. The shape of the lysimeter was that of a large cylinder, 1.30 m long with a diameter of 0.85 m. The volume of the lysimeter was:

$$\text{volume} = \pi r^2 h$$
$$\text{volume} = \pi(0.85/2)^2\, 1.30 = 0.737\ m^3.$$

Let θ_s = saturated water content = 0.40 m³/m³. This value was known from previous measurements.

One pore volume = volume × θ_s = 0.737 m³ × 0.40 m³/m³ = 0.295 m³.
    0.295 m³ × (1 × 10⁶ cm³)/(1 m³) = 295,000 cm³ = 295 liters.

After 7 days of drainage, 70 liters had been collected. Thus, (70 L)/(295 L) = 0.24 or about 1/4 pore volume.

To get one pore volume, Vogeler et al. (2001) needed to collect drainage water for one month. The experiment was run for two months to get two pore volumes.

Another example of calculating pore volumes is given by Singh and Kanwar (1991). They did a column study and calculated the value of one pore volume for each column by multiplying total porosity by the total volume of the soil column. Soil porosity for each column was estimated by using bulk density and particle density (Equation 7.2). Note that in either case [the experiment by Vogeler et al. (2001) or the experiment by Singh and Kanwar (1991)], the whole pore volume is used to calculate pore volume. Vogeler et al. (2001) used the saturated water content (all pores filled with water); Singh and Kanwar used the total porosity.

In the theoretical development of pore volume (Section III), the water-filled porosity, α, is used to calculate a pore volume. In practice, we do not know which pores have air and which pores have water, and hence, which pores are carrying the solute. So in our calculations in this section (above), the water-filled porosity becomes the porosity or the saturated water content. However, we note that the level of saturation affects the shape of the breakthrough curve. We see this by comparing Fig. 12.7, the breakthrough curve for a saturated soil, with Fig. 12.8, the breakthrough curve for an unsaturated soil. As one pore volume is attained, the slope in Fig. 12.7 is steeper than in Fig. 12.8.

V. PORE VOLUMES BASED ON LENGTH UNITS

Jury et al. (1991, pp. 224–225) calculate pore volumes by multiplying the length of a column by the water content. They give the formula

$$d_{wb} = J_w\, t_b = J_w L/V = L\theta, \tag{12.7}$$

where d_{wb} is the drainage water (cm) evolved at the breakthrough time $d_{wb} = J_w t_b$, J_w is the soil water flux (cm/sec), t_b is the breakthrough time (sec), L is the length of the soil column (cm), and V is the solute velocity (cm/sec) (Figs. 12.12 and 12.13). (We will assume this velocity is the average pore velocity that Kirkham and Powers define; the definition is given in a preceding section.) Jury et al. (1991) say that the value $L\theta$ is the volume of water per unit area held in the wetted soil pores of the column during transport. For this reason $d_{wb} = L\theta$ is called a pore volume, and it requires approximately one pore volume of water to move a mobile solute through a soil column (Jury et al., 1991, p. 225).

Why do Jury et al. (1991) use a length instead of a volume in getting a pore volume? A pore volume is a calculation of the equivalent amount of transmitted water in depth units (where the area has been taken out, as with evapotranspiration, where we use the units of mm). So in the case of Kirkham and Powers (1972), they deal with soil in which the water is being transmitted through the water-filled porosity. The calculation is turning the water that is being transmitted (i.e., a θ) into a volume, by multiplying by the soil's volume. But if one is dealing with areas, one can divide through by an area, as we do when we turn a [volumetric] water content (m^3/m^3)

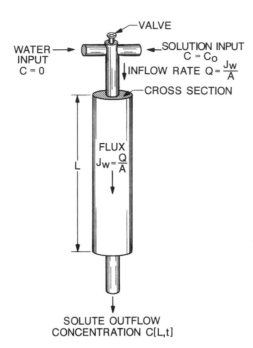

FIG. 12.12 Schematic diagram of a soil column outflow experiment, where solute is added at $t = 0$. (From Jury, W.A., Gardner, W.R., and Gardner, W.H., *Soil Physics*, 5th ed., p. 224, ©1991, John Wiley & Sons: New York. This material is used by permission of John Wiley & Sons, Inc.)

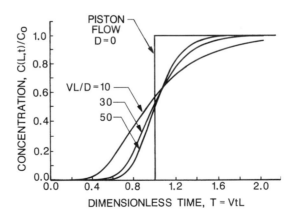

FIG. 12.13 Outflow concentration versus time for a step change in solute input at $t = 0$. D = dispersion coefficient and it has units of length^2/time. If $D = 0$, there is no dispersion. Curves correspond to different values of D ($V = 2$ cm day^{-1}, $L = 30$ cm). (From Jury, W.A., Gardner, W.R., and Gardner, W.H., *Soil Physics*, 5th ed., p. 224, ©1991, John Wiley & Sons: New York. This material is used by permission of John Wiley & Sons, Inc.)

into a depth of storage water (mm) (B.E. Clothier, personal communication, February 25, 1999).

VI. MISCIBLE DISPLACEMENT

Pore volumes are analyzed in miscible displacement studies. Miscible displacement is the process that occurs when one fluid mixes with and displaces another fluid. Leaching of salts from a soil is an example, because the added water mixes with and displaces the soil solution. A pioneer in the application of miscible displacement techniques to soil science is D.R. Nielsen. (For a biography of Nielsen, see the Appendix, Section VIII.) In a key paper, Nielsen et al. (1965) showed that chloride movement in soil depends upon the method of water application. They found that intermittently ponding the soil with 2-inch (5-cm) increments of water was more efficient in leaching applied chloride from the soil surface than continuous ponding or leaching with 6-inch (15-cm) increments. This finding has important applications in salinity management.

For a mathematical discussion of miscible displacement, the interested reader is referred to Kirkham and Powers (1972; see their Chapter 8).

VII. RELATION BETWEEN MOBILE WATER CONTENT AND PORE VOLUME

As noted in the first paragraph of this chapter, calculation of pore volumes does not tell us about the mobility of a solute, as we determined in Section IX, Chapter 11. The question arises, "How does one relate mobile water content to pore volumes?" Or, in other words, "How does the ratio of c^*/c_m, needed to determine mobility, relate to C/C_o on the ordinate in breakthrough curves?" The answer to this question is tricky. If we know that the soil wets to θ_o (the soil water content under a tension infiltrometer), then we can calculate the nonpreferential pore volume using this θ_o. But if our solute comes through earlier (i.e., a smaller pore volume), then not all the pore volume could have been active. So we could define an active pore volume, which we could directly relate to a mobile (volume) fraction, θ_m (B.E. Clothier, personal communication, February 25, 1999).

VIII. APPENDIX: BIOGRAPHY OF DONALD NIELSEN

Donald Rodney Nielsen, soil and water science educator, was born in Phoenix, Arizona, on October 10, 1931. He got his B.S. degree in

agricultural chemistry and soils at the University of Arizona in 1953; his
M.S. degree in soil microbiology at the University of Arizona in 1954; and
his Ph.D. in soil physics at Iowa State University in 1958. His career has
been spent at the University of California, Davis, where he started as an
assistant professor in 1958, moved to associate professor in 1963, and then
to professor in 1968. He was the director of the Kearney Foundation of
Soil Science from 1970–1975; associate dean, 1970–1980; director of the
Food Protection and Toxicology Center, 1974–1975; chairman of the
Department of Land, Air, and Water Resources, 1975–1977; executive
associate dean of the College of Agricultural and Environmental Sciences,
1986–1989; and chairman of the Department of Agronomy and Range
Science, 1989–1991 (Marquis Who's Who, 1994). In his administrative
duties, he emphasized the important links between agriculture and environmental science.

Nielsen has been a pioneer in three areas of soil-science research: link-
ing theory to field measurements of water movement, miscible displacement (Nielsen and Biggar, 1962), and geostatistics. One of his first papers
with colleagues on geostatistics (Nielsen et al., 1973) became a citation
classic (Institute for Scientific Information, 1983). He is co-author of a
book on soil hydrology (Kutílek and Nielsen, 1994) and a book on spatial
and temporal statistics (Nielsen and Wendroth, 2003). He has edited several books, including a major compendium on nitrogen (Nielsen and
MacDonald, 1978).

Nielsen has had an outstanding career not only in research and administration, but also in teaching and serving as editor on important journals.
Nielsen was an associate editor of *Water Resources Research* from when it
was established in 1965 until 1986, and he was its editor-in-chief from
1986–1989. He has taught 15 different courses dealing with soil physics,
water science, and irrigation. He has guided 17 students through to the
M.S. degree and 20 students through to the Ph.D., and they are leaders in
the field now. Seventy-five scientists from around the world have spent
leaves with him.

He has taught workshops at numerous locations around the world,
including the International Atomic Energy Agency in Vienna, Austria, and
the famous International Centre for Theoretical Physics, established by the
Nobel Prize-winning Pakistani physicist, Abdus Salam, in Trieste, Italy.

Nielsen has won many awards. He is Fellow of the American Society of
Agronomy, the Soil Science Society of America, and the American
Geophysical Union. He has been president of the Soil Science Society of
America, the American Society of Agronomy, and the Hydrology Section

of the American Geophysical Union. He was on the National Research Council's Board on Agriculture. He received an honorary doctor of science degree from Ghent State University in Belgium and received the M. King Hubbert Award of the National Ground Water Association. He was made an honorary member of the European Geophysical Society and in 2001 he received the Horton Medal from the American Geophysical Union for outstanding contributions to the geophysical aspects of hydrology.

Nielsen married Joanne Joyce Locke on September 26, 1953. They have three daughters and two sons.

REFERENCES

Institute for Scientific Information. (1983). This Week's Citation Classic: Nielsen, D.R., Biggar, J.W., and Erh, K.T. Spatial variabilty of field-measured soil-water properties. *Hilgardia* 42:215–259, 1973. *Current Contents, Agr, Biol, Environ Sci* 14(1); 16.

Iskandar, I.K., and Kirkham, M.B. (Eds.). (2001). *Trace Elements in Soil. Bioavailability, Flux, and Transfer*. Lewis Publishers: Boca Raton, Florida.

Jury, W.A., Gardner, W.R., and Gardner, W.H. (1991). *Soil Physics, 5th ed*. Wiley: New York.

Kirkham, Don, and Powers, W.L. (1972). *Advanced Soil Physics*. Wiley-Interscience: New York.

Kutílek, M., and Nielsen, D.R. (1994). *Soil Hydrology*. Catena Verlag: Cremlingen-Destedt, Germany.

Marquis Who's Who. (1994). *Who's Who in America*. 48th ed., p. 2541. Marquis Who's Who: New Providence, New Jersey.

Nielsen, D.R., and Biggar, J.W. (1962). Miscible displacement. III. Theoretical considerations. *Soil Sci Soc Amer Proc* 26; 216–221.

Nielsen, D.R., and MacDonald, J.G. (Eds.). (1978). *Nitrogen in the Environment*. Two Volumes. Vol. 1. *Nitrogen Behavior in Field Soil*. 526 pp. Vol. 2. *Soil-Plant-Nitrogen Relationships*. Academic Press: New York.

Nielsen, D.R., and Wendroth, O. (2003). *Spatial and Temporal Statistics*. Catena Verlag: Reiskirchen, Germany.

Nielsen, D.R., Miller, R.J., and Biggar, J.W. (1965). Chloride displacement in Panoche clay loam in relation to water movement and distribution. *Water Resources Res* 1; 63–73.

Nielsen, D.R., Biggar, J.W., and Erh, K.T. (1973). Spatial variability of field-measured soil-water properties. *Hilgardia* 42, 215–259.

Singh, P., and Kanwar, R.S. (1991). Preferential solute transport through macropores in large undisturbed saturated soil columns. *J Environ Quality* 20; 295–300.

Soil Science Society of America. (1997). *Glossary of Soil Science Terms*. Soil Science Society of America: Madison, Wisconsin.

Vogeler, I., Green, S.R., Clothier, B.E., Kirkham, M.B., and Robinson, B.H. (2001). Contaminant transport in the root zone. In *Trace Elements in Soil. Bioavailability, Flux, and Transfer* (Iskandar, I.K., and Kirkham, M.B., Eds.), pp. 175–197. Lewis: Boca Raton, Florida.

13

Time Domain Reflectometry to Measure Volumetric Soil Water Content

In Chapter 4 we learned how to measure the matric potential energy of water in soil with a tensiometer. Here we learn how to measure the water content of soil using time domain reflectometry. This method is the most widely used one, aside from the gravimetric method, to determine soil water content.

Time domain reflectometry (TDR) makes use of the dielectric constant, ε, of water to determine the volumetric water content of soil. We are going to see how the dielectric constant of a soil sample depends on the amount of water in it. We measure it and then use an empirical relation, which equates the volumetric water content to the dielectric constant.

I. DEFINITIONS

The dielectric constant of a medium is defined by ε in the following equation (Weast, 1964, p. F-37):

$$F = (QQ')/(\varepsilon r^2) \tag{13.1}$$

where F is the force of attraction between two charges Q and Q' separated by a distance r in a uniform medium. The dielectric constant of a material is the ratio of the capacitance of a capacitor with the material between the plates to the capacitance with a vacuum between the plates (Shortley and Williams, 1971, p. 519). It is dimensionless. The dielectric constant for a vacuum = 1 exactly.

Before we look at the TDR method, let us first review some basic definitions from physics (Shortley and Williams, 1971, p. 243). A *cycle* is one complete execution of a periodic motion. The *period* of a periodic motion is the *time T* required for the completion of a cycle. The *frequency* of a periodic motion is the number f of cycles completed per unit time (units = cycles per second).

Thus, it is seen that the period T of a periodic motion is the reciprocal of the frequency f, that is $T = 1/f$ and $f = 1/T$. For example, a pendulum with a *period T* = 1/5 s has a *frequency* of 5 cycles per second or $f = 5$ s^{-1} = 5 Hz.

Frequency is measured in a unit called the *hertz* (Hz), which corresponds to *one cycle per second*, or 1 Hz = 1 s^{-1}. The unit is named after Heinrich Rudolph Hertz (1857–1894), who was a German physicist. (See the Appendix, Section IX, for a biography of Hertz.) Hertzian waves are radio waves or other electromagnetic radiation resulting from the oscillation of electricity in a conductor. Hertz was the first to demonstrate the production and reception of radio waves.

Mechanical wave motion has a single nonrepeated disturbance, called a *pulse*, which is initiated at the source and then travels away from the source through the medium (Shortley and Williams, 1971, p. 415). Another important type of wave motion is the *regular wave train* or *continuous wave*. In this type of wave, a regular succession of pulses is initiated at the source and transmitted through the medium. Thus, if a floating block of wood is pushed up and down regularly on a water surface, a regular train of waves will be propagated outward. The simplest type of regular wave train is a sinusoidal wave motion, which is illustrated in Fig. 13.1. Part **A** of this figure shows one end of a long stretched string attached to a weight supported by a spring. The weight is arranged so that it can move freely in the vertical "ways" of a frame. If the weight is pulled downward a distance A and then released, the weight will move in the vertical direction with simple harmonic motion of a certain period T. Because the end of the string is attached to the weight, the oscillating weight acts as a source of a sinusoidal transverse wave that travels to the right along the string in the manner indicated by the curves of Fig. 13.1(**B**). These curves show successive "snapshots"

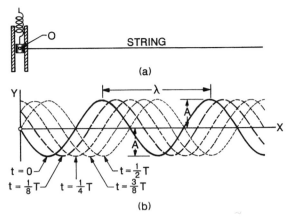

FIG. 13.1 Production and propagation of a sinusoidal transverse wave in a long string shown in part (**A**). Part (**B**) shows the motion of the string during one half-cycle of oscillation of O, shown in part (**A**), from $Y = 0$ to $+A$ and back to 0. (From *Elements of Physics*, 5th ed., by Shortley, G., and Williams, D., p. 416, ©1971. Reprinted by permission of Pearson Education, Inc.: Upper Saddle River, New Jersey.)

of the shape of the string during one half-cycle, after the motion has been well established. The distance between adjacent crests or adjacent troughs in such a wave is called the *wavelength*; in the figure the wavelength is denoted by λ. Each time the particle O attached to the weight makes a complete oscillation, the wave moves a distance λ in the X-direction. Hence the wave speed, v, and the wavelength are related by the equation

$$v = \lambda / T, \tag{13.2}$$

where T is the period of oscillation and λ = wavelength. In terms of the frequency $f = 1/T$, this equation can be written as

$$v = f\lambda \tag{13.3}$$

or

$$\lambda = v/f. \tag{13.4}$$

II. DIELECTRIC CONSTANT, FREQUENCY DOMAIN, AND TIME DOMAIN

We recall the physical properties of water and remember that the dielectric constant varies with temperature (Weast, 1964, p. E-36):

Temperature	
°C	ε
0	88.00
10	84.11
20	80.36
30	76.75
40	73.28
50	69.94
100	55.33

One can obtain information about a dielectric in either the frequency domain or the time domain. In the frequency domain, a number of measurements over a wide frequency range is required for complete characterization of the dielectric, which is time consuming and requires a considerable investment in instrumentation. However, one can obtain the same information over a wide frequency range in only a fraction of a second by making the measurement not in the frequency domain but in the time domain. In time domain reflectometry, a pulse is used that simultaneously contains all the frequencies of interest (Fellner-Feldegg, 1969). In the frequency range of 1 megaHertz (MHz) to 1 gigaHertz (GHz), the dielectric constant is not strongly frequency dependent (mega = 10^6 and giga = 10^9). Figure 13.2 shows the wavelength and frequency of commonly used devices, such as radios, televisions, and cellular phones (Clark, 1994).

In passing, we note that the safety of electromagnetic waves, especially those associated with cellular telephones, is still in doubt. A link between brain cancer and electromagnetic fields has been found in some studies (Bishop, 1995). Experiments have shown that radio waves at about the same power as that emitted by today's cellular phones can break down the binding of calcium to the surface of cells. Calcium is essential for virtually all living processes, including enzyme action and cell growth. Data showed that the breakdown occurred at 145 MHz, the frequency at which ham radios operate, and at 450 MHz, the frequency used by security guards' radio phones. European cellular systems operate at 450 MHz (Clark, 1994).

III. THEORY FOR USE OF THE DIELECTRIC CONSTANT TO MEASURE SOIL WATER CONTENT

Most of the solid components of soil have dielectric constants in the range of 2 to 7, and that of air is effectively 1 (ε of air = 1.000590). Thus

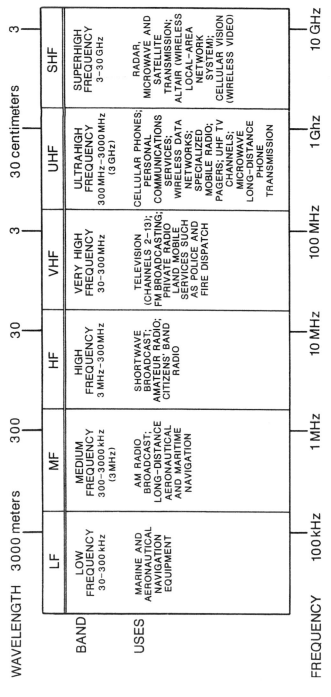

FIG. 13.2 Wavelengths, frequencies, and their uses. (From Clark, D., 1994. Reprinted by permission of The Wall Street Journal, ©1994, Dow Jones & Company, Inc. All Rights Reserved Worldwide.)

a measure of the dielectric constant of soil is a good measure of the water content of the soil.

Here is a brief outline of what we are going to do to use TDR to get the volumetric water content of soil. We are going to measure a travel time, and by knowing the length of the rods (waveguides) in the soil, we are going to get a velocity (velocity = length/time). We are going to relate this velocity to the dielectric constant. And then we will relate the dielectric constant to volumetric water content.

The TDR technique measures the velocity of propagation of a high-frequency signal (1 MHz to 1 GHz). The velocity of propagation is as follows:

$$V = c/(K')^{1/2}, \tag{13.5}$$

where

V = is the velocity of propagation in the soil

c = is the propagation velocity of light in free space; $c = 3 \times 10^8$ m/s

K' = dielectric constant of the soil.

By determining the travel time, t, of the pulse traveling in the transmission line or wave guide of length L, one can get the velocity as L/t.

Equation 13.5 can be rearranged to give the apparent dielectric constant as

$$K_a = [(ct)/L]^2 \tag{13.6}$$

where K_a is the apparent dielectric constant. However, we need to add a "2" to the denominator in Equation 13.6, because the line length is the distance *traveled,* but commercial cable testers measure the length down and the echo (reflection). Hence, the distance *measured* is two times the line length. So we have

$$K_a = [(ct)/2L]^2. \tag{13.7}$$

The relationship in Equation 13.5 is approximate, so in Equations 13.6 and 13.7 we use K_a, the apparent dielectric constant, instead of K' (Topp and Davis, 1982).

Commercial TDR cable testers reduce the transfer time to an apparent probe length, l_a (Fig. 13.3 from Clothier et al., 1994), so that

$$K_a = [(ct)/2L]^2 = (l_a/Lv_p)^2 \tag{13.8}$$

where v_p is the relative velocity setting of the instrument (Vogeler et al., 1996; see their Equation 10). It is relative so it is unitless (v = the Greek letter nu.) The reason the 2 appears in the first equation of Equation 13.8 is that the echo must travel down and back along the rods of

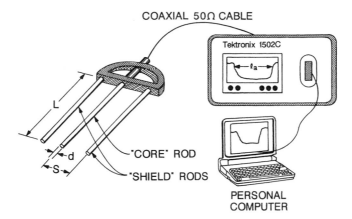

COAXIAL 50Ω CABLE

Tektronix 1502C

ℓ_a

L

d

S

"CORE" ROD

"SHIELD" RODS

PERSONAL
COMPUTER

FIG. 13.3 Schematic diagram of the TDR system with the three-rod probe, the cable tester, and the computer that controls, acquires, and analyzes the data. (From Clothier et al., 1994. Reprinted by permission of Brent E. Clothier.)

length L, as noted above. On the right-hand side of Equation 13.8, the travel time is normalized to the length that is found relative to the true length. The length is $2l_a$ over $2L$, so the 2's cancel (B.E. Clothier, personal communication, March 6, 1997). The commercial cable tester made by Tektronix (Model 1502C, Wilsonville, Oregon) does not display $2l_a$ because it is used as a cable tester to find breaks in a cable. So it knows it is measuring 2 times the length, and saves the operator the hassle by dividing by two before it puts the trace on the screen (Fig. 13.3).

Experimental results (Topp et al., 1980) have given the following relation between volumetric water content and the dielectric constant:

$$\theta_v = -0.053 + 0.0292\,K_a - 5.5 \times 10^{-4}\,K_a^2 + 4.3 \times 10^{-6}\,K_a^3 \quad (13.9)$$

Equation 13.9 has been shown to hold for many different types of soil. The relationship between volumetric water content (θ) and the dielectric constant (K_a) is essentially independent of soil texture, porosity, and salt content. However, if a soil is high in organic matter, Equation 13.9 does not hold and a separate calibration equation needs to be determined. Herkelrath et al. (1991), who studied organic soil, found that the equation of Topp et al., Equation (13.9), predicted values of soil water content that were 30% too low. Topp et al. (1980) also reported a similar shift in their calibration for an organic soil. The soil cores of Herkelrath et al. (1991) had a large fraction of organics: 12.6% carbon.

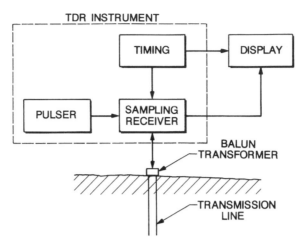

FIG. 13.6 A block diagram of a TDR instrument and its display units. (From Topp, G.C., Soil water content. In *Soil Sampling and Methods of Analysis*, (M.R. Carter, Ed.), p. 545, ©1993, CRC Press: Boca Raton, Florida.)

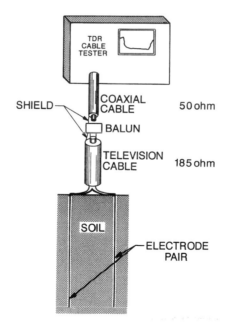

FIG. 13.7 The basic TDR circuit for use in determining soil moisture. (From Herkelrath, W.H., Hamburg, S.P, and Murphy, F. Automatic, real-time monitoring of soil moisture in a remote field area with time domain reflectometry. *Water Resources Research* 27(5); 857, 864, 1991. Copyright ©1991, American Geophysical Union. Reproduced by permission of the American Geophysical Union.)

The coaxial cable in Fig. 13.7 is 50 ohm. The coaxial cable is connected to an 185-ohm shielded television cable and a balun to provide a "balanced line" (Herkelrath et al., 1991). *Impedance* is the apparent resistance in an alternating electrical current corresponding to the true resistance in a direct current (*Webster's New World Dictionary of the American Language*, 1959). If one is using three or more pronged probes, they simulate the coaxial line and require no impedance matching transformer (Clothier et al., 1994).

Third, we need cables for connecting the TDR instrument and soil probes (Topp, 1993). Cable combinations between the TDR instrument and soil probes are determined by the type of probes used (two-pronged, three-pronged, or more).

Fourth, we need tools for installation of soil probes. Three procedures can be used for insertion. First, for short probes (in most soils, except the most resistant), we can insert the probes by hand, take a reading, and remove them and move on to the next spot. This way we can easily take many readings and quickly get a "feel" for spatial variability. Second, for longer probes, it is necessary to make "pre-holes" with a dummy probe. This could also be done to obtain repeated measurements in space, although once the longer probes are inserted we tend to leave them in place connected to a multiplexer, or with caps on the coaxial connector if single measurements are to be made. Third, B.E. Clothier and S.R. Green in New Zealand have made "direct-wired" probes that they insert horizontally into a face of a pit that they have dug. The probes are inserted horizontally, the hole backfilled, and the probes remain in place underneath undisturbed soil during the summer. One has to be careful when removing them at the end of the experiment, for it is easy to put a spade right through the connector cable when exhuming them (B.E. Clothier, personal communication, February 23, 1994).

And, fifth, if we are automating measurements, we need a multiplexer. Baker and Allmaras (1990) describe a system for automated measurements using a multiplexer.

VI. PRACTICAL INFORMATION WHEN USING TDR TO MEASURE SOIL WATER CONTENT

Following are some notes on the use of the TDR technique.

1. Rods normally range in length from 100 mm to 1 m. The shortest depth that I have seen documented in the literature is 50 mm (Mallants et al.,

1996). Probes shorter than 50 mm do not give good traces (B.E. Clothier and M.B Kirkham, personal observations, January–April, 1991). Probes longer than 1 m are difficult to insert into the soil without them bending. Sometimes it is difficult even to insert the probes to this depth (e.g., in the caliche soils of Texas; Todd A. Vagts, personal communication, February 11, 2000). Miller and Buchan (1996) in New Zealand describe the challenges of inserting rods at depth in a silt loam soil overlying unweathered greywacke gravels and stones with a sand matrix.

However, the fact that TDR probes measure soil water content only to 1 m does not mean that we do not need to measure deeper than 1 m. Neutron probes can measure to a depth of 3 meters or more. Even though TDR is replacing the use of neutron probes, because no danger from radioactivity is involved with TDR, we cannot abandon neutron probes. They provide the only method that can be used to get soil water content at deep depths. In semi-arid regions like Kansas, it is important to measure 2 to 3 m below the surface of the soil, to determine maximum depth of water depletion by roots. Miller and Buchan (1996) report the widespread use of neutron probes in Australia and South Africa to schedule irrigations.

2. The rods allow flexibility in determining water content. The spacing and geometry can be changed. Probes can be inserted horizontally or vertically. The ability to insert probes horizontally allows calculation of the velocities of both the wet front and solute front in a soil (Duwig et al., 1997).

3. There is some heating with the TDR method, but given the power levels involved, it is minuscule (B.E. Clothier, personal communication, February 22, 1994).

4. The accuracy of the method is ± 0.01 m^3 m^{-3} (Topp, 1993). For comparison, Song et al. (1998) found that the dual-probe heat-pulse technique monitored soil water content within 0.03 m^3 m^{-3} and changes in soil water content within 0.01 m^3 m^{-3}.

5. The magnitude of reflected signals, after the first one for soil water content, can be used to determine electrical conductivity of the soil (Topp, 1993). The exact relation has yet to be established. The degree to which the signal is "lost" (attenuated) after all the multiple reflections have died away is due to the soil's electrical conductivity, which is, in some large part, due to the salt content of the solution (B.E. Clothier, personal communication, January 20, 1994).

6. Simultaneous measurements of soil water content and electrical conductivity by using TDR have been done (Dalton et al., 1984; Dasberg and Dalton, 1985; Zegelin et al., 1989; Nadler et al., 1991), and the method is being used to determine solute transport (Lundin and Johnsson, 1994; Ward et al., 1994; Vogeler et al., 1996; Duwig et al., 1997).

VII. EXAMPLE OF USING TDR TO DETERMINE ROOT WATER UPTAKE

Many papers have been published that analyze the TDR technique (e.g., Heimovaara and Bouten, 1990) and its use for routine measurements of soil water content (e.g., Grantz et al., 1990), and the literature is growing rapidly. TDR permits observations of the changing pattern of water content in the soil that occurs as a result of root water uptake. Here we present only one example in which a kiwifruit vine was studied (Clothier and Green, 1994). After an initial irrigation, the soil water content was uniform across the root zone of the kiwifruit vine; also, the water uptake was quite uniform (Fig. 13.8, top). Beginning in the tenth week of 1992, just one half of the vine's root zone, the southern half, was wetted by a sprinkler irrigation. Following this differential irrigation of the root zone, the flow of water in the "wet" southern root increased, but the flux in the "dry" northern root was about halved. Thus, the vine quickly switched its pattern of uptake away from the drier parts of its root zone.

Of greater interest, however, was the depthwise pattern of root uptake observed on the wet side. The preference for near-surface water uptake can be seen (Fig. 13.8, bottom). The vine continued to extract water in the densely rooted region surrounding its base, but the shift in uptake to the surface roots on the wet southern side was remarkable

The results show that greater efficiency in irrigation water might be obtained by applying small amounts of water, more frequently. A small amount of irrigation water would be rapidly used by active, near-surface roots. This would then eliminate drainage of irrigation water into the lower regions of the root zone, where draining water passes by inactive roots and goes to greater depth. Such observations are made possible by using TDR.

VIII. HYDROSENSE™

Campbell Scientific, Inc. sells the HydroSense™ to measure soil water content (Tanner, 1999). It consists of a sensor with two parallel rods. The HydroSense™ is often referred to as a quasi-TDR device. Its frequency is

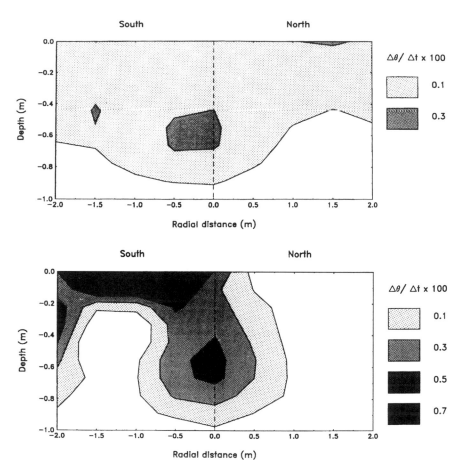

FIG. 13.8 Measurement by time domain reflectometry of the changing spatial pattern of soil water content in the root zone of a kiwifruit vine growing near Palmerston North, New Zealand. The upper figure depicts the average rate of water content change over the four-week period 11 February–9 March 1992. The lower figure shows the change that occurred over the two weeks following irrigation of just the south side on 10 March. Rate of water extraction $\Delta\theta/\theta t$ is given in units of $m^3 \ m^{-3} \ s^{-1}$. The vine is located at the center. (From Clothier, B.E., and Green, S.R., Rootzone processes and the efficient use of irrigation water, *Agricultural Water Management* 25; 1–12, ©1994, Elsevier Science: Amsterdam. Reprinted by permission of Elsevier, Amsterdam.)

not in the GHz range of true TDR, but it is a good instrument once an electrical conductivity has been calibrated (B.E. Clothier, personal communication, March 6, 2000). The HydroSense™ has the advantages of being easily portable (hand-held) and cheap. Because it measures both water and solute concentration, one can track tracers with it.

The product brochure describes the instrument as follows (Tanner, 1999): The HydroSense™ "consists of an electronic circuit encapsulated in epoxy. Replaceable rods are 5 mm in diameter and are available in 12- and 20-cm lengths. A measurement is made by fully inserting the rods into the soil and pressing the READ button." The display gives volumetric water content in percent. The instrument also can be set in the water deficit mode in which the user sets "lower and upper water content references by taking measurements under those conditions and storing the values in memory. Once reference values are stored, subsequent measurements provide a display of the relative water content and the water deficit."

To calibrate the probe, mix up some soil of known water content in the laboratory and insert the probes into it. If one chooses 2-3 water contents, one can get a feel as to which electrical conductivity setting to use in the Hydrosense™ to get the right water content result (B.E. Clothier, personal communication, March 3, 2000).

IX. APPENDIX: BIOGRAPHY OF HEINRICH HERTZ

Heinrich Rudolf Hertz (1857–1894), a German physicist, was born on February 22, 1857, at Hamburg. After leaving the gymnasium, he studied civil engineering, but at the age of 20, he came to a turning-point in his career (Cajori, 1929, p. 258) and abandoned engineering in favor of physics. He went to Berlin, and worked under Hermann Ludwig Ferdinand von Helmholtz (German physiologist and physicist, 1821–1894), advancing rapidly to become his assistant by 1880. In 1883 he became a private docent (official but unpaid lecturer) at Kiel. There he began the studies of Maxwell's electromagnetic theory (James C. Maxwell, Scottish physicist, 1831–1879), which resulted in the discoveries—between 1885 and 1889, while he was professor of physics in the Polytechnic in Karlsruhe, Germany (Preece, 1971)—that made Hertz's name famous. It was there that he performed his memorable experiments on electromagnetic waves.

In 1888 Hertz found means of detecting the presence of electromagnetic waves arising from a Leyden jar (Cajori, 1929, p. 259). A Leyden jar, named after the Dutch city of Leiden in The Netherlands, where it was invented, is a glass jar coated outside and inside with tin foil and having a metallic rod connecting with the inner lining and passing through the lid (*Webster's New World Dictionary of the American Language*, 1959). It acts as a condenser for static electricity (Fig. 13.9). During the oscillatory discharge of a Leyden jar, electromagnetic waves radiate into space. Such a wave is called "electromagnetic," because it has two components: an electric

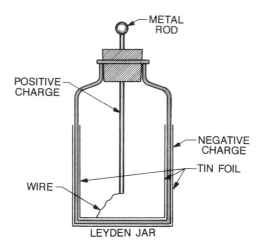

FIG. 13.9 Leyden jar. (From *Webster's New World Dictionary of the American Language: College Edition.* Copyright ©1959, The World Publishing Company: Cleveland and New York. All rights reserved. Rights now owned by Wiley. Reproduced here by permission of Wiley Publishing, Inc., Indianapolis, Indiana.)

wave and a magnetic wave. Hertz was able to observe each separately, an accomplishment that Maxwell had feared would never be realized (Cajori, 1929, p. 259).

In 1889 Hertz was appointed to succeed Rudolf Julius Emanuel Clausius (German physicist, who made important contributions to molecular physics; 1822–1888) as professor of physics at the Univeristy of Bonn. Thus, at the age of 32, he occupied a position attained much later in life by most men of his time. There he continued his researches on the discharge of electricity in rarefied gases, only just missing the discovery of the X-rays described by Wilhelm Konrad Röntgen (German physicist, 1845–1923, who received the Nobel Prize in physics in 1901) a few years later. There Hertz wrote his treatise *Principles of Mechanics.* In 1892 a chronic blood poisoning began to undermine his health, and, after a long illness he died in the prime of life on January 1, 1894, in Bonn. By his premature death, science lost one of its most promising disciples (Preece, 1971). For a book that describes Hertz's experiments and production of electromagnetic waves, see Buchwald (1994).

X. APPENDIX: BIOGRAPHY OF SERGEI SCHELKUNOFF

Sergei A. Schelkunoff, inventor and expert on electromagnetism, was born in Samara, Russia. He researched the coaxial cable now widely used for television transmission (Lambert, 1992). He was a student at the University of Moscow when he was caught up in the tumult of World War I and the Bolshevik Revolution. Drafted and trained as a Russian Army officer in 1917, he fought and worked his way across Siberia into Manchuria and on to Japan before landing in Seattle in 1921. He learned English and worked his way through school, earning both bachelor's and master's degrees in mathematics from the State College of Washington, now the University of Washington, and a doctorate from Columbia University in New York City in 1928.

He went to work for Western Electric's laboratories and its successor, Bell Labs, and in his 35 years there, he became assistant director of mathematical research and assistant vice president for university relations. The government granted him 15 patents for radio antennas, resonators, and wavelength guides. In 1935, he and three colleagues reported that the newly developed coaxial cable could transmit television or up to 200 telephone circuits. He specialized in coaxial's frequency, impedance, attenuation, coupling, shielding, circuit, and field characteristics. He published four books and dozens of papers in scientific journals, and also taught for five years at Columbia University, where he retired in 1965. The Institute of Radio Engineers awarded him a prize for his contributions to radio wave transmission theory, and the Franklin Institute awarded him a medal for his communication and reconnaissance research. He died of a heart ailment at age 95 on May 2, 1992. He had no immediate survivors. His wife of 51 years, the former Jean Kennedy, died in 1979.

REFERENCES

Baker, J.M., and Allmaras, R.R. (1990). System for automating and multiplexing soil moisture measurement by time-domain reflectometry. *Soil Sci Soc Amer J* 54; 1–6.

Bishop, J.E. (1995). Link between EMF, brain cancer is suggested in major new study. However, no added risk of leukemia is found, unlike previous work. *Wall Street Journal* Wednesday, January 11, 1995; p. B4.

Buchwald, J.Z. (1994). *The Creation of Scientific Effects: Heinrich Hertz and Electric Waves.* University of Chicago Press: Chicago, Illinois.

Cajori, F. (1929). *A History of Physics.* Macmillan: New York.

Clark, D. (1994). Squeeze play. With usable frequencies scarce, companies try to ease the crunch. *Wall Street Journal* Friday, February 11, 1994; p. R16.

Clothier, B.E., and Green, S.R. (1994). Rootzone processes and the efficient use of irrigation water. *Agr Water Manage* 25; 1–12.

Clothier, B., Gaudet, J.-P., Angulo, R., and Green, S. (1994). Application of the TDR (time domain reflectometry) method to measure water content and concentration of solutes in soils. In *French Society of Thermal Engineers, One-Day Workshop, 9 February 1994, Paris, France, on Methodologies for Porous-Media Measurement.* French Society of Thermal Engineers, Paris. (In French; English translation by M.B. Kirkham).

Dalton, F.N., Herkelrath, W.N., Rawlins, S.L, and Rhoades, J.D. (1984). Time domain reflectometry: Simultaneous measurement of soil water content and electrical conductivity with a single probe. *Science* 224; 989–990.

Dasberg, S., and Dalton, F.N. (1985). Time domain reflectometry field measurements of soil water content and electrical conductivity. *Soil Sci Soc Amer J* 49; 293–297.

Duwig, C., Vogeler, I., Clothier, B.E., and Green, S.R. (1997). Nitrate leaching to groundwater under mustard growing on soil from a coral atoll. In *Nutritional Requirements of Horticultural Crops* (Currie, L.D., and Loganathan, P., Eds.), pp. 36–43. Occasional Report No. 10. Fertilizer and Lime Research Centre, Massey University: Palmerston North, New Zealand.

Fellner-Feldegg, H. (1969). The measurement of dielectrics in the time domain. *J Phys Chem* 73; 616–623.

Grantz, D.A., Perry, M.H., and Meinzer, F.C. (1990). Using time-domain reflectometry to measure soil water in Hawaiian sugarcane. *Agronomy J* 82; 144–146.

Heimovaara, T.J., and Bouten, W. (1990). A computer-controlled 36-channel time domain reflectometry system for monitoring soil water contents. *Water Resources Res* 26; 2311–2316.

Herkelrath, W.H., Hamburg, S.P., and Murphy, F. (1991). Automatic, real-time monitoring of soil moisture in a remote field area with time domain reflectometry. *Water Resources Res* 27; 857–864.

Lambert, B. (1992). S.A. Schelkunoff, 95, researcher and developer of coaxial cable. *New York Times*, May 17, 1992; 23Y.

Lorrain, P., and Corson, D.R. (1979). *Electromagnetism. Principles and Applications.* W.H. Freeman and Co: San Francisco.

Lundin, L.-C., and Johnsson, H. (1994). Ion dynamics of a freezing soil monitored in situ by time domain reflectometry. *Water Resources Res.* 30; 3471–3478.

Mallants, D., Vanclooster, M., Toride, N., Vanderborght, J., van Genuchten, M.T., and Feyen, J. (1996). Comparison of three methods to calibrate TDR for monitoring solute movement in undisturbed soil. *Soil Sci Soc Amer J* 60; 747–754.

Miller, B., and Buchan, G. (1996). TDR vs. neutron probe—how do they compare? *WISPAS. A Newsletter about Water in the Soil-Plant-Atmosphere System* 65, 8–9. The Horticultural and Food Research Institute: Palmerston North, New Zealand.

Nadler, A., Dasberg, S., and Lapid, I. (1991). Time domain reflectometry measurements of water content and electrical conductivity of layered soil columns. *Soil Sci Soc Amer J* 55; 938–943.

Preece, W.E. (General Ed.). (1971). Hertz, Heinrich Rudolf. *Encyclopaedia Britiannica* 11; 456.

Shortley, G., and Williams, D. (1971). *Elements of Physics.* 5th ed. Prentice-Hall: Englewood Cliffs, New Jersey.

Song, Y., Ham, J.M., Kirkham, M.B., and Kluitenberg, G.J. (1998). Measuring soil water content under turfgrass using the dual-probe heat-pulse technique. *J Amer Soc Hort Sci* 123; 937–941.

Tanner, B. (Executive Ed.). (1999). HydroSense™ now available. *The Campbell Update. A Newsletter for the Customers of Campbell Scientific, Inc.* 10(1); 7. Campbell Scientific, Inc.: Logan, Utah.

Topp, G. C. (1993). Soil water content. In *Soil Sampling and Methods of Analysis* (Carter, M.R., Ed.), pp. 541–557. Lewis Publishers: Boca Raton, Florida.

Topp, G.C., and Davis, J.L. (1982). Measurement of soil water content using time domain reflectometry. In *Canadian Hydrology Symposium: 82*, pp. 269–287. Associate Committee on Hydrology National Research Council of Canada: Fredericton, New Brunswick.

Topp, G.C., Davis, J.L., and Annan, A.P. (1980). Electromagnetic determination of soil water content: Measurements in coaxial transmission lines. *Water Resources Res* 16; 574–582.

Vogeler, I., Clothier, B.E., Green, S.R., Scotter, D.R., and Tillman, R.W. (1996). Characterizing water and solute movement by time domain reflectometry and disk permeametry. *Soil Sci Soc Amer J* 60; 5–12.

Ward, A.L., Kachanoski, R.G., and Elrick, D.E. (1994). Laboratory measurements of solute transport using time domain reflectometry. *Soil Sci Soc Amer J* 58; 1031–1039.

Weast, R.C. (1964). *Handbook of Chemistry and Physics*. 45th ed. Chemical Rubber Co.: Cleveland, Ohio.

Webster's New World Dictionary of the American Language. (1959). College ed. World Publishing: Cleveland and New York.

Zegelin, S.J., White, I., and Jenkins, D.R. (1989). Improved field probes for soil water content and electrical conductivity measurement using time domain reflectometry. *Water Resources Res* 25; 2367–2376.

14

Root Anatomy and Poiseuille's Law for Water Flow in Roots

We now turn to water movement in plant roots. We shall apply Poiseuille's law for the flow through the xylem. But first let us review root anatomy and the cell types that make up the xylem.

I. ROOT ANATOMY

A. The Four Regions of an Elongating Root

Elongating roots usually possess four regions: the root cap, the meristematic region, the region of cell elongation, and the region of differentiation and maturation (Fig. 14.1). But these regions are not always clearly delimited (Kramer, 1983, p. 122). The root cap is composed of loosely arranged cells and is usually well defined. Because it has no direct connection with the vascular system, it probably has no role in absorption. It is said to be the site of perception of the gravitation stimulus, but this is debatable.

The meristematic region typically consists of numerous small, compactly arranged, thin-walled cells almost completely filled with cytoplasm. Relatively little water or salt is absorbed through this region, largely because of the high resistance to movement through the cytoplasm and the lack of a conducting system (Kramer, 1983, p. 122).

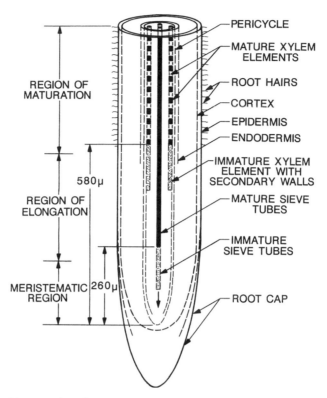

FIG. 14.1   Diagram of a tobacco root tip showing relative order of maturation of various tissue. The distance from the tip at which the various tissues differentiate and mature depends on the kind of root and the rate of growth. (From Kramer, P.J., *Water Relations of Plants*, p. 123, ©1983, Academic Press: New York. Reprinted by permission of Academic Press.)

Usually there is a zone of rapid cell elongation and expansion a few tenths of a millimeter behind the root apex. It is difficult to indicate a definite zone of differentiation because various types of cells and tissues are differentiated at different distances behind the root apex (Kramer, 1983, p. 122). Typically, sieve tubes of the phloem differentiate before the xylem elements (Esau, 1965, p. 498). As the newly enlarged, thin-walled cells at the base of the zone of enlargement cease to elongate, they become differentiated into the epidermis, cortex and stele, which constitute the primary structures of roots.

B. Root Hairs

Root hairs appear when the epidermis differentiates. The epidermis has specialized cells that are root hair cells. Much attention has been given to root hairs because of their presumed importance as absorbing surfaces. The epidermis is usually composed of relatively thin-walled, elongated cells that form a compact layer covering the exterior of young roots. Sometimes a second compact layer, the hypodermis, lies beneath the epidermis. In some plants, including citrus and conifers, root hairs can arise not only from the epidermis, but also from the layer of cells beneath the epidermis, or even from deeper in the cortex (Kramer, 1983, p. 125).

C. Dicotyledonous Roots

The arrangement of the principal tissues in a dicotyledonous root is shown in Fig. 14.2. The conductive tissues form a solid mass in the center, instead of being dispersed in bundles around the periphery of the pith, as in stems of most herbaceous, dicotyledonous plants (Kramer, 1983, p. 124). (We will study stem anatomy in Chapter 18.) The primary xylem in the roots of dicots usually consists of two to several strands extending radially outward from the center, with the primary phloem located between them. The outermost layer of the stele is the pericycle. Its cells retain their ability to divide, and they give rise not only to branch roots but also to the cork cambium, if secondary growth occurs. The endodermis usually consists of a single layer of cells and forms the inner layer of the cortex. The endodermis has the casparian strip, which is a bandlike wall formation within primary

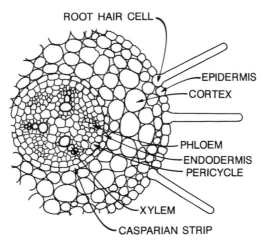

FIG. 14.2 A dicotyledonous root (a squash root). Note the central stele of vascular tissue. (From Kramer, P.J., *Water Relations of Plants*, p. 12, ©1983, Academic Press: New York. Reprinted by permission of Academic Press.)

walls that contains suberin and lignin. It occurs on the radial and transverse walls in the endodermis (Esau, 1977, p. 504).

D. Monocotyledonous Root

In monocotyledonous plants, a variable number of xylem vessels are arranged in a circle around a pith (Fig. 14.3). However, in some monocotyledonous roots such as wheat (Fig. 14.4), a single vessel occupies the center and is separated by nontracheary elements from other vessels (Esau, 1965, p. 496). The large central xylem vessel is part of the metaxylem. The metaxylem is part of the primary xylem that differentiates after the protoxylem and before the secondary xylem, if any secondary xylem is formed in a given species. Protoxylem is the first formed element of the xylem in a plant organ. It is the first part of the primary xylem. The protoxylem gets crushed as the metaxylem develops. Although most monocotyledons lack secondary growth from a vascular cambium, they can undergo a type of "secondary growth" by an intense and protracted thickening growth. Large trees can result, such as palm trees (Esau, 1965, p. 400), which are monocotyledons.

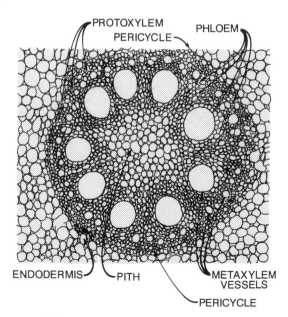

FIG. 14.3 A monocotyledonous root (a corn root). Note the central pith. (From Esau, K., *Plant Anatomy*, 2nd ed., p. 713, ©1965, John Wiley & Sons, Inc.: New York. This material is used by permission of John Wiley & Sons, Inc.)

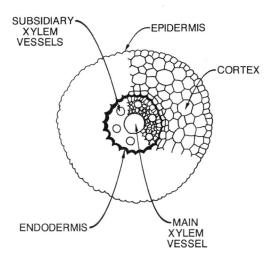

FIG. 14.4   A monocotyledonous root (a wheat root). Note the central large xylem vessel, part of the metaxylem. (From Richards, R.A., and Passioura, J.B., p. 250, ©1981, Crop Science Society of America: Madison, Wisconsin. Reprinted by permission of the Crop Science Society of America.)

E. Movement of Water and Solutes Across the Root

The structure of the root is of particular interest with regard to the movement of water and the dissolved salts in it from the absorbing cells to the conducting tissues, and their release from the living cells of the vascular cylinder into the nonliving tracheary elements. Figure 14.5 illustrates the pathway of the soil solution in the wheat root (Esau, 1965, pp. 516–517). The arrows indicate the direction of movement in certain selected cells. The living cells among these are stippled. The most notable features of this pathway are: 1) the presence of abundant intercellular spaces in the cortex, 2) the lack of such spaces in the vascular cylinder, and 3) the presence of a specialized endodermis between the two systems. The endodermis between the two distinct systems (cortex and vascular cylinder) acts as a barrier that facilitates the development of hydrostatic pressure in the vascular cylinder by preventing a leakage of solutes from the vascular cylinder into the cortex.

F. Endodermis

Let us consider the endodermis and its casparian strip in more detail. A prominent feature of the primary structure of most roots is the endodermis, the inner layer of cells of the cortex which separates it from the

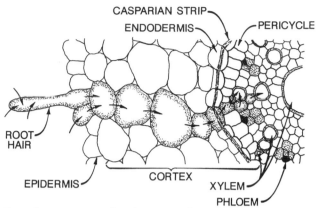

FIG. 14.5 Part of a transection of a wheat root, illustrating the kinds of cells that may be traversed by water and salts absorbed from the soil before they reach the tracheary elements of the xylem. Arrows indicate the direction of movement through a selected series of cells. Among these, the living cells are partly stippled. The casparian strip in the endodermis is shown as though exposed in surface views of end walls. (From Esau, K., *Plant Anatomy*, 2nd ed., p. 517, ©1965, John Wiley & Sons, Inc.: New York. This material is used by permission of John Wiley & Sons, Inc.)

stele. The endodermis is not part of the stele. The stele comprises the vascular system (xylem and phloem) and the associated ground tissue (pericycle; interfascicular regions, and pith, if it occurs). Early in the development of the endodermis, suberin (a fatty substance) is deposited in bands on the transverse walls and radial walls in the longitudinal direction, forming the casparian strip (Fig. 14.6). It renders them relatively impermeable to water and presents a barrier to inward movement of water and solutes in the apoplast (Kramer, 1983, p. 128). The apoplast is the supposedly dead part of the plant tissue, including the cell walls. The symplast or symplasm is the continuum of communicating cytoplasm, which is created by the intercellular connections. Plasmodesmata are fine, cytoplasmic threads that pass from a protoplast through a cell wall directly into the protoplast of a second cell (Nobel, 1974, p. 37).

The endodermis characterized by casparian strips is almost universally present in roots (Esau, 1965, p. 489). The strip is formed during the early ontogeny of the cell and is a part of the primary wall. It varies in width and is often much narrower than the wall in which it occurs. The incrustation of the cell wall by the material constituting the casparian strip presumably blocks the submicroscopic capillaries in the wall and hinders the movement of substances through the walls (Esau, 1965, p. 517). Moreover, the

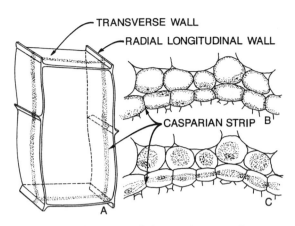

FIG. 14.6 Endodermal cells. A, entire cell showing location of casparian strip. B and C, effect of treatment with alcohol on cells of endodermis and of parenchyma. B, cells before treatment; C, after. The casparian strip is seen only in sectional views in B, C. (From Esau, K., *Plant Anatomy*, 2nd ed., p. 489, ©1965, John Wiley & Sons, Inc.: New York. This material is used by permission of John Wiley & Sons, Inc.)

cytoplasm of the endodermal cells is relatively firmly attached to the casparian strip, so that it does not readily separate from the strip when the tissue is subjected to the effects of plasmolytic or other agents normally causing a contraction of protoplasts (Fig. 14.6). Thus the casparian strip appears to form a barrier at which the soil solution is forced to pass through the selectively permeable cytoplasm (the symplasm) rather than through the cell wall (apoplast).

G. Cell Types in Xylem Tissue

Xylem is a tissue that is comprised of four cell types (Table 14.1): tracheids and vessel members, which make up the tracheary elements, fibers, and parenchyma cells (Esau, 1977, p. 103). The tracheary elements are the most highly specialized cells of the xylem and are concerned with the conduction of water and substances in water. They are nonliving cells at maturity. They have lignified walls with secondary thickenings and a variety of pits.

The two kinds of tracheary cells, the tracheids and the vessel members, differ from each other in that the tracheid is an imperforate cell whereas the vessel member has perforations, one or more at each end and sometimes also on a side wall (Esau, 1977, p. 106). The longitudinal series of vessel members interconnected through their perforations are called xylem vessels

TABLE 14.1   The four cell types in xylem tissue: tracheids, vessel members, fibers, and parenchyma cells. (From Esau, K., *Anatomy of Seed Plants*, 2nd ed., p. 106, ©1977, John Wiley & Sons, Inc.: New York. This material is used by permission of John Wiley & Sons, Inc.)

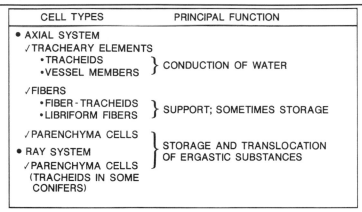

or simply, vessels. The perforated part of a wall of a vessel member is called the perforation plate. A plate may be simple, with only one perforation, or multiperforate, with more than one perforation.

The fibers are long cells with secondary, commonly lignified, walls (Esau, 1977, p. 108). The walls vary in thickness but are usually thicker than the walls of tracheids in the same wood (wood is secondary xylem). Two principal types of xylem fibers are recognized, the fiber-tracheids and the libriform fibers. If both occur in the same wood, the libriform fiber is longer and has thicker walls than the fiber-tracheid. Fibers give support to the xylem.

Parenchyma cells of the primary xylem occur in interfascicular regions and are considered to be part of the ground tissue (Esau, 1977, pp. 112–113). In the root, the primary xylem forms a core with parenchyma (as in some monocotyledonous roots) or a core without parenchyma (as in dicotyledonous roots). In secondary xylem, the parenchyma cells make up the axial and ray parenchyma. These parenchyma cells store starch, oils, and many other ergastic substances. Ergastic subtances are products of protoplasts such as starch grains, fat globules, crystals, and fluids. They occur in the cytoplasm, organelles, vacuoles, and cell walls (Esau, 1977, p. 509).

We can list the characteristics of tracheids, vessel members, fibers, and parenchyma cells as follows:

TRACHEIDS

- More primitive (in angiosperms, gymnosperms, and lower vascular plants)
- Tapered ends
- Long cells
- Thin cells
- Lignified secondary cell walls
- No protoplasts at maturity (dead)

VESSEL MEMBERS

- Present only in angiosperms (more evolved than tracheids)
- Short cells
- Broad cells
- Flat ends
- Perforated end walls with perforation plates
- Lignified secondary cell walls
- No protoplasts at maturity (dead)
- Several together form a continuous tube, which is called a xylem vessel
- A xylem vessel is a low-resistance circuit
- Width varies from 10 to 800 μm; an average diameter is about 40 μm
- Length of a xylem vessel varies from few hundred microns (μm) to a few millimeters. Long and wide xylem vessels occur in tropical plants such as vines (e.g., kiwifruit) and rattans (climbing palms). Fisher and colleagues (2002) of the Fairchild Tropical Garden in Miami, Florida, found a vessel in a rattan that was 3 m long and 532 μm in diameter.

FIBERS

- For structural support
- Long cells
- Thin cells
- Heavily lignified cell walls
- No protoplasts at maturity (dead)

PARENCHYMA CELLS

- For storage
- For lateral movement of water and solutes into and out of conducting cells (the ray system in secondary xylem) (Table 14.1).

II. POISEUILLE'S LAW

Poiseuille, a French physiologist, discovered the law on velocity of flow of a liquid through a capillary tube. (For a biography of Poiseuille, see the Appendix, Section VI.) He found that the volume of fluid moving in unit time along a cylinder is proportional to the fourth power of its radius and that the movement depends on the drop in pressure. Poiseuille's law applies to cylindrical, capillary tubes. Even though the soil can be considered to consist of cylindrical tubes, Poiseuille's law is usually not applied to water movement in soil. Darcy's law is used. As noted in Table 7.1, Darcy's law is a linear-flow law (Poiseuille's law is not), and Darcy's law applies to an area-averaged section of the soil. Darcy's law does not consider water movement at the small scale of a capillary tube.

Poiseuille's law states (Weast, 1964, p. F-62):

$$v = (\pi p r^4)/(8X\eta), \tag{14.1}$$

where

v = volume (cm^3) escaping per second
p = difference of pressure at the ends of the tube (dyne/cm^2)
r = radius of the tube (cm)
X = length of tube (cm)
η = coefficient of viscosity (poise or dyne-seconds per cm^2).

The volume will be given in cm^3 per sec if X and r are in cm, p in dyne/cm^2, and η in poises or dyne-seconds per cm^2.

We take a moment here to consider viscosity (Weast, 1964, p. F-62). All fluids possess a definite resistance to change of form and many solids show a gradual yielding of forces tending to change their form. This property, a sort of internal friction, is called viscosity. It is expressed in dyne-seconds per cm^2 or poises. The unit is named after Poiseuille. Remember that poise also can be expressed as gram cm^{-1} s^{-1} if we replace dyne by (gram-cm)/s^2 using Newton's law, $F = ma$.

We present the Poiseuille equation in the form that Nobel (1974, p. 392; 1983, p. 494; 1991, p. 508) gives:

$$v = -(\pi r^4/8\eta)(\partial P/\partial X), \tag{14.2}$$

where

v = rate of volume movement (e.g., in units of cm^3/sec)
r = radius of the capillary tube (cm)
η = viscosity of the solution (poise)
$-(\partial P/\partial X)$ = the negative gradient of the hydrostatic pressure.

Hydrostatics (construed as singular) is the branch of physics having to do with the pressure and equilibrium of water and other liquids. Statics is the branch of mechanics dealing with bodies, masses, or forces at rest or in equilibrium. Henceforth, instead of saying "hydrostatic pressure," we simply will say "pressure." Because positive flow occurs in the direction of decreasing pressure [$(\partial P/\partial X)$ is less than 0], the minus sign is necessary in Equation 14.2.

In Equation 14.2 we use partial derivatives. When a quantity is a function of more than one independent variable, it is necessary to use partial derivatives when discussing differentials or derivatives. For example, if $u = f(x,y)$ the partial derivative of u with respect to x, $(\partial u/\partial x)_y$, is the rate with which u changes with a change in x at a constant value of y. A subscript is used on the partial derivative when it is important to emphasize which variable is held constant (Daniels and Alberty, 1966, p. 740). See Appendix 6 of Nobel (1974, pp. 438–440) for a discussion of partial derivatives. Note that v depends both on P and X.

In Poiseuille's law, we are concerned with the volume flowing per unit time and area, often represented by J_v, and called *flux*. For flow in a cylinder of radius r, and hence area πr^2, J_v is

$$v/\pi r^2 = J_v = - (r^2/8\eta)(\partial P/\partial X), \qquad (14.3)$$

where J_v has units of length/time. This is the form of Poiseuille's law that we shall use in our calculations (see Section IV).

III. ASSUMPTIONS OF POISEUILLE'S LAW

Before we use Poiseuille's law, we need to consider the assumptions used in deriving it. These assumptions are important to know so we can apply the law correctly.

Poiseuille's law assumes two things. First, it assumes that the fluid in the cylinder moves in layers or laminas, each layer gliding over the adjacent one. In laminar flow, two particles of water moving will describe paths (streamlines) that will never cross each other (Kirkham, 1961, p. 47). Such laminar movement occurs only if flow is slow enough to meet a criterion deduced by Reynolds in 1883 (Nobel, 1974, p. 392) called the Reynolds number, Re, which is the dimensionless quantity

$$Re = (\rho J_v r)/ \eta. \qquad (14.4)$$

The symbols in Equation 14.4 are the same as those defined above and ρ is the solution density. (Reynolds did not call the number "the Reynolds number"; his students started to refer to it as "the Reynolds's number." It is usually called the Reynolds number and the possessive form is not used. For a biography of Reynolds, see the Appendix, Section VII.) *Re* must be less than 2000 to have laminar flow. Otherwise, a transition to turbulent flow occurs and Equation 14.3 is no longer valid. In turbulent flow, whirls and eddies develop. An eddy is a current of water or air running contrary to the main current, especially, a small whirlpool. For an article on turbulence, see Moin and Kim (1997). Turbulence is still one of the great unsolved problems of classical physics (Nelkin, 1992). The number 2000 is not exact and other numbers in this range are cited in the literature as the Reynolds number at which turbulence takes over.

Second, Poiseuille's law assumes that the fluid in Poiseuille (laminar) flow is actually stationary at the wall of the cylinder. The velocity of solution flow increases in a parabolic fashion to a maximum value in the center of the tube. Thus the flux in Equation 14.3 is actually the mean flow averaged over the entire cross section of cylinder of radius *r*.

Poiseuille's law requires advanced mathematics for its proof. For a proof, see Childs (1969, pp. 194–196). For a discussion of Poiseuille velocity distribution in a circular tube, see Bird et al. (1960, pp. 123–130).

IV. CALCULATIONS OF FLOW BASED ON POISEUILLE'S LAW

Following the analysis of Nobel (1974, pp. 391–395; 1983, pp. 493–498; 1991, pp. 508–513), we now shall use Poiseuille's law to estimate the pressure gradient in different parts of the pathway that water takes when it goes from outside the root and through the apoplast to the endodermis, where it is forced into the living part of the plant (symplasm) because of the casparian strip. The region of a plant made up of cell walls and the hollow xylem vessels is part of the apoplast. Water and the solutes that it contains can move fairly readily in the apoplast. But they must cross a membrane to enter the symplast, the living part of the cells (Nobel, 1974, p. 395). The role of the apoplast in water transport is not fully understood and is an area of active investigation (Steudle and Frensch, 1996; Schreiber et al., 1999).

Nobel first calculates the pressure gradient that occurs in the xylem vessels of diffuse-porous wood. Before we continue with his analysis, we review wood anatomy. There are two types of wood (secondary xylem): diffuse-porous wood and ring-porous wood (Esau, 1965, 1977). In diffuse-porous wood, the xylem members have more or less equal diameters in the spring and summer wood. Esau (1977, p. 508) defines diffuse-porous wood as secondary xylem in which the pores (vessels) are distributed fairly uniformly throughout a growth layer or change in size gradually from early to late wood. Examples of diffuse-porous wood are *Acer* (maple), *Betula* (birch), and *Liriodendron* (tulip tree). In ring-porous wood, the vessel members are large in diameter in the spring wood and are small in diameter in the summer wood. Esau (1977, p. 524) defines ring-porous wood as secondary xylem in which the pores (vessels) of the early wood are distinctly larger than those of the late wood and form a well-defined zone or ring in a cross section of wood. Examples of ring-porous wood are *Castanea* (chestnut), *Fraxinus* (ash), *Robinia* (locust), and some *Quercus* (oak).

We note that Nobel's analysis is for wood, which can occur in woody stems or woody roots. His calculations for Poiseuille-law flow, therefore, are not confined to roots.

Nobel assumes, based on experimental data, that the velocity of sap ascent in the xylem of a transpiring tree with diffuse-porous wood is 0.1 cm/s. This is the value for J_v. For a tree with ring-porous wood, he assumes J_v is 10 times faster (1.0 cm/s). He assumes that the radius of the vessel member in the diffuse-porous wood is 20 μm and the radius of the vessel member in the ring-porous wood is 100 μm. Nobel also assumes that the xylem sap is a dilute aqueous solution, so the volume flow (flux) (J_v) is essentially the same as the volume of water flow, and the viscosity of the solution is the same as that for water.

We now use Equation 14.3 to solve for the pressure gradient in a tree with diffuse-porous wood.

$$J_v = - (r^2/8\eta)\,(\partial P/\partial X)$$

$$0.1 \text{ cm/s} = -(20 \times 10^{-4})^2 / 8(0.010 \text{ dyne-s} / \text{cm}^2)(\partial P/\partial X)$$

or

$$(\partial P/\partial X) = -2 \times 10^3 \text{ dynes/cm}^3.$$

We know that 1 bar = 1×10^6 dynes/cm^2. Therefore -2×10^3 dynes/cm^3 = -2×10^{-3} bar/cm or -0.2 bar/m. Remember this number. We will come back to it.

We digress to calculate Reynolds number for water movement in diffuse-porous and ring-porous wood. We use a density for the sap in the xylem of 1 gram/cm^3 (same as water at 20°C). We now change the units of density to dyne-s^2/cm^4. Remember: $F = ma$; 1 dyne = 1 gram × 1 cm/s^2; so 1 gram = dyne-s^2/cm. Substituting into the formula for density, ρ, we get units for density of dyne-s^2/cm^4. We find the Reynolds number, Re, for diffuse-porous wood with a radius of 20 μm for the vessel members and a velocity of sap flow of 0.1 cm/s to be

$$Re = (\rho\, J_v\, r)/\eta$$

$$Re = [(1 \text{ dyne s}^2/\text{cm}^4)(0.1 \text{ cm/s})(2 \times 10^{-3} \text{ cm})]/0.010 \text{ dyne-s/cm}^2$$

$$= 0.02 \text{ (unitless)}.$$

For a ring-porous tree with a radius of 100 μm for the vessel members and a velocity of sap flow of 1 cm/s, Re is

$$[(1 \text{ dyne-s}^2/\text{cm}^4)(1 \text{ cm/s})(0.01 \text{ cm})]/0.010 \text{ dyne-s/cm}^2 = 1 \text{ (unitless)}.$$

The value of $Re = 1$ for the ring-porous tree is still far less than the value of 2000, where turbulence generally starts. Therefore, we can be assured that Poiseuille's law applies even to plants with large diameter vessel members.

Let us get back to the main problem of determining the pressure gradient in different parts of the pathway that water takes as it crosses a root. We have calculated the pressure gradient that occurs in the xylem vessels. Now let us calculate the pressure drop in the cell walls, again following Nobel's (1974, 1983, 1991) analysis.

The interfibrillar spaces in a cell wall have diameters of about 10 nm. Let us assume that the average radius of these interstices is 5 nm. We are forgetting tortuosity and we are assuming that the cell walls are tubes where Poiseuille's law applies. Let us assume the same J_v that we used previously ($J_v = 0.1$ cm/s). The pressure gradient in the cell walls is:

$$J_v = -(r^2/8\eta)(\partial P/\partial X)$$

$$0.1 \text{ cm/s} = -(5 \times 10^{-7})^2/8(0.010 \text{ dyne-s/cm}^2)(\partial P/\partial X)$$

or

$$(\partial P/\partial X) = -3.2 \times 10^{10} \text{ dynes/cm}^3 = -3.2 \times 10^6 \text{ bars/m}.$$

A $(\partial P/\partial X)$ of only -0.2 bar/m is needed for the same J_v in the vessel member having a radius of 20 μm. Thus, the $(\partial P/\partial X)$ for Poiseuille flow through the small interstices of a cell wall is over 10^7 times greater than for the same flux through the lumen of the vessel member. Because of the tremendous pressure gradients required to force water through the small interstices available for solution conduction in the cell wall, a solution cannot flow rapidly enough up a tree in the cell walls, as has been suggested, to account for the observed rates of water movement (Nobel, 1974, p. 394).

Now let us calculate the pressure gradient for the cell membrane. The Poiseuille law no longer applies, because we have no capillary tubes. We use the following equation for flux, J_v, through the plasmalemma (Nobel, 1974, p. 395 and p. 144):

$$J_v = L_p (\Delta P - \sigma \Delta \pi), \qquad (14.5)$$

where

L_p = hydraulic conductivity coefficient of the cell membrane

ΔP = difference in pressure across the membrane

σ = reflection coefficient. This is a unitless number and varies between 0 and 1. If $\sigma = 1$, all solutes are reflected from the membrane and no solute gets across it. If $\sigma = 0$, all solutes can cross the membrane.

$\Delta \pi$ = osmotic pressure difference across the membrane.

We are considering a dilute solution. The xylem sap is very dilute and we can consider it to be like water. So we shall consider that $\Delta \pi = 0$ and the last term in Equation 14.5 drops out.

So we have $J_v = L_p (\Delta P)$. Let us assume we are still studying plants with a J_v of 0.1 cm/s. A reasonable value to assume for L_p is 1×10^{-5} cm/s-bar (Nobel, 1974, p. 395). So,

$$\Delta P = [0.1 \text{ cm/s}]/1 \times 10^{-5} \text{ cm/s-bar} = 10,000 \text{ bar} = 1 \times 10^4 \text{ bar}.$$

Now let us consider a vessel member before the plasmalemma has broken down. (This is probably an unrealistic situation, because in the mature vessel member, the cell membrane is broken down. But we continue with the analysis of Nobel, 1974, p. 395, and compare flux through the vessel member, cell wall, and cell membrane.) Let us assume that the lumen in our vessel member is 1000 μm long and has a diameter of 20 μm and that the cell wall is 5 μm thick all the way around the cell. As water moves from below and up through this rectangular cell, it passes through the cell wall (5 μm), then the cell membrane, then goes into the lumen that is 1000 μm long, and then passes another membrane and finally another cell wall (5 μm). With these distances, we now can compare the pressure drop across the membrane, cell wall, and lumen. For the membrane, it is -2×10^4 bar. This value includes both ends of the cell (the water traverses the membrane two times—going into the cell and coming out of the cell).

Now let us consider the cell walls. Again, the water moves through one cell wall at the proximal end of the cell and another cell wall through the distal end. Taking the value for $(\partial P/\partial X)$, -3.2×10^6 bars/m, and multiplying it by 10×10^{-6} m (each end of the cell has a cell wall 5 μm thick for a total cell wall thickness that the water must traverse of 10 μm), we get

$$-3.2 \times 10^6 \text{ bar/m} \times 10 \times 10^{-6} \text{ m} = -30 \text{ bars (rounding off)}.$$

For the lumen, we found $(\partial P/\partial X) = -0.2$ bar/m. If we multiply this by the length of the lumen, 1000 μm or 1×10^{-3} m, we get: -0.2 bar/m $\times 1 \times 10^{-3}$ m $= -2 \times 10^4$ bar.

Comparing:
membranes: -2×10^4 bar
cell walls: -30 bar
lumen: -2×10^{-4} bar

The main barriers to water transport are the cell membranes. The interstices of the cell walls provide a much easier pathway for solution flow, while a hollow xylem vessel presents the least obstacle. The evolution of xylem, in particular the vessel members in the angiosperms, provides a plant with a tube well suited for moving water over long distances.

V. AGRONOMIC APPLICATIONS OF POISEUILLE'S LAW

One of the first important agronomic uses of Poiseuille's law was in a pioneering paper by Passioura (1972) in Australia where wheat plants face *terminal drought,* as they do in other semi-arid areas. This is a term that is used when plants are grown in dryland. If they are not irrigated, the crops often use up water stored in the soil by the time they reach flowering, and then no water is available for flowering, grain fill, and the remainder of the life cycle. So the drought at the end part of the life cycle is "terminal."

Passioura (1972) suggested that, when wheat is growing predominantly on stored water, it is an advantage for the plants to have root systems of high hydraulic resistance, so that they will conserve water during early growth and thus have more water available while filling their grain. The xylem of the seminal roots in wheat is dominated by one large metaxylem element (vessel member), the diameter of which (about 50 μm in diameter on average) probably determines the amount of water flowing through the wheat plant, and, thus, indirectly, its hydraulic resistance. (Note: Poiseuille's law says nothing about resistance—only the flux in a capillary tube as it is related to the pressure drop.) Passioura (1972) suggested that the resistance to flow in the wheat root could be increased in one of two ways: 1) by decreasing the size of the central metaxylem element, or 2) by reducing the number of seminal roots.

Passioura (1972) chose to reduce the number of seminal roots. He forced wheat plants to grow on one seminal root. The number of seminal roots that different cultivars (varieties) of wheat produce is under genetic control. Percival (1921) reported that the wheat plant may produce up to eight seminal roots. (Note: The words *cultivar* and *variety* are used interchangeably. A "cultivar" is a "cultivated variety," i.e., a native variety cultivated for specific characteristics.)

Passioura (1972) grew wheat in two columns of soil in a greenhouse. The wheat plants in one column had only one seminal root because he cut off all the others. The wheat plants in the other column had their natural number of seminal roots (three seminal roots). He irrigated at the beginning of the experiment when the plants were young and then let the plants reach terminal drought. He applied Poiseuille's law to the central large metaxylem element for both treatments. His measurements and calculations showed:

	J_v (measured)	$\partial P/\partial X$ (press. gradient) (calculated)
1 seminal root	800 mm/s	$\mid 0.1$ bar/mm$\mid$
3 seminal roots	250 mm/s	$\mid 0.03$ bar/mm$\mid$

The single-rooted plants had double the available water at anthesis and produced double the grain yield. The plants with the one seminal root had a large pressure gradient. [Note, for comparison, Nobel (1974, p. 393) found for a diffuse-porous tree with a J_v of 0.1 cm/s the pressure drop was -0.2 bar/m or -0.0002 bar/mm.] However, the plant with the one root could not sustain the large pressure drop and closed its stomata. Passioura (1972) concluded that it might be possible to conserve water by growing wheat plants with a single seminal root, and it may be possible to breed high root resistance into existing cultivars by breeding for smaller vessels. Passioura's (1972) classic paper showed that Poiseuille's law can be used to calculate the pressure drop in crops and that this value can be used to breed for drought-resistant varieties.

This idea was carried forward when Richards and Passioura (1981a, 1981b) screened the world's wheat collection for the two factors that control resistance to water movement in the wheat root: number of axes (number of seminal roots) and central metaxylem vessel diameter. They screened about 1000 accessions in the collection, and they found that the number of axes varied betweeen 3 and 5 and the central metaxylem vessel diameter varied between 35 to 75 μm. Because there was greater variability in the vessel diameter than in the number of axes, they said the diameter was more important in determining resistance than the number of axes. One should breed for diameter, not for number of axes.

The wheat breeding program, as suggested by Passioura (breeding for small central metaxylem element), has been applied in Australia (Graeme L. Hammer, personal communication, February 12, 1986). However, many factors other than metaxylem vessel diameter and number of seminal roots determine the amount of water lost by a wheat plant. For example, even if the seminal roots are cut (or the wheat is bred for a small number of them), or if a small central metaxylem element diameter in the axes is bred for, the wheat plant develops adventitious roots. The large number of adventitious roots swamps the change in number of seminal axes or metaxylem vessel diameter. Nevertheless, the application of Poiseuille's law to a breeding program is important. Perhaps only a small amount of water can be saved through this breeding method. However, if applied over a large land area (dryland wheat is planted on huge acreages), then a large amount of water could be saved.

Meyer and Alston (1978) saw that the diameter of root vessels of wheat increased with depth. On average, a plant with metaxylem vessels that change in diameter from 30 μm at the soil surface to 45 μm at one meter depth, carries five times the water at one meter depth as at the surface. (Note: One metaxylem vessel would not be 1 meter long.) This adaptation (wheat vessel diameter at depth is wider) might be valuable evolutionarily. If the soil surface is dry (as it often is under dryland conditions), then the vessel at depth, where more water might be, can have a higher flow rate and allow survival. But, at any one level in soil, we would like the diameter of the metaxylem vessels of a drought-resistant variety to be smaller in diameter than the metaxylem vessels of a drought-sensitive variety. This would allow water conservation in the drought-resistant variety.

Meyer et al. (1978) grew wheat that depended for moisture on subsoil water extracted below 45 cm. The plants had 1, 3, or 5 seminal roots. The size of the metaxylem vessel in the seminal axes affected water uptake. Using Poiseuille's law, they calculated that the flow rate at a subcrown potential of −15 bars would be increased by 30% in a plant with three axes without changing the root length, if the radius of each metaxylem vessel were increased by 3 μm.

Meyer and Ritchie (1980) also applied Poiseuille's law to sorghum. They suggested that cultivars with small vessels might have higher resistances and, therefore, be better adapted to dry conditions because they use less water.

The main limitations in the use of the Poiseuille equation to calculate pressure drop in the xylem (and indirectly resistance to water movement) are the inability to measure xylem vessel radii sufficiently accurately, lack of knowledge about effects of growth conditions on longitudinal variations in vessel radii, lack of knowledge about numbers of vessels that carry water from one layer to another in field crops, and inability to account for water exchange along the vessel length (Klepper and Taylor, 1979, p. 61). The most serious of these limitations is that associated with vessel radii, especially in view of the fact that vessel radii enter the Poiseuille equation as r^4. However, with improved scanning electron microscopes (e.g., Cooper and Cass, 2001), vessel member characteristics are being described with better resolution than in the past. Most investigators measure vessel diameter at one depth within the profile (Klepper and Taylor, 1979). To understand the hydraulics of the xylem vessels, their diameters at different depths in the root zone need to be known.

VI. APPENDIX: BIOGRAPHY OF J.L.M. POISEUILLE

Jean Leonard Marie Poiseuille was a French physiologist born in Paris, France, on April 22, 1799. He got his M.D. in 1828 and practiced medicine in Paris. He received the Gold Medal of the French Academy of Sciences. He was the author of "Sur la force du coeur aortique" (1828) ["On the force of the aorta of the heart"]. (The aorta is the main artery of the body; it carries blood from the left ventricle of the heart to all organs except the lungs.) Poiseuille also wrote "Le Mouvement des liquides dans des tubes de petits diamètres" (1844) ["The movement of liquids in tubes of small diameters"]. In 1828, he was the first to use the mercury manometer for measurement of blood pressure, and this is still the most accurate method to measure blood pressure. In 1843, he discovered the law on velocity of flow of a liquid through a capillary tube, and in 1846, he studied flow of viscous liquids. He invented the hemodynamometer for measuring blood pressure inside arteries and also invented the viscosimeter. He died in Paris on December 26, 1869 (Debus, 1968).

VII. APPENDIX: BIOGRAPHY OF OSBORNE REYNOLDS

Osborne Reynolds (1842–1912), an English engineer and physicist, is best known for his work in the fields of hydraulics and hydrodynamics. He was born in Belfast (seaport and capital of Northern Ireland) on August 23, 1842. Gaining early workshop experience and graduating at Queens'

College, Cambridge, England, in 1867, he became in 1868 the first professor of engineering at Owens College, Manchester, England. He was elected a fellow of the Royal Society in 1877 and a Royal medalist in 1888. He retired in 1905 and died at Watchet, Somerset, England on February 21, 1912 (Priestley, 1971).

Reynolds's studies of condensation and the transfer of heat between solids and fluids brought radical revision in boiler and condenser design, and his work on turbine pumps laid the foundation for their rapid development. A fundamentalist among engineers, he formulated the theory of lubrication (1886), and, in his classical paper on the law of resistance in parallel channels (1883), he investigated the transition from smooth, or laminar, to turbulent flow. He later (1889) developed the mathematical framework that became standard in turbulence work. His name is perpetuated in the "Reynolds stress," or drag exerted between adjacent layers of fluid due to turbulent motion, and in the "Reynolds number," which provides a criterion for correct modeling in many fluid flow experiments. He developed corresponding criteria for wave and tidal motions in rivers and estuaries. Among his other work was the explanation of the radiometer and an early absolute determination of the mechanical equivalent of heat. Reynolds's *Scientific Papers* were published in three volumes (1900–1903) (Priestley, 1971).

REFERENCES

Bird, R.B., Stewart, W.E., and Lightfoot, E.N. (1960). *Transport Phenomena.* Wiley: New York.

Childs, E.C. (1969). *An Introduction to the Physical Basis of Soil Water Phenomena.* Wiley-Interscience, John Wiley and Sons: London.

Cooper, R.L., and Cass, D.D. (2001). Comparative evaluation of vessel elements in *Salix* spp. (Salicaceae) endemic to the Athabasca sand dunes of northern Saskatchewan, Canada. *Amer J Bot* 88; 583–587.

Daniels, F., and Alberty, R.A. (1966). *Physical Chemistry.* 3rd ed. Wiley: New York.

Debus, A.G. (Ed.). (1968). Poiseuille, Jean Leonard Marie. *In World Who's Who in Science. A Biographical Dictionary of* "Notable" *Scientists from Antiquity to the Present.* 1st ed., p. 1357. A.N. Marquis Company, Marquis-Who's Who: Chicago, Illinois.

Esau, K. (1965). *Plant Anatomy.* 2nd ed. Wiley: New York.

Esau, K. (1977). *Anatomy of Seed Plants.* 2nd ed. Wiley: New York.

Fisher, J.B., Tan, H.T.W., and Toh, L.P.L. (2002). Xylem of rattans: vessel dimensions in climbing palms. *Amer J Bot* 89; 196–202.

Kirkham, D. (1961). Lectures on Agricultural Drainage. Institute of Land Reclamation, College of Agriculture, Alexandria University: Alexandria, Egypt. (Copy in the Iowa State University Library, Ames, Iowa.)

Klepper, B., and Taylor, H.M. (1979). Limitations to current models describing water uptake by plant root systems. In *The Soil-Root Interface* (Harley, J.L., and Scott Russell, R., Eds.), pp. 51–65. Academic Press: London.

Kramer, P.J. (1983). *Water Relations of Plants.* Academic Press: New York.

Meyer, W.S., and Alston, A.M. (1978). Wheat responses to seminal root geometry and subsoil water. *Agronomy J* 70; 981–986.

Meyer, W.S., and Ritchie, J.T. (1980). Resistance to water flow in the sorghum plant. *Plant Physiol* 65; 33–39.

Meyer, W.S., Greacen, E.L., and Alston, A.M. (1978). Resistance to water flow in the seminal roots of wheat. *J Exp Bot* 29; 1451–1461.

Moin, P., and Kim, J. (1997). Tackling turbulence with supercomputers. *Sci Amer* 276; 62–68 + cover.

Nelkin, M. (1992). In what sense is turbulence an unsolved problem? *Science* 255; 566–570.

Nobel, P.S. (1974). *Introduction to Biophysical Plant Physiology*. W.H. Freeman: San Francisco.

Nobel, P.S. (1983). *Biophysical Plant Physiology and Ecology*. W.H. Freeman: San Francisco.

Nobel, P.S. (1991). *Physicochemical and Environmental Plant Physiology*. Academic Press: San Diego.

Passioura, J.B. (1972). The effect of root geometry on the yield of wheat. *Australian J Agr Res* 23; 745–752.

Percival, J. (1921). *The Wheat Plant*. E.P. Dutton: New York.

Priestley, C.H.B. (1971). Reynolds. *Encyclopaedia Britannica* 19; 252.

Richards, R.A., and Passioura, J.B. (1981a). Seminal root morphology and water use of wheat. I. Environmental effects. *Crop Sci* 21; 249–252.

Richards, R.A., and Passioura, J.B. (1981b). Seminal root morphology and water use of wheat. II. Genetic variation. *Crop Sci* 21; 253–255.

Schreiber, L., Hartmann, K., Skrabs, M., and Zeier, J. (1999). Apoplastic barriers in roots: Chemical composition of endodermal and hypodermal cell walls. *J Exp Bot* 50; 1267–1280.

Steudle, E., and Frensch, J. (1996). Water transport in plants: Role of the apoplast. *Plant Soil* 187; 67–79.

Weast, R.C. (1964). *Handbook of Chemistry and Physics*. 45th ed. Chemical Rubber Co.: Cleveland, Ohio.

15

Gardner's Equation for Water Movement to Plant Roots

W.R. Gardner was one of the first people to model analytically water movement to plant roots. His solution is widely cited today and known by any plant or soil scientist modeling water movement to plant roots. Therefore, it is important for us to understand this basic work. (For a biography of Gardner, see the Appendix, Section X.)

I. DESCRIPTION OF THE EQUATION

In his paper, Gardner (1960) solved the flow equation to determine water movement to a plant root. The flow equation for a single root in an infinite, two-dimensional medium is

$$\partial\theta/t = (1/r)(\partial/\partial r)[rD(\partial\theta/\partial r)] \qquad (15.1)$$

where θ is the water content of the soil on a volumetric basis, D is the diffusivity, t is the time, and r is the radial distance from the axis of the root. The solution to Equation 15.1, subject to boundary conditions that Gardner (1960) defines, is given by Carslaw and Jaeger (1959) (Gardner cites a 1947 reprint of the first edition of Carslaw and Jaeger published in 1946):

$$\tau - \tau_o = (q/4\pi k)[ln(4Dt/a^2) - \gamma], \qquad (15.2)$$

where $\gamma = 0.57722\ldots$ is Euler's constant, a is the radius of the root, k is the unsaturated hydraulic conductivity of the soil, q is the rate of water uptake by the root, D is the diffusivity of the soil, τ is the soil suction, and τ_o is the suction in the soil for initial conditions. We are interested in the difference in suction required at the boundary between the plant root and the (bulk) soil to maintain a constant rate of water movement to the plant. (We will refer to the soil at some distance from the root as the "bulk" soil.) The solution to Equation 15.2 assumes a constant k and D. Because root diameters are small, it is possible to consider the root as a line source, or in this case, sink, of strength q per unit length for which the solution of Equation 15.1 also is given by Carslaw and Jaeger (1959):

$$\tau - \tau_o = (q/4\pi k)[ln(4Dt/r^2) - \gamma]. \tag{15.3}$$

When $r = a$, Equations 15.2 and 15.3 are the same.

Gardner pointed out that study of Equation 15.3 shows that a large part of the water being taken up by the roots comes from some distance from the roots. It is instructive to compare Equation 15.2 with the steady-state solution for flow in a hollow cylinder:

$$\tau - \tau_o = (q/4\pi k)[ln(b^2/a^2)], \tag{15.4}$$

where τ_o is now the suction at the outer radius of the cylinder $r = b$ and τ is the suction at the inner radius $r = a$. If we take $b = 2(Dt)^{1/2}$, Equation 15.4 becomes identical with Equation 15.2 except for the constant term γ. γ is relatively small compared with the logarithmic term, so that the distribution of suction in the transient case is not very different from that in the steady-state case, with all the water coming from a distance $b = 2(Dt)^{1/2}$. The maximum radius b is limited by the density of roots and can be taken as one-half the average distance between neighboring roots.

Using the modern terminology of matric potential instead of the old terminology of soil suction, Equation 15.4 becomes (Baver et al., 1972, p. 404):

$$\psi_b - \psi_a = (q/4\pi k)[ln(b^2/a^2)], \tag{15.5}$$

where ψ_b (bars or MPa) is the matric potential midway between roots and is that which could be measured by any finite sized measuring device, ψ_a (bars or MPa) is the matric potential at the plant root-soil boundary, q is the volume of water taken up per unit length of root per unit time (cm^3/cm/s or mL/cm/day), and k is the unsaturated hydraulic conductivity of the soil (cm^2/s-bar). Gardner (1960) neglects any contact resistance at the soil-root interface. Figure 15.1 from Clothier and Green (1997) provides a picture of Gardner's model.

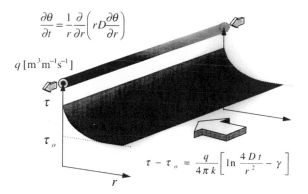

FIG. 15.1 For the gravity-free, linearized form of the Richards equation with a radial coordinate (equation at top of figure), Gardner (1960) solved for the field of suction surrounding a root of sink strength q, in terms of the suction τ, distance r, and time t, given the soil's hydraulic characteristics (the unsaturated hydraulic conductivity k and diffusivity D). γ is Euler's constant. (From Clothier, B.E., and Green, S.R., Roots: The big movers of water and chemical in soil. *Soil Science* 162(8); 534–543, ©1997. This material is used by permission of Lippincott Williams & Wilkins, A Wolters Kluwer Company: Hagerstown, Maryland; and Brent E. Clothier.)

II. ASSUMPTIONS

Gardner (1960) made several assumptions in deriving Equation 15.4 (or Equation 15.5), as follows:

1. The roots are infinitely long cylinders and a distance $2b$ apart.
2. The roots have a uniform radius = a.
3. There is uniform water absorption along the root. [Kramer (1969, p. 179) reports that, as expected, the highest rate of water entry into roots is found in root hairs and unsuberized roots and the lowest rate of water entry into roots occurs in suberized woody roots. However, even suberized roots do have a low rate of water entry. They are not completely impermeable to water.]
4. Water moves in a radial direction only (gravity-free movement).
5. There is a uniform value for the initial soil water content, θ_o, and it corresponds to an initial matric potential in the bulk soil, ψ_o for ψ_b.

III. VALUES FOR THE RATE OF WATER UPTAKE

What might values of q be? The uptake of water by a young root 1 mm in diameter is usually 0.1–0.5 mL/day per cm of root length (e.g., see Nobel, 1974, p. 389). If we take an intermediate value of 0.3 mL/day per cm of root, this corresponds to:

q = (0.3 cm³/day)(1 day/86,400 s) = 3.5×10^{-6} cm³/s per cm of root.

As an aside, we can compare this q to values of flux (J_v) reported by Nobel (1974, p. 393) for transpiration rates of diffuse-porous and ring-porous wood. This q of 3.5×10^{-6} cm^3/s occurs over a root surface area of $2\pi r l$, which is the surface area of a cylinder with radius r and length l. So, we have

$$(2\pi)(0.05\text{cm})(1 \text{ cm}) = 0.31 \text{ cm}^2.$$

Thus J_v or flux at the root surface is

$$(3.5 \times 10^{-6}\text{cm}^3/s)/(0.31 \text{ cm}^2) = 1.1 \times 10^{-5} \text{ cm/s}.$$

Nobel assumed a J_v for diffuse-porous wood of 0.1 cm/s and for ring-porous wood of 1 cm/s. The J_v for the root is four orders of magnitude less than that for the diffuse-porous tree and five orders of magnitude less than that for the ring-porous tree. We note that the q (or J_v) being considered by Gardner is that to a root, which is much less than the total flux that accumulates in the shoot from numerous roots penetrating the soil and funneling all the water into one stem (Fig. 15.2). Water absorption by plant roots occurs at a slower rate than movement of water through the open vertical tubes of the xylem vessels in the stems of plants.

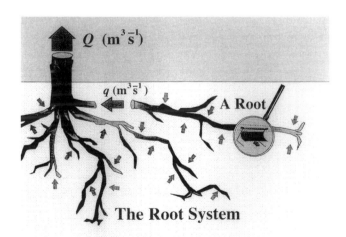

FIG. 15.2 Gardner's (1960) model of uptake describes the local field of flow in the soil that is required to supply active root segments with an internal flux of sap flow q (magnifying glass). The anastomosis of these spatially distributed fluxes in the roots forms the plant's total transpiration Q. (From Clothier, B.E., and Green, S.R., Roots: The big movers of water and chemical in soil. *Soil Science* 162(8); 534–543, ©1997. This material is used by permission of Lippincott Williams & Wilkins, A Wolters Kluwer Company: Hagerstown, Maryland; and Brent E. Clothier.)

IV. EXAMPLES

Let us follow an example given by Nobel (1974, p. 389). Let us assume that at $b = 1.05$ cm, $\psi_b = -2$ bars, which we can measure with a device such as a thermocouple psychrometer. (Note: We could not measure two bars tension with a tensiometer because it is beyond the tensiometer range.) Let us assume that $k = 1 \times 10^{-6}$ cm^2/s bar, which might apply to a loam of moderately low water content.

Using Equation 15.5:

$$-2 \text{ bars} - (\psi_a) = \{3.5 \times 10^{-6} \text{ cm}^3/\text{cm-s})/(4\pi)(1 \times 10^{-6} \text{ cm}^2/\text{s-bar})\}$$
$$\{\ln[(1.05 \text{ cm})^2/(0.05 \text{ cm})^2]\}$$

Solving, we get $\psi_a = -3.7$ bars or about -4 bars. Thus, the matric potential drops about 2 bars across a distance of 1.05 cm in the soil next to the root.

We can consider an osmotic (solute) component in the soil that is constant throughout the soil. As we saw in Chapter 5 (Equation 5.3), the equation for total water potential, ψ, is

$$\psi = \psi_m + \psi_s + \psi_g + \psi_p,$$

where ψ_m is the matric potential, ψ_s is the osmotic (solute) potential, ψ_g is the gravitational potential, and ψ_p is the pressure potential. But the last two terms on the right-hand side drop out because water is moving laterally in the soil (gravity stays the same) and we are dealing with unsaturated conditions, so $\psi_p = 0$. The ψ_s is additive to the ψ_m. Let us assume $\psi_s = -1$ bar. For ψ_a, we have:

$$\psi_m + \psi_s = -4 \text{ bars} + (-1 \text{ bar}) = -5 \text{ bars}.$$

For ψ_b, we have:

$$\psi_m + \psi_s = -2 \text{ bars} + (-1 \text{ bar}) = -3 \text{ bars}.$$

The difference in matric potential between the root-soil boundary and the bulk soil at 1.05 cm away from the root remains the same (2 bars) as in the nonsaline case.

V. EFFECT OF WET AND DRY SOIL

Gardner (1960) applied his equation to different situations. Figure 15.3 shows the matric potential at the root (ψ_a) as a function of the distance from the root, when the matric potential in the bulk soil (ψ_b) is -5 or -15 bars and the rate of uptake is 0.1 mL/cm/day. At -5 bars matric potential, the gradient is very small, except right at the root, so that the matric potential

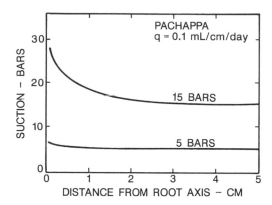

FIG. 15.3 A solution to Gardner's equation. It shows the suction at the plant root as a function of the distance from the root when the suction in the bulk soil is 5 or 15 bars. In current terminology, suction (positive value) is now called matric potential (negative value). (From Gardner, W.R., Dynamic aspects of water availability to plants. *Soil Science* 89(4); 63–73, ©1960. This material is used by permission of Lippincott Williams & Wilkins, A Wolters Kluwer Company: Hagerstown, Maryland; and Wilford R. Gardner.)

is virtually uniform throughout the soil. When $\psi_b = -15$ bars, a large gradient is required for the same q, because of the lower hydraulic conductivity.

VI. EFFECT OF ROOT RADIUS

Figure 15.4 shows the effect of root radius. On the ordinate is the relative value of $\Delta\tau$ (difference in matric potential between the bulk soil and soil-

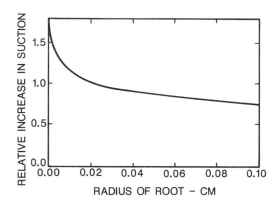

FIG. 15.4 A solution to Gardner's equation. It shows the relative increase in suction at the plant root as a function of the root radius. (From Gardner, W.R., Dynamic aspects of water availability to plants. *Soil Science* 89(4); 63–73, ©1960. This material is used by permission of Lippincott Williams & Wilkins, A Wolters Kluwer Company: Hagerstown, Maryland; and Wilford R. Gardner.)

root boundary). The scale on the ordinate is arbitrary because the actual matric potential would depend on the initial matric potential, q, and k. The figure shows that root radius is not extremeley important. A tenfold increase in root radius can bring about only approximately a twofold decrease in the difference in matric potential between the bulk soil and soil-root boundary.

VII. COMPARISON OF MATRIC POTENTIAL AT ROOT AND IN SOIL FOR DIFFERENT RATES OF WATER UPTAKE

Figure 15.5 shows suction at the plant root plotted as a function of the suction in the bulk soil for a Pachappa sandy loam for three different rates of water uptake. Except for unusual circumstances, the lowest rate indicated (0.05 cm^2/day) is probably more nearly that which occurs in nature and may, in fact, be high for a fully developed root system. The value of $q = 0.1$ cm^2/day is consistent with data of Ogata et al. (1960) for alfalfa. The figure shows that the suction at the root does not exceed the bulk soil suction appreciably until the suction difference is a few bars. When the bulk soil suction is as high as 15 bars, the suction at the root must be 30 bars or 40 bars to maintain a given value of q. The figure shows that, over a short distance, there is little difference in the suction between the root and bulk soil. But as the soil becomes dry, the difference becomes large in order to maintain a constant q. (Figure 15.5 from the original 1960 paper has been reproduced by Baver et al., 1972, p. 405.)

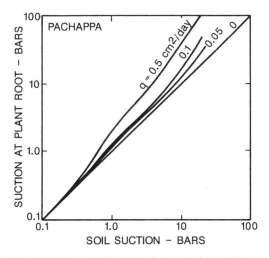

FIG. 15.5 Suction at the root-soil boundary as a function of the bulk soil suction for different rates of water uptake q. (From Gardner, W.R., Dynamic aspects of water availability to plants. *Soil Science* 89(4); 63–73, ©1960. This material is used by permission of Lippincott Williams & Wilkins, A Wolters Kluwer Company: Hagerstown, Maryland; and Wilford R. Gardner.)

The main point of Fig. 15.5 is that over short distances there is little difference in suction between the root and soil in the vicinity of the root until the soil becomes very dry. Even then, the flux tends to decrease markedly due to plant wilting. Gardner's analysis shows that in many cases one can probably assume that the roots are at very nearly the same potential as the surrounding soil (Baver et al., 1972, p. 404).

VIII. EFFECT OF ROOT DISTRIBUTION ON WILTING

The rate of uptake of water per unit length of root is proportional to the total transpiration rate, and inversely proportional to the length of the root system (Gardner, 1960, p. 68). Assuming a given rate of transpiration, the more extensive the root system the lower is the rate of uptake per unit length of root. The uptake rate q thus depends on the transpiration rate and the extent of the root system. To study the effect of transpiration rate and extent of root system on the wilting of a plant, Gardner (1960) assumed that wilting occurs when the suction in the plant root is above some value, say 20 bars. In Fig. 15.6 the average soil suction that results in a suction of 20 bars in the plant root is plotted as a function of q. This is, then, a plot of the soil suction at the wilting point as a function of q. For low rates of uptake, the wilting suction of the soil is very nearly the suction in the plant. For high values of q, large differences between the two are possible (Gardner, 1960).

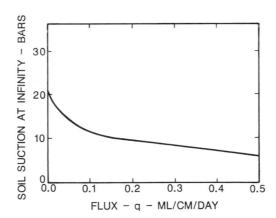

FIG. 15.6 The soil suction at a distance from the root, when suction at the plant root is 20 bars, plotted as a function of water uptake rate. (From Gardner, W.R., Dynamic aspects of water availability to plants. *Soil Science* 89(4); 63–73, ©1960. This material is used by permission of Lippincott Williams & Wilkins, A Wolters Kluwer Company: Hagerstown, Maryland; and Wilford R. Gardner.)

IX. FINAL COMMENT

Gardner said, when his 1960 paper became a citation classic (Institute for Scientific Information, 1985), "I believe this paper has been so frequently cited because the approach is essentially the same that all computer models of plant water uptake now follow." Clothier and Green (1997) discuss the incorporation of Gardner's theoretical ideas into the many simulation models of root zone function that have followed.

X. APPENDIX: BIOGRAPHY OF WILFORD GARDNER

Wilford Robert Gardner, a physicist and educator, was born October 19, 1925, in Logan, Utah, the son of Robert and Nellie (Barker) Gardner. After serving in the U.S. Army during World War II (1943–1946) (Marquis Who's Who, 2003), he got his B.S. degree at Utah State University in 1949 and his M.S. degree at Iowa State University in 1951 under the direction of Don Kirkham. They did the pioneering work on the neutron probe, which is now widely used to monitor soil water content (Gardner and Kirkham, 1952). Gardner got his Ph.D. degree at Iowa State University in 1953 under the guidance of Gordon C. Danielson. Gardner's uncle, Willard Gardner, brother to his father Robert, was the famous soil physicist at Utah State College (as it was known then). Willard Gardner is considered to be the "Father of Soil Physics in America." Gardner's cousin, Walter H. Gardner, the son of Willard, is also a soil physicist located at Washington State University in Pullman.

After obtaining his Ph.D., Wilford joined the U.S. Salinity Laboratory in Riverside, California. There, Gardner did seminal work deriving equations for flow of water under unsaturated conditions (Gardner, 1958, Gardner and Mayhugh, 1958), which are used worldwide by mathematical soil physicists today. This theoretical work has resulted in a soil being named after him, the "Gardner soil" (Kirkham and Powers, 1972, p. 275). At the U.S. Salinity Laboratory he also wrote his important paper on water movement to a plant root (Gardner, 1960).

In 1966 Gardner moved to the University of Wisconsin, where he was a professor in the Soil Science Department. His students there included M.B. Kirkham, Frank N. Dalton, David B. Lesczynski (who died of a heart attack on August 4, 2001, at the age of 55 years), William A. Jury, William N. Herkelrath, John H. Knight, Don F. Yule, and K. John McAneney. In 1980 he moved from the University of Wisconsin to become head of the Department of Soil and Water Science at the University of Arizona, Tucson.

In 1987 he accepted the position of dean of the College of Natural Resources at the University of California, Berkeley. In addition to being dean, he was associate director of the California Agricultural Experiment Station and professor of soil physics in the Departments of Plant & Soil Biology and Forestry & Resource Management. He was one of the few scientists who was both an outstanding researcher and an able administrator. In 1994, he became dean emeritus, and since 1995 has been adjunct professor at Utah State University.

Gardner has received many recognitions. In 1959, he was NSF Senior Fellow and in 1971–1972, he was a Fulbright Fellow. In 1972, he was recipient of an Honorary Faculty Award at the University of Ghent, Belgium. In 1983, he was elected to the National Academy of Sciences. He received the Centennial Alumnus Award at Utah State University in 1986. He is Fellow of the American Association for the Advancement of Science, the American Society of Agronomy, and the Soil Science Society of America. He was President of the Soil Science Society of America in 1990 and received its Research Award in 1962. He is an Honorary Member of the International Union of Soil Science (Marquis Who's Who, 2003). In 2002 he received an honorary doctor's degree from The Ohio State University.

Gardner married Marjorie Louise Cole, the granddaughter of Willard Gardner, on June 9, 1949. They have three children, Patricia, Robert, and Caroline.

REFERENCES

Baver, L.D., Gardner, W.H., and Gardner, W.R. (1972). *Soil Physics*. 4th ed. Wiley: New York.

Carslaw, H.S., and Jaeger, J.C. (1959). *Conduction of Heat in Solids*. 2nd ed. Clarendon Press: Oxford. (First ed., 1946).

Clothier, B.E., and Green, S.R. (1997). Roots: The big movers of water and chemical in soil. *Soil Sci* 162; 534–543.

Gardner, W.R. (1958). Some steady-state solutions of the unsaturated moisture flow equation with application to evaporation from a water table. *Soil Sci* 84; 228–232.

Gardner, W.R. (1960). Dynamic aspects of water availability to plants. *Soil Sci* 89; 63–73.

Gardner, W.R., and Kirkham, D. (1952). Determination of soil moisture by neutron scattering. *Soil Sci* 73; 391–401.

Gardner, W.R., and Mayhugh, M.S. (1958). Solutions and tests of the diffusion equation for the movement of water in soil. *Soil Sci Soc Amer Proc* 22; 197–201.

Institute for Scientific Information. (1985). Gardner, W.R. Dynamic aspects of water availability to plants. Soil Sci 89:63–73. This Week's Citation Classic. *Current Contents, Agr, Biol, Environ Sci* 16(35); 20.

Kirkham, D., and Powers, W.L. (1972). *Advanced Soil Physics*. Wiley: New York.

Kramer, P.J. (1969). *Plant and Soil Water Relationships: A Modern Synthesis*. McGraw-Hill Book Co.: New York.

Marquis Who's Who. (2003). *Who's Who in America*. 57th ed. Marquis Who's Who: New Providence, New Jersey.

Nobel, P.S. (1974). *Introduction to Biophysical Plant Physiology*. W.H. Freeman: San Francisco.

Ogata, G., Richards, L.A., and Gardner, W.R. (1960). Transpiration of alfalfa determined from soil water content changes. *Soil Sci* 89; 179–182.

16

Measurement of Water Potential with Thermocouple Psychrometers

Thermocouple hygrometers are generally accepted as the standard for measurement of plant-water potential (Oosterhuis et al., 1983; Savage, Cass, and de Jager, 1983). They measure vapor pressure in a small chamber by using either a psychrometric (wet bulb/dry bulb) or dew-point technique. Because both are used, the more general term thermocouple hygrometer is sometimes used rather than thermocouple psychrometer (Campbell and Campbell, 1974). However, most people call the instruments thermocouple psychrometers.

We use thermocouple psychrometers to determine water potential by measuring relative humidity. But we actually are measuring a temperature depression (either the wet-bulb or the dew-point temperature depression). To relate the temperature depression to relative humidity, we use the psychrometric equation. Let us look at these points now in detail.

I. RELATION BETWEEN WATER POTENTIAL AND RELATIVE HUMIDITY

The use of thermocouple psychrometers to measure water potential is based on a sound physical-chemical foundation. A definite, quantitative relation exists between water potential of a sample and the relative vapor pressure above it (Barrs, 1968, p. 281; Rawlins, 1972; Savage and Cass, 1984), as follows:

$$\psi = [(RT)/V_w°)] \: ln \: (e/e°), \qquad\qquad (16.1)$$

where ψ = water potential, R = ideal gas constant, T = absolute temperature (°K), $V_w°$ = molar volume of pure water, e = partial pressure of water vapor in air, $e°$ = saturated vapor pressure, and $e/e°$ = relative humidity. Equation 16.1 is called the Kelvin equation (Rawlins, 1972).

Except for T, which is always in °K, units vary according to values used. If ψ is expressed in bars, then R = 83.2 cm³-bar/mole-degree, $V_w°$ = 18.048 cm³/mole at 20°C, e and $e°$ = bars (or millibars). Other values of R are: 0.0821 L-atm/mole-degree; 0.0832 L-bar/mole-degree; and 82.1 cm³-atm/mole-degree. We remember that °K = °C + 273.16. Absolute zero is at −273.16 (or −459.69°F) and it is the temperature at which a gas would show no pressure if the general law for gases would hold for all temperatures (Weast, 1964, p. F-29). Absolute zero is the hypothetical point at which a substance would have no molecular motion and no heat (*Webster's New World Dictionary of the American Language*, 1959). The Kelvin scale of temperature measured is in degrees centigrade from absolute zero and is named after William Thomson, Baron Kelvin. (For a biography of William Thomson, see the Appendix, Section XI.) In 1967 the 13th General Conference on Weights and Measures adopted the unit *kelvin* (K) as its standard for temperature, making it one of the seven base units of Le Système International d'Unités (SI system; see Chapter 2, Section II).

A measurement of relative vapor pressure, or of some related property, gives the water potential of the sample directly, provided that the sample and the space in the chamber have first come to equilibrium. Suitable electrical transducers are thermocouple psychrometers (Barrs, 1968, p. 281) or, using the more general term, thermocouple hygrometers. A hygrometer is an instrument for measuring humidity or the moisture in the air. A psychrometer is a type of hygrometer in which the humidity is measured with wet and dry bulb thermometers. The initial combining form of the word is *psychro-*, which comes from the Greek word *psychros*, meaning "cold" (*Webster's New World Dictionary of the American Language*, 1959).

II. THERMOELECTRIC EFFECTS

Before discussing thermocouple psychrometers used in plant-water measurements, let us review the thermoelectric effects on which they are based.

Figure 16.1 shows an electric circuit of two metals formed into two junctions. If a temperature difference exists between the two junctions, an electric current will flow between them (Barrs, 1968, p. 287). This is the Seebeck effect, named after Johann Seebeck in Berlin, who discovered it in 1821 (Shortley and Williams, 1971, p. 578). Holding the two junctions at different temperatures causes a current when no other source of electromotive force (emf) is present.

If both junctions are initially at the same temperature, then, by passing an electric current through them, one junction will cool and the other will heat (Barrs, 1968, p. 287). This is the Peltier effect, named after Jean Charles Athanase Peltier in Paris, who described the phenomenon about 1834 (Shortley and Williams, 1971, p. 577). (For a biography of Peltier, see the Appendix, Section IX.) The rates of heat generation and absorption are proportional to the current. When the current is reversed, the roles of the two junctions are reversed. Although the Peltier current tends to heat the reference junction, while the free junction is cooled, the rise in temperature is negligible due to the rapid outflow of heat along the massive copper wires attached to junction A (Fig. 16.1) or the junction at the top of the thermocouple psychrometer (Fig. 16.2, right).

The Peltier effect can give only a small degree of cooling, but it is of interest biologically. For example, if the osmotic pressure of a solution is 10 atm, depression of the dew-point temperature is about 0.124°C at 25°C, which corresponds to a relative humidity of 99.3%. For a dew-point

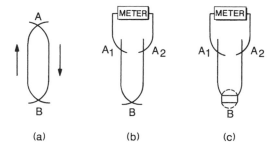

(a)   (b)   (c)

FIG. 16.1 Thermoelectric effects used in thermocouple psychrometry. (A) The Seebeck effect; current flows due to a temperature difference between junctions A and B; (B) Measurement of temperature difference between A and B. B may initially be cooled by the Peltier effect (Spanner psychrometer); (C) Maintenance of permanently wet junction at B (Richards and Ogata psychrometer). (From Barrs, H.D. Determination of water deficits in plant tissues. In *Water Deficits and Plant Growth. Vol. 1. Development, Control, and Measurement*, T.T. Kozlowski, ed., pp. 235–368, ©1968, Academic Press: New York. Reprinted by permission of Academic Press.)

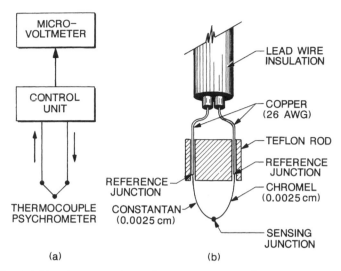

FIG. 16.2 A Peltier thermocouple psychrometer system used to measure water potential consisting of (**A**) a microvoltmeter, control unit, and the thermocouple psychrometer; (**B**) a single-junction Peltier thermocouple psychrometer illustrated in detail. (From van Haveren, B.P., and Brown, R.W. The properties and behavior of water in the soil-plant-atmosphere continuum. In *Psychrometry in Water Relations Research*, R.W. Brown and B.P. van Haveren, eds., pp. 1–27, ©1972. Utah Agricultural Experiment Station, Logan, Utah. Reprinted by permission of the Utah Agricultural Experiment Station.)

depression of 1°C, osmotic pressure would be 80 atm (94.3% relative humidity) (Spanner, 1951).

III. JOULE HEATING

The temperature changes associated with the Peltier effect appear in addition to increases in temperature resulting from the normal joule heating, which we now review. (For a biography of Joule, see the Appendix, Section X.) The work W in joules done in transferring in a circuit a charge of q coulombs between two terminals having a potential difference V volts is

$$W = qV = It(V) = IVt, \tag{16.2}$$

[Remember that I (amperes) = q (coulombs)/t (s).] Because $V = IR$, where R is resistance, $IVt = I(IR)t = I^2Rt$. Thus the electrical energy in joules converted into heat in a conductor of resistance R ohms carrying a current I amperes is

$$W = I^2Rt, \tag{16.3}$$

which is called Joule's law of heating (Schaum, 1961, p. 153). Because 1 joule = 0.239 calories, the heat H in calories developed in the conductor is

$$H = 0.239I^2Rt. \tag{16.4}$$

By use of low-resistance pieces of metal, it is possible, in spite of Joule heating, to get one of the junctions to cool below room temperature (Shortley and Williams, 1971, p. 578). The maximum degree of cooling due to the Peltier effect is limited by Joule heating.

The value of a thermocouple, when used as a thermometer (measurement of temperature difference between two junctions), depends on the fact that the net emf developed is directly related to the temperature difference between the junctions. For small temperature differences, it is approximately proportional to the temperature difference (Shortley and Williams, 1971, p. 578).

The thermoelectric effect is not an unmixed blessing. In any electrical apparatus in which the circuits contain different metals or even different grades of the same metal, temperature differences arising from any cause will set up small "thermal emfs" and "thermal currents," as they are called. Even when a piece of equipment is constructed of a single grade of metal, small thermal emfs exist if there are temperature differences between different portions of the equipment. These emfs appear as a result of a phenomenon known as the Thomson effect. If a copper rod is heated at one end and cooled at the other, a difference of potential is observed between the ends. This Thomson difference of potential arises from a temperature dependence of the density of free electrons in the metal (Shortley and Williams, 1971, p. 579).

IV. THERMOELECTRIC POWER

The thermoelectric power of thermocouples varies. One of the most commonly used thermocouples in plant-water measurements is constantan-chromel, because it is commercially available. Its thermoelectric power is 60 μV/°C. Bismuth-bismuth + 5% tin is preferred by some workers, because its thermoelectric power (126 μV/°C) is twice as high as that of constantan-chromel (Barrs, 1968, p. 289). A further point in favor of the bismuth-bismuth + 5% tin thermocouple is that the maximum Peltier cooling possible is 4.9°C as against 1.5°C for constantan-chromel. There is a maximum Peltier cooling (dew-point temperature). If the temperature depression is greater than the dew-point-temperature depression, it is not possible to condense dew on the wet junction. With most biological systems, this is not a limitation, because such low potentials usually are not encountered. For example, if one uses constantan-chromel, the lower useful limit is about −65 bars (Barrs, 1968, p. 291), which is below the potentials measured in most plants.

V. RELATIONSHIP BETWEEN VAPOR PRESSURE AND TEMPERATURE

As stated earlier, we obtain water potential by observations of vapor pressure. The changes in temperature measured are actually minute. The relation between vapor pressure and temperature is as follows:

$$e = e_w^o - \gamma(T_A - T_W),$$ (16.5)

where e = partial pressure of water vapor in air; $e_w^{\;\circ}$ = saturated vapor pressure at the wet-bulb temperature; T_A = dry-bulb temperature (air temperature); T_W = wet-bulb temperature; γ = psychrometric constant, taken to be 0.658/°C at 20°C and 1000 mb pressure (Monteith, 1973, p. 221). (γ is 0.655 at 15°C; 0.662 at 25°C; 0.665 at 30°C; 0.668 at 35°C; and 0.671 at 40°C—all at 1000 mb.) The interested reader can study Monteith (1973, pp. 171–173) for an explanation of the psychrometric constant and the basis for Equation 16.5.

The Smithsonian Meteorological Tables (List, 1951) give exact values of saturation vapor pressure over water in metric and English units. In the metric units, values for saturation vapor pressure over water in millibars are given for temperatures ranging from −50.0 to 120.0°C, in tenths of degree increments (List, 1951, pp. 351–353, Table 94). Rigorous expressions for the dependence of saturation vapor pressure on temperature are obtained by integrating the Clausius-Clapeyron equation (Monteith and Unsworth, 1990, p. 8). Many authors have proposed simpler equations for estimating the saturation vapor pressure of water at different temperatures. Perhaps the most useful form is the Tetens (1930) formula (Ham, 2004). Values of saturation vapor pressure from the Tetens formula are within 1 Pa of the exact values from −5 to 35°C (Monteith and Unsworth, 1990, p. 10). The Tetens formula, as given by Murray (1967; see his Equation 6), is

$$e_s = 6.1078 \exp[a(T - 273.16)/(T - b)]$$ (16.6)

where a = 17.2693882 and b = 35.86.

Buck (1981) also gives a form of the Tetens formula, and most microclimatologists now use the Buck formula (Jay M. Ham, Department of Agronomy, Kansas State University, personal communication, December

30, 2003). Ham (2004) compares the coefficients in the Murray (1967) and Buck (1981) formulas.

To avoid the necessity of knowing coefficients, we can use Equation 16.5 and Table 94 from List (1951) to determine the partial pressure of water vapor in air (e), as well as relative humidity (RH), and vapor pressure deficit (VPD). Brown and van Haveren (1972, pp. 266–277) also give tables of saturated vapor pressure over water both in mb and mm Hg. Or one can get values from Monteith (1973, pp. 222–223) or Monteith and Unsworth (1990, p. 269), who give saturation vapor pressure in mb and kPa, respectively, at temperatures ranging from −5 to 45°C. Our method is simple. All we need is Equation 16.5 and Table 94 to determine e, RH, and VPD.

A. Sample Problem

Assume that the dry-bulb temperature is 27.2°C and the wet-bulb temperature is 23.9°C. Find e, relative humidity, vapor pressure deficit, and dew-point temperature ($T_{\text{dew-point.}}$). Let $\gamma = 0.66/°C$.

Solution: From Table 94 in List (1951), we find at 27.2°C, $e° = 36.070$ mb ($e°$ = saturated vapor pressure at the dry-bulb temperature); at 23.9°C, $e_w° = 29.652$ mb ($e_w°$ = saturated vapor pressure at wet-bulb temperature).

Putting the known values in Equation 16.5, we get:

$$e = 29.652 - [(0.66/°C)(27.2°C - 23.9°C)] = 27.474 \text{ mb.}$$
$$\text{RH} = e/e° = 27.474 \text{ mb}/36.070 \text{ mb} = 0.76 \text{ or } 76\%.$$
$$\text{VPD} = e° - e = 36.070 \text{ mb} - 27.474 \text{ mb} = 8.596 \text{ mb.}$$
$$T_{dew-pt.} = 22.6°C.$$

(Read Table 94 backwards; temperature for saturated vapor pressure of 27.474 mb = 22.6°C.)

VI. CALIBRATION

Each thermocouple psychrometer is calibrated to yield an answer in units such as bars, megaPascals, or atmospheres. Calibration solutions are often NaCl because water potentials of sodium-chloride solutions at different molalities have been published by Lang (1967) and have been reproduced (e.g., see Barrs, 1968, p. 288; Brown and van Haveren, 1972, pp. 304–305). A table of water potentials of potassium-chloride solutions also has been published (Rawlins and Campbell, 1986). Care must be taken that the filter paper soaked with the salt solution is exposed in the same way as subsequent samples to minimize effects of changed geometry (Barrs, 1968, p. 294).

The calibrating solution is put on filter paper in the thermocouple psychrometer chamber and the sample is equilibrated. Solutions take less time for equilibration than do plant samples; a solution may take an hour or less for equilibration, depending upon the concentration. After equilibration, a cooling current (e.g., 3 milliamps for 15 s) is passed through the thermocouple psychrometer. A microvoltmeter is used to measure the μvolt output of the different salt solutions. Salt solutions varying from 0.05 molal NaCl (−2.3 bars at 25°C) to 1 molal NaCl (−46.4 bars at 25°C) cover the range of interest when measuring plant-water potential. Most plants are severely wilted well above −46.4 bars. After calibration, the plant tissue is put in the chamber, using the same geometry for the sample as was used for the filter paper. The tissue is equilibrated. This usually takes 2–3 hours. A cooling current again is passed through the thermocouple psychrometer, and the μvolt output is recorded. From the calibration curve, the water potential of the tissue is determined.

VII. IMPORTANCE OF ISOTHERMAL CONDITIONS WHEN MAKING MEASUREMENTS

The measurements of plant-water potential made with thermocouples must be done under isothermal conditions. Rawlins and Dalton (1967) [summarized by Savage and Cass (1984)] point out four ways in which temperature affects the measurements:

1. Through the relationship between water potential and relative humidity (Equation 16.1);
2. Through the temperature dependence of the relationship between wet-bulb depression and vapor pressure (Equation 16.5);
3. Through differences in temperature between the sensing junction of the thermocouple and the sample (arising, for example, from respiration by the tissue);
4. Through changes of temperature within the cavity formed by the thermocouple psychrometer and sample, which will alter the relative humidity of the air in the cavity, if water vapor cannot be exchanged with the surrounding system.

In the first two cases (temperature dependence of water potential on relative humidity and temperature dependence of wet-bulb depression on relative humidity), the errors are relatively small, and are about 0.3%/°C and 2%/°C, respectively (Rawlins and Dalton, 1967). These errors can be reduced by following suitable calibration procedures (Savage and Cass,

1984). Heat-of-respiration effects are corrected by reading the psychrometer with its free junction first dry and then wet (Barrs, 1968, pp. 302–303 and 311–312). By using thermocouple walls with adequate conductivity, temperature effects due to number 4 above can be minimized and will not be significant (Rawlins and Dalton, 1967).

There are several other sources of error, in addition to those caused by temperature. They include resistance to diffusion of water vapor into or out of a leaf, adsorption of water on the walls of the container, effects of excision of leaves from plants, and surface contamination. Errors resulting from low tissue permeability are probably negligible, unless leaves are heavily cutinized (Barrs, 1968, p. 311). Adsorption errors can be minimized by using Teflon for the sample container. Cut-edge effects can be decreased by using samples with a small cut-edge-to-volume ratio (Barrs, 1968, p. 307–308; Nelsen et al., 1978). Extraneous dust and soil can be easily washed off. But water potential apparently cannot be measured reliably in salt-extruding species such as cotton (Barrs, 1968, p. 309).

VIII. TYPES OF THERMOCOUPLE PSYCHROMETERS

Four types of instruments with thermocouples are in use to measure water potential of plants:

1. Isopiestic thermocouple psychrometer (Boyer, 1972a, 1972b) (Fig. 16.3) in which solutions of varying concentrations are put manually on the wet junction of the thermocouple psychrometer. The isopiestic solution is the solution that has the same vapor pressure as that of the tissue and produces no thermocouple output (the "null point").
2. The Peltier thermocouple psychrometer (Fig. 16.4, left) in which the wet junction is cooled, using the Peltier effect, to the dew-point temperature. The junction then quickly rises to the wet-bulb temperature. This thermocouple psychrometer is also called the Spanner thermocouple psychrometer, named after Spanner (1951), who described the instrument. The degree of cooling is a function of the water potential of the tissue.
3. The thermocouple psychrometer in which a drop of water is put manually on the wet junction (Fig. 16.4, right), instead of having a cooling current form the drop of water, as is done with the Peltier thermocouple psychrometer. This instrument is known as the Richards-and-Ogata thermocouple psychrometer or just the Richards thermocouple psychrometer, named after Richards and Ogata (1958), who developed the psychrometer. As with the Spanner thermocouple psychrometer, the degree of cooling is a function of the water potential of the tissue.

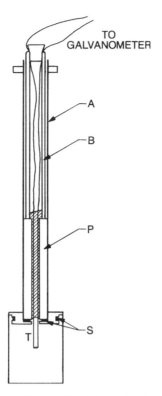

FIG. 16.3 Thermocouple for making isopiestic determinations. The psychrometer chamber and barrel are made of brass and are submerged in a constant-temperature water bath. Key to symbols: barrel, **A**; plexiglas tube, **B**; plunger heat sink, **P**; diagrammatic representation of O-ring seal (coated with stopcock grease) for chamber and seal of stopcock grease for plunger, **S**; thermocouple with ring junction, **T**. (From Boyer, J.S. Use of isopiestic technique in thermocouple psychrometry. I. Theory. In *Psychrometry in Water Relations Research*, R.W. Brown and B.P. van Haveren, eds., pp. 51–55, ©1972. Utah Agricultural Experiment Station: Logan, Utah. Reprinted by permission of the Utah Agricultural Experiment Station.)

    The Richards-and-Ogata thermocouple psychrometer is similar to the isopiestic one in that both require a liquid to be placed on the wet junction. The liquid is sucrose in Boyer's isopiestic method (1972a, p. 53, legend to his Fig. 16.3) and water in the Richards-and-Ogata thermocouple psychrometer.

4. The dew-point hygrometer (Neumann and Thurtell, 1972) (Fig. 16.5) in which the wet junction is cooled to the dew point and stays at the dew point for the measurement. This method uses the Peltier effect to cool the wet junction, but differs from the Peltier thermocouple psychrometer in that it detects the dew-point depression rather than the wet-bulb depression (Neumann and Thurtell, 1972). Neumann and Thurtell

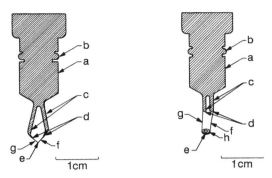

FIG. 16.4   Left, Silhouette of Spanner-type thermocouple psychrometer; **Right**, silhouette of Richards-and-Ogata-type thermocouple psychrometer. Symbols: a, brass mount; b, O-ring seal; c, twin core, PVC-covered copper flex, bared in this region; d, reference junction; e, free junction; f, chromel-P 0.001 inch (0.0254 mm) diameter; g, constantan 0.001 inch (0.0254 mm) diameter; h, silver cylinder. (From Barrs, H.D., Determination of water deficits in plant tissues. In *Water Deficits and Plant Growth. Vol. 1. Development, Control*, and *Measurement*, T.T. Kozlowski, ed., pp. 235–368, ©1968, Academic Press: New York. Reprinted by permission of Academic Press.)

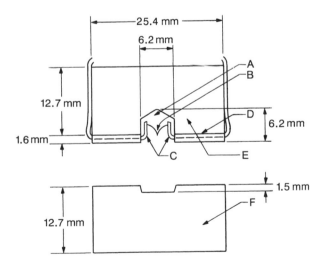

FIG. 16.5   Cross-section view of leaf hygrometer. A, cavity; B, thermocouple; C, posts supporting thermocouple; D, faceplate; E, cavity containing section; F, base plate. (From Neumann, H.H., and G.W. Thurtell. A Peltier cooled thermocouple dewpoint hygrometer for *in situ* measurement of water potentials. In *Psychrometry in Water Relations Research*, R.W. Brown and B.P. van Haveren, eds., pp. 103–112, ©1972. Utah Agricultural Experiment Station: Logan, Utah. Reprinted by permission of the Utah Agricultural Experiment Station.)

(1972) reported that a dew-point measurement is preferable to a wet-bulb measurement in determination of water potential, because with a dew-point measurement, no net water exchange occurs at the wet junction, allowing the measurement to be made without disturbing the vapor equilibrium within the chamber.

The dew-point technique is similar to the isopiestic procedure (Boyer, 1972a, 1972b). Both methods adjust the vapor pressure of the droplet on the wet junction until it is in equilibrium with the vapor within the chamber. In the isopiestic technique, the vapor pressure of the droplet is adjusted by changing the osmotic potential. With the dew-point hygrometer, the vapor pressure is regulated by controlling the temperature of the droplet (Neumann and Thurtell, 1972).

Dew-point hygrometers were designed to minimize the need for temperature control (Neumann and Thurtell, 1972; Campbell and Campbell, 1974). Internal temperature gradients between the sample and the thermocouple, however, must be small (Neumann and Thurtell, 1972; Shackel, 1984). Savage, Cass, and de Jager (1981a, 1983) reported that dew-point hygrometers were less sensitive to temperature than thermocouple psychrometers, but the accuracy of dew-point hygrometers was dependent upon the correct setting of the dew-point cooling coefficient. Measurements made with thermocouple psychrometers (isopiestic, Spanner, Richards-and-Ogata) require careful temperature control (Rauscher and Smith, 1978; Bristow and de Jager, 1980; Slack and Riggle, 1980; Bruckler, 1984).

In situ hygrometers, which measure the water potential of intact plants, have been extensively studied (Savage et al., 1979; Savage, Cass, and de Jager, 1981a, 1981b, 1982, 1983; McBurney and Costigan, 1982; Savage and Cass, 1984). They can be used either in the psychrometric or dew-point mode. Several designs have been described (Neumann and Thurtell, 1972; Michel, 1977, 1979; Brown and McDonough, 1977), including one that is commercially available (Campbell and Campbell, 1974) (Figs. 16.6, 16.7) and one that measures water potential of the soil (McAneney et al., 1979). The advantage of an in situ measurement is that tissue does not need to be excised to determine water potential, which avoids errors due to cutting (Campbell and Campbell, 1974; Baughn and Tanner, 1976b; Nelsen et al., 1978; Savage, Cass, and Wiebe, 1984).

Measurements obtained with in situ hygrometers have been compared to those obtained with other methods. Baughn and Tanner (1976a) found that readings of water potential of plants in a greenhouse, made with a

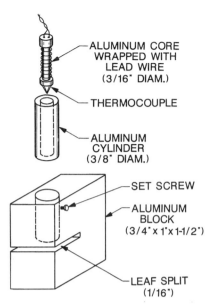

FIG. 16.6 Expanded view of *in-situ* leaf hygrometer. (From Campbell, G.S., and Campbell, M.D., ©1974. American Society of Agronomy: Madison, Wisconsin. Reprinted by permission of the American Society of Agronomy.)

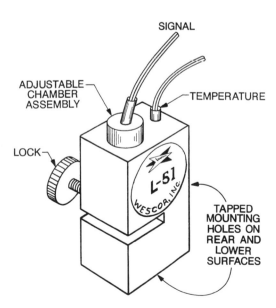

FIG. 16.7 A commercially available *in-situ* hygrometer. (From Wescor, Inc., Logan, Utah, brochure. Reprinted by permission of Wescor, Inc., Logan, Utah.)

pressure chamber, did not agree with those made with an in situ hygrometer. The pressure chamber gave a lower (drier) water potential than the hygrometer in the high potential range and a higher potential than the hygrometer in the dry potential range. Under field conditions, however, water potentials measured with a hygrometer were about 0.2 MPa (2 bars) greater (less negative) than those measured with a pressure chamber (Brown and Tanner, 1981). Brown and Tanner (1981) felt that the higher potential obtained with the hygrometer was caused by its covering the leaf and decreasing transpiration. In contrast, Savage, Wiebe, and Cass (1983) found that the water potential of field-grown plants, obtained with an in situ hygrometer, agreed well with that measured by using a pressure chamber.

In all these experiments (Baughn and Tanner, 1976a; Brown and Tanner, 1981; Savage, Wiebe, and Cass, 1983), the cuticle of the leaf was abraded to obtain rapid vapor equilibrium between the leaf and the hygrometer. Savage, Wiebe, and Cass (1984) showed that the water potential was dependent upon the amount of abrasion. Coarse abrasion resulted in deep cavities in the epidermis and large variability in readings of water potential. Turner et al. (1984) said that in situ hygrometers should not be used on plants with thick cuticles because, even after abrasion, the instruments gave inaccurate values of water potential.

Oosterhuis et al. (1983) compared measurements of water potential obtained by using three instruments: an in situ hygrometer, a pressure chamber, and screen-caged psychrometers, described by Brown and Bartos (1982) (Fig. 16.8). They found that measurements made with the in situ hygrometer gave reliable, nondestructive measurements of water potential, if precautions were followed. These included thermal insulation of the aluminum housing, careful positioning of the hygrometer to the leaf to minimize shading, and allowing adequate time for vapor equilibrium.

The Spanner and dew-point methods are more popular than the isopiestic and Richards-and-Ogata methods. The isopiestic technique takes more time than the other three techniques. It is not readily adaptable to automatic measurements (Boyer, 1972a). The solutions must be put quickly on the wet junction to minimize evaporation, which changes their potentials. The chamber holding the thermocouple is perturbed each time a new solution is introduced into it. With the Spanner, Richards-and-Ogata, and dew-point methods, the chamber remains closed after the sample is placed in it. Instruments commercially available utilize the Spanner or dew-point methods (Figs. 16.9, 16.10). Boyer's laboratory uses the isopiestic procedure (Matthews et al., 1984), as does the laboratory of Robert E. Sharp of the University of Missouri (personal communication, December 3, 1994).

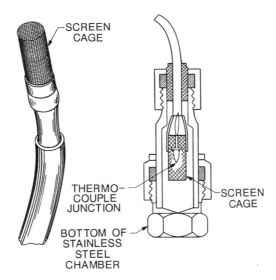

FIG. 16.8   Screen cage psychrometer. (From a figure in a brochure from J.R.D. Merrill Specialty Equipment, Logan, Utah.)

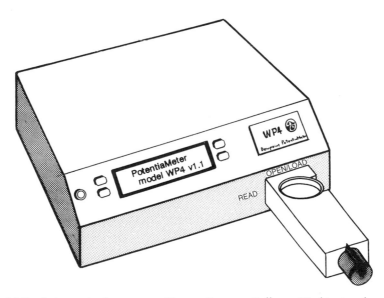

FIG. 16.9   A dew-point hygrometer. (From a Decagon, Pullman, Washington, brochure. Reprinted by permission of Decagon Devices, Inc., Pullman, Washington.)

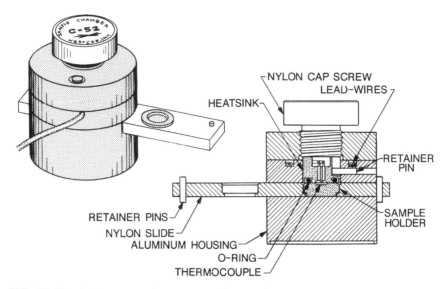

FIG. 16.10   An instrument that operates either as a hygrometer or psychrometer. It measures water potential of small samples in the laboratory or in the field without requiring a constant temperature bath. The large aluminum casing insulates the sample from temperature changes. A thermocouple in an internal chamber functions either as a psychrometer (wet-bulb depression method) or a hygrometer (dew-point depression method), depending on the type of readout equipment employed. (From a Wescor, Inc., Logan, Utah, brochure. Reprinted by permission of Wescor, Inc., Logan, Utah.)

All four techniques require precise electrical measurements and expensive equipment: for example, voltmeters that can read in the microvolt range. Measurements of water potential using thermocouple psychrometers was not possible until about the early 1960s, when microvoltmeters came on the market. Therefore, these methods are not available to some plant physiologists (Barrs, 1968, p. 281). Plant-water potential, however, can be determined, using vapor-phase techniques, with simple equipment (Barrs, 1968, pp. 280–285). Zyalalov (1977) described a method to measure water potential that used only salt solutions, capillary tubes, a weighing bottle with a greased ring to hold the capillary tubes, and a ruler. Learning how to take proper measurements with thermocouple psychrometers requires much training (several months) and skill. Many precautions are required for accurate measurements (Brown and Oosterhuis, 1992). Most people do not have the time or dedication to learn how to use thermocouple psychrometers, which is why the pressure chamber is the most popular method to measure water potential (see Chapter 17).

IX. APPENDIX: BIOGRAPHY OF J.C.A. PELTIER

Jean Charles Athanase Peltier (1785–1845), French physicist, was born at Ham (Somme), France, on February 22, 1785. He was originally a clockmaker, but retired at about the age of 30 to devote himself to experimental and scientific observations. He is best known for his discovery (1834) that an electric current produces, according to direction, either heat or cold at the junction of two dissimilar metals in a circuit. This is called the Peltier effect. Peltier is also remembered for introducing the concept of electrostatic induction. His papers, which are numerous, are devoted in great part to atmospheric electricity, waterspouts, the polarization of skylight, the temperature of water in the spheroidal state, and the boiling point at great elevations. There are also a few papers devoted to points of natural history. He died in Paris on October 27, 1845 (McKie, 1971b).

X. APPENDIX: BIOGRAPHY OF JAMES PRESCOTT JOULE

James Prescott Joule (1818–1889) was an English physicist who established the principle of the interconvertibility of the various forms of energy (i.e., the first law of thermodynamics), and whose name was given to an energy unit, the "joule" (McKie, 1971a). He was born at Salford, Lancashire, on Christmas Eve, 1818, into a famous brewing family and spent some of his early years working for the firm (Hughes, 1989). Eventually his scientific interests predominated. The requirements of brewing technology and the accountancy needed to run a business helped to mold his scientific attitudes. A spinal weakness at birth turned him into a hunchback, and this shy and unassertive man was always sensitive about his public appearances (Hughes, 1989). Science at the time of Joule was changing from being the affair of the gentleman devotee to being the occupation of the full-time professional, ensconced in the university laboratory. Joule was in the first category. Almost all of his research was carried out in his laboratory at home and at his own expense. His reticence often meant that his discoveries were attributed to more verbose and flamboyant researchers. Joule's main interest lay in exact measurement and his special genius showed itself at its best in the invention of methods for obtaining greater accuracy in quantitative experiments. He was systematic and hardworking (Hughes, 1989).

Joule found that the heat generated by the flow of electricity was proportional to the electrical resistance multiplied by the square of the current. His experimental skills firmly established the law of conservation of energy. We take this law for granted now, but in Joule's time the complete conversion of heat into work or work into heat was not conceivable (Hughes, 1989).

Except for some instruction from John Dalton (1766–1844; English chemist and physicist and originator of the atomic theory), Joule was self-taught in science. He early realized the importance of accurate measurement. In a long series of experiments, he studied the quantitative relation between electrical, mechanical, and chemical effects, and was thus led to his great discovery. Joule announced in 1843 his determination of the amount of work required to produce a unit of heat. This is called the mechanical equivalent of heat. He used several methods to show this. The best-known method produced heat from friction in water by means of paddles rotating under the action of a falling weight. His paddle-wheel experiment, which showed that any fluid could be heated merely by agitating it, is famous. Because of this simple fact, the water that has dropped the 49 m over Niagara Falls is 0.11°C higher in temperature than the water at the top of the falls (Hughes, 1989). In 1853, with W. Thomson (later Lord Kelvin; see next section), he researched the work done in compressing gases and the thermal changes gases undergo when forced under pressure through small apertures. Joule's *Scientific Papers* were collected and published in two volumes by the Physical Society of London (1885–1887). Joule died at Sale, Cheshire, on October 11, 1889 (McKie, 1971a).

XI. APPENDIX: BIOGRAPHY OF WILLIAM THOMSON, BARON KELVIN

Baron William Thomson Kelvin (1824–1907) was a British physicist, who discovered the second law of thermodynamics and was an inventor of telegraphic and scientific instruments. He was born in Belfast, Ireland, on June 26, 1824. He was first educated by his father, but at the age of 11 he entered the University of Glasgow, Scotland, where his father was professor of mathematics. Leaving Glasgow without taking a degree, in 1841 he entered Peterhouse, Cambridge, and in 1845 took his degree as second wrangler. Wranglers were mathematically brilliant boys who competed to get the top prize in mathematics at Cambridge University.

At that time (1845) there were few facilities for the study of experimental science in Great Britain. On his father's advice (Gooding, 1990), Thomson traveled to Paris to learn experimental methods in the laboratory of Henri Victor Regnault (1810–1878; French chemist and physicist noted for his work on the properties of gases), who was then engaged in his classical researches on the thermal properties of steam. In 1846 Thomson accepted the chair of natural philosophy at the University of Glasgow, which he filled for 53 years (Preece, 1971). Within four years of his appointment as professor at the age of 22, Thomson established the century's most successful applied physics laboratory, remembered for its compasses and precision instruments. It also was known for another innovation: using laboratory instruction to teach experimental practice and habits of accuracy and precision (Gooding, 1990). This was Britain's first teaching laboratory. It harnessed the skills of a large corps of students to produce intellectual capital, which Thomson invested in new ventures (Gooding, 1990).

In 1847 Thomson first met James Prescott Joule, whose views of the nature of heat strongly influenced Thomson's mind. In 1848 Thomson proposed his absolute scale of temperature, which is independent of the properties of any particular thermometric substance, and in 1851 he presented to the Royal Society of Edinburgh a paper on the dynamical theory of heat. It was in this paper that the principle of the dissipation of energy, briefly summarized in the second law of thermodynamics, was first stated.

Although his contributions to thermodynamics may properly be regarded as his most important scientific work, it is in the field of electricity, especially in its applicaion to submarine telegraphy, that Lord Kelvin is best known (Preece, 1971). The compass went through a process of complete reconstruction in his hands, a process that enabled both the permanent and the temporary magnetism of the ship to be readily compensated, while the weight of the 10-inch (25-cm) card (the dial of a compass) was reduced to one-seventeenth of that of the standard card previously in use (Preece, 1971). Thomson also invented his sounding apparatus, whereby soundings can be taken in shallows and in deep water. Thomson's tide gauge, tidal harmonic analyzer, and tide predicter are famous. He developed tables to simplify the method for determining the position of a ship at sea. The firm of Kelvin and White, in which he was a partner, was formed to manufacture his inventions (Preece, 1971).

In 1866 Thomson was knighted for helping to engineer that year the first successful trans-Atlantic cable (Gooding, 1990). He was raised to the peerage in 1892 with the title of Baron Kelvin of Largs. Thomson took the name Kelvin from the river that flows past the University of Glasgow. In 1890 he became president of the Royal Society and in 1902 received the Order of Merit. In 1904 he was elected chancellor of the University of Glasgow. Thomson published more than 300 original papers bearing upon nearly every branch of physical science. Thomson's extraordinary productivity shows how effectively his father taught him to expend his energy in highly efficient ways (Gooding, 1990), typical of the Scottish attitude of never wasting anything, even time. He wrote, "When you can measure what you are speaking about ... you know something about it, and when you cannot measure it ... your knowledge is of a meagre and unsatisfactory kind...." (Gooding, 1990). Kelvin died on December 17, 1907, at his residence, Netherhall, near Largs, Scotland, and was buried in Westminster Abbey (Preece, 1971).

REFERENCES

Barrs, H.D. (1968). Determination of water deficits in plant tissues. In *Water Deficits and Plant Growth. Vol. 1. Development, Control, and Measurement* (Kozlowski, T.T., Ed.), pp. 235–368. Academic Press: New York.

Baughn, J.W., and Tanner, C.B. (1976a). Leaf water potential: Comparison of pressure chamber and *in situ* hygrometer on five herbaceous species. *Crop Sci* 16; 181–184.

Baughn, J.W., and Tanner, C.B. (1976b). Excision effects on leaf water potential of five herbaceous species. *Crop Sci* 16; 184–190.

Boyer, J.S. (1972a). Use of isopiestic technique in thermocouple psychrometry. I. Theory. In *Psychrometry in Water Relations Research* (Brown, R.W., and van Haveren, B.P., Eds.), pp. 51–55. Utah Agricultural Experiment Station: Logan, Utah.

Boyer, J.S. (1972b). Use of isopiestic technique in thermocouple psychrometry. II. Construction. In *Psychrometry in Water Relations Research* (Brown, R.W., and van Haveren, B.P., Eds.), pp. 98–102. Utah Agricultural Experiment Station: Logan, Utah.

Bristow, K.L., and de Jager, J.M. (1980). Leaf water potential measurements using a strip chart recorder with the leaf psychrometer. *Agr Meteorol* 22; 149–152.

Brown, P.W., and Tanner, C.B. (1981). Alfalfa water potential measurement: A comparison of the pressure chamber and leaf dew-point hygrometer. *Crop Sci* 21; 240–244.

Brown, R.W., and van Haveren, B.P. (Eds.). (1972). *Psychrometry in Water Relations Research*. Utah Agr Exp Sta: Logan, Utah.

Brown, R.W., and McDonough, W.T. (1977). Thermocouple psychrometer for *in situ* leaf water potential determinations. *Plant Soil* 48; 5–10.

Brown, R.W., and Bartos, D.L. (1982). *A calibration model for screen-caged Peltier thermocouple psychrometers*. United States Department of Agriculture, Forest Service Research Paper INT-293. Intermountain Forest and Range Experiment Station: Ogden, Utah.

Brown, R.W., and Oosterhuis, D.M. (1992). Measuring plant and soil water potentials with thermocouple psychrometers: Some concerns. *Agronomy J.* 84; 78–86.

Bruckler, L. (1984). Utilisation des micropsychrometres pour la mesure du potential hydrique du sol en laboratoire et *in situ*. *Agronomie* 4; 171–182. (In French, English sum.)

Buck, A.L. (1981). New equation for computing vapor pressure and enhancement factor. *J Appl Meteorol* 20; 1527–1532.

Campbell, G.S., and Campbell, M.D. (1974). Evaluation of a thermocouple hygrometer for measuring leaf water potential *in situ*. *Agronomy J* 66; 24–27.

Gooding, D. (1990). First lord of British science. Book review of "Energy and Empire: A Biographical Study of Lord Kelvin" by Crosbie Smith and M. Norton Wise. Cambridge University Press, 1989, 866 pp. *Nature* 344; 900.

Ham, J.M. 2004. Appendix. In Hatfield, J.L., and Baker, J.M. (Eds.). Micrometeorology in Agricultural Systems. American Society of Agronomy, Crop Science Society of America, and Soil Science Society of America, Madison, Wisconsin (In press).

Ham, J.M. (2004). Appendix. In *Microclimatology in Agricultural Systems* (Hatfield, J.L., Ed.). American Society of Agronomy, Soil Science Society of America, Crop Science Society of America: Madison, Wisconsin (In press).

Hughes, D.W. (1989). The northern light. Book review of "James Joule: A Biography" by Donald S.L. Cardwell. Manchester University Press, 1989, 333 pp. *Nature* 340; 686.

Lang, A.R.G. (1967). Osmotic coefficients and water potentials of sodium chloride solutions from 0 to 40°C. *Australian J Chem* 20; 2017–2023.

List, R.J. (1951). Hygrometric and psychrometric tables. *Smithsonian Misc. Collections*, Vol. 114. Smithsonian Meteorological Tables, 6th Rev Ed., Smithsonian Institution: Washington DC.

Matthews, M.A., van Volkenburgh, E., and Boyer, J.S. (1984). Acclimation of leaf growth to low water potentials in sunflower. *Plant Cell Environ* 7; 199–206.

McAneney, K.J., Tanner, C.B., and Gardner, W.R. (1979). An *in situ* dewpoint hygrometer for soil water potential measurement. *Soil Sci Soc Amer J* 43; 641–645.

McBurney, T., and Costigan, P.A. (1982). Measurement of stem water potential of young plants using a hygrometer attached to the stem. *J Exp Bot* 33; 426–431.

McKie, D. (1971a). Joule, James Prescott. *Encyclopaedia Britannica* 13; 93.

McKie, D. (1971b). Peltier, Jean Charles Athanase. *Encyclopaedia Britannica* 17; 542.

Michel, B.E. (1977). A miniature stem thermocouple hygrometer. *Plant Physiol* 60; 645–647.

Michel, B.E. (1979). Correction of thermal gradient errors in stem thermocouple hygrometers. *Plant Physiol* 63; 221–224.

Monteith, J.L. (1973). *Principles of Environmental Physics*. American Elsevier: New York.

Monteith, J.L., and Unsworth, M.H. (1990). *Principles of Environmental Physics*. 2nd ed. Edward Arnold: London.

Murray, F.W. (1967). On the computation of saturation vapor pressure. *J Appl Meteorol* 6; 203–204.

Nelsen, C.E., Safir, G.R., and A.D. Hanson, A.D. (1978). Water potential in excised leaf tissue. Comparison of a commercial dew point hygrometer and a thermocouple psychrometer on soybean, wheat, and barley. *Plant Physiol* 61; 131–133.

Neumann, H.H., and Thurtell, G.W. (1972). A Peltier cooled thermocouple dewpoint hygrometer for *in situ* measurement of water potentials. In *Psychrometry in Water Relations Research* (Brown, R.W., and van Haveren, B.P., Eds.), pp. 103–112. Utah Agricultural Experiment Station,: Logan, Utah.

Oosterhuis, D.M., Savage, M.J., and Walker, S. (1983). Field use of *in situ* leaf psychrometers for monitoring water potential of a soybean crop. *Field Crops Res* 7; 237–248.

Preece, W.E. (General Ed.). (1971). Kelvin, William Thomson. *Encyclopaedia Britannica* 13; 275.

Rauscher, H.M., and Smith, D.W. (1978). A comparison of two water potential predictors for use in plant and soil thermocouple psychrometry. *Plant Soil* 49; 679–683.

Rawlins, S.L. (1972). Theory of thermocouple psychrometers for measuring plant and soil water potential. In *Psychrometry in Water Relations Research* (Brown, R.W., and van Haveren, B.P, Eds.), pp. 43–50. Utah Agricultural Experiment Station: Logan, Utah.

Rawlins, S.L., and Campbell, G.S. (1986). Water potential: Thermocouple psychrometry. In *Methods of Soil Analysis. Part 1. Physical and Mineralogical Methods* (Klute, A., Ed.). 2nd ed., pp. 597–618. American Society of Agronomy, Soil Science of America: Madison, Wisconsin.

Rawlins, S.L. and F.N. Dalton. 1967. Psychrometric measurement of soil water potential without precise temperature control. *Soil Sci Soc Amer Proc* 31; 297–301.

Richards, L.A., and Ogata, G. (1958). Thermocouple for vapor pressure measurements in biological and soil systems at high humidity. *Science* 128; 1089–1090.

Savage, M.J., and Cass, A. (1984). Measurement of water potential using *in situ* thermocouple hygrometers. *Advance Agron* 37; 73–126.

Savage, M.J., de Jager, J.M., and Cass, A. (1979). Calibration of thermocouple hygrometers using the psychrometric technique. *Agrochemophysica* 11; 51–56.

Savage, M.J., Cass, A., and de Jager, J.M. (1981a). Calibration of thermocouple hygrometers. *Irrigation Sci* 2; 113–125.

Savage, M.J., Cass, A., and de Jager, J.M. (1981b). Measurement of water potential using thermocouple hygrometers. *South Afr J Sci* 77; 24–27.

Savage, M.J., Cass, A., and de Jager, J.M. (1982). An accurate temperature correction model for thermocouple hygrometers. *Plant Physiol* 69; 526–530.

Savage, M.J., Cass, A., and de Jager, J.M. (1983). Statistical assessment of some errors in thermocouple hygrometric water potential measurement. *Agr Meteorol* 30; 83–97.

Savage, M.J., Cass, A., and Wiebe, H.H. (1984). Effect of excision on leaf water potential. *J Exp Bot* 35; 204–208.

Savage, M.J., Wiebe, H.H., and Cass, A. (1983). *In situ* field measurement of leaf water potential using thermocouple psychrometers. *Plant Physiol* 73; 609–613.

Savage, M.J., Wiebe, H.H., and Cass, A. (1984). Effect of cuticular abrasion on thermocouple psychrometric *in situ* measurement of leaf water potential. *J Exp Bot* 35; 36–42.

Schaum, D. (1961). *Theory and Problems of College Physics*. 6th ed. Schaum: New York.

Shackel, K.A. (1984). Theoretical and experimental errors for *in situ* measurements of plant water potential. *Plant Physiol* 75; 766–722.

Shortley, G., and Williams, D. (1971). *Elements of Physics*. 5th Ed. Prentice-Hall: Englewood Cliffs, New Jersey.

Slack, D.C., and Riggle, F.R. (1980). Effects of Joule heating on thermocouple psychrometer water potential determinations. *Trans Amer Soc Agr Eng* 23; 877–833.

Spanner, D.C. (1951). The Peltier effect and its use in the measurement of suction pressure. *J Exp Bot* 11; 145–168.

Tetens, O. 1930. Über einige meteorologische Begriffe. *Zeitschrift Geophysik* 6; 297–309.

Turner, N.C., Spurway, R.A., and Schulze, E.-D. (1984). Comparison of water potentials measured by *in situ* psychrometry and pressure chamber in morphologically different species. *Plant Physiol* 74; 316–319.

van Haveren, B.P., and Brown, R.W. (1972). The properties and behavior of water in the soil-plant-atmosphere continuum. In *Psychrometry in Water Relations Research* (Brown, R.W., and van Haveren, B.P., Eds.), pp. 1–27. Utah Agricultural Experiment Station: Logan, Utah.

Weast, R.C. (Ed.) (1964). *Handbook of Chemistry and Physics*. 45th edition. Chemical Rubber Co: Cleveland, Ohio.

Webster's New World Dictionary of the American Language. (1959). College Ed. World Publishing: Cleveland and New York.

Zyalalov, A.A. (1977). Indices of the state of water and determination of depression of the water chemical potential. *Soviet Plant Physiol* 24(Part 2, No. 6); 1004–1007.

17

Measurement of Water Potential with Pressure Chambers

The pressure chamber described by Scholander and colleagues (1964, 1965) is the most popular method used to measure water potential of plants. (For a biography of Scholander, see the Appendix, Section V.) The method consists of increasing the pressure around a leafy shoot until sap from the xylem appears at the cut end of the shoot, which extends outside of the chamber and is exposed to atmospheric pressure (Figs. 17.1 and 17.2). The pressure necessary to retain this condition represents the negative pressure existing in the intact stem. It is felt that the amount of pressure necessary to force water out of the leaf cells into the xylem is a function of the water potential of the leaf cells (Boyer, 1967).

I. COMPARISON OF MEASUREMENTS MADE WITH THE PRESSURE CHAMBER AND THE THERMOCOUPLE PSYCHROMETER

For accurate measurements, one should compare measurements made with a thermocouple psychrometer with those made with a pressure chamber before assuming that the pressure chamber is giving valid measurements of water potential. Because the thermocouple-psychrometer method is based on sound physics using the Kelvin equation (Rawlins, 1972), measurements

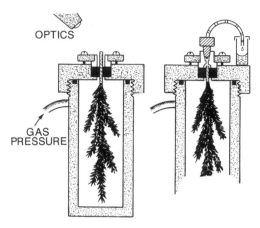

FIG. 17.1   Pressure chamber for measurement of sap presure in the xylem of a twig. Left: direct observation; right: stepwise sap extrusion and pressure measurement to obtain a pressure-volume curve. (Reprinted with permission from Scholander, P.R., Hammel, H.T., Bradstreet, E.D., and Hemmingsen, E.A., Sap pressure in vascular plants. *Science* 148; 339–346, ©1965, American Association for the Advancement of Science.)

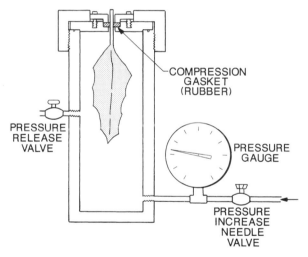

FIG. 17.2   Diagrammatic cross section through a pressure chamber for measurement of leaf water potential by pressure equilibration. (From Kramer, P.J., *Water Relations of Plants*, p. 385, ©1983, Academic Press: New York. Reprinted by permission of Academic Press.)

made with thermocouple psychrometers are the standard ones. But relatively few comparisons exist in the literature. Most people take for granted that the pressure chamber is giving an accurate measurement of water potential and most people use the pressure chamber when measuring plant water potential. It has the advantages of relative simplicity and provision of pressure-volume curves to estimate osmotic potential and turgor potential (see Chapter 18).

Boyer (1967) was one of the first to compare measurements made with thermocouple psychrometers to those made with a pressure chamber. (For a biography of Boyer, see the Appendix, Section VI.) He estimated leaf water potentials from the sum of the balancing pressure measured with a pressure chamber and the osmotic potential of the xylem sap in leafy shoots or leaves of yew (*Taxus cuspidata* Sieb. & Zucc.), rhododendron (*Rhododendron roseum* Rehd.), and sunflower (*Helianthus annuus* L.). Measurements made with the pressure chamber were within ±2 bars of the psychrometric measurements with sunflower and yew (Figs. 17.3 and 17.4). In rhododendron, water potentials measured with the pressure chamber plus xylem sap were 2.5 bars less negative to 4 bars more negative than the psychrometric measurements (Fig. 17.5). As we shall see when we discuss the ascent of sap in plants (Chapter 19), xylem sap is very dilute. Boyer (1967) found xylem sap in yew, rhododendron, and sunflower to have a solute potential of about −0.5 bar (Fig. 17.6). Only when plants got very stressed (e.g., when the rhododendron leaves were at −30 bars) was the xylem sap about −2.0 bars. So the solute potential of the sap was usually

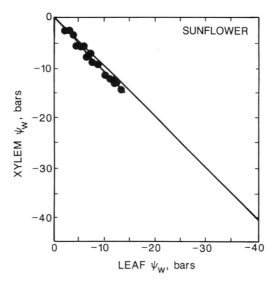

FIG. 17.3 Xylem and leaf water potentials in sunflower. The equipotential values are represented by the diagonal line. Each point represents a single determination. (From Boyer, J.S., Leaf water potentials measured with a pressure chamber. *Plant Physiology* 42; 133–137, ©1967, American Society of Plant Physiologists. Reprinted by permission of the American Society of Plant Biologists, Rockville, Maryland.)

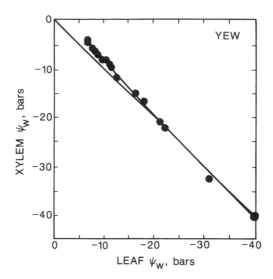

FIG. 17.4 Xylem and leaf water potentials in yew. The equipotential values are represented by the diagonal line. Each point represents a single determination. (From Boyer, J.S., Leaf water potentials measured with a pressure chamber. *Plant Physiology* 42; 133–137, ©1967, American Society of Plant Physiologists. Reprinted by permission of the American Society of Plant Biologists, Rockville, Maryland.)

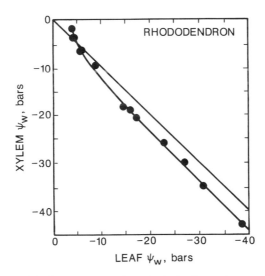

FIG. 17.5 Xylem and leaf water potentials in rhododendron. The equipotential values are represented by the diagonal line. Each point represents a single determination. (From Boyer, J.S., Leaf water potentials measured with a pressure chamber. *Plant Physiology* 42; 133–137, ©1967, American Society of Plant Physiologists. Reprinted by permission of the American Society of Plant Biologists, Rockville, Maryland.)

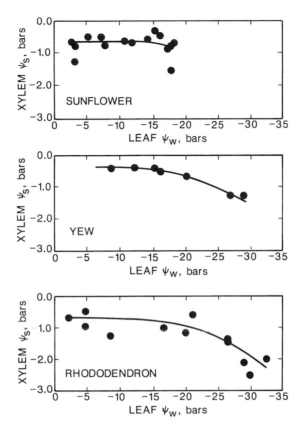

FIG. 17.6 Xylem osmotic potentials (xylem ψ_s) measured at various leaf water potentials in sunflower, yew, and rhododendron. Each point represents a single determination. (From Boyer, J.S., Leaf water potentials measured with a pressure chamber. *Plant Physiology* 42; 133–137, ©1967, American Society of Plant Physiologists. Reprinted by permission of the American Society of Plant Biologists, Rockville, Maryland.)

within the error of comparison (±2 bars). When making measurements with pressure chambers, the osmotic potential of the xylem sap is ignored, and it is assumed that the balancing pressure is the water potential of the leaves.

Gandar and Tanner (1975) compared potato (*Solanum tuberosum* L.) leaf and tuber water potentials measured with both a pressure chamber and thermocouple psychrometers. They used soil psychrometers to measure the tuber water potential. They bored holes in the tuber and put the soil psychrometer in the hole. For leaves drier than −3 bars, the pressure chamber gave estimates of water potential that were zero to three bars drier than potentials measured using thermocouple psychrometers. Pressure chamber readings ranged ±2.5 bars from the psychrometric value for leaves wetter than −3 bars. The psychrometric measurement usually was drier than the pressure chamber when leaves were sampled in the evening. With tubers, water potential measurements using the in situ soil psychrometers and the pressure chamber agreed to within one bar, except in tubers drier than −7 bars, in which there were discrepancies of ±2.5 bars. However, if the interval between psychrometer insertion and water potential measurement was longer than 24 hours, serious errors arose in the psychrometer measurements, apparently from suberization of tissues surrounding the psychrometers that prevented vapor equilibrium.

II. ADVANTAGES AND DISADVANTAGES OF THE PRESSURE CHAMBER

The Scholander pressure chamber is commercially available (Figs. 17.7 and 17.8), and it is widely used (Cochard et al., 2001) because of its many advantages. They include simplicity, comparative speed of measurement, and fair portability (Oosterhuis et al., 1983). Even though thermocouple psychrometers appear to provide more accurate measurements than pressure chambers (Millar, 1982) and are based on sound theory, they are not

FIG. 17.7   A commercially available pressure chamber. The pressure chamber is designed for either laboratory or field use. A safety valve on the lug cover ensures that pressure can be applied to the chamber only when the cover is properly and completely secured. The gas tank is an accessory and is not shown. (From a PMS Instrument Company, Corvallis, Oregon, brochure. Reprinted by permission of PMS Instrument Company.)

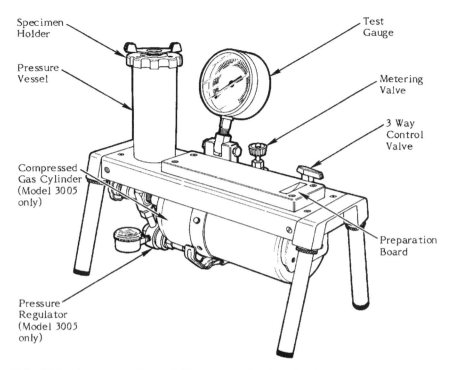

Specimen Holder

Pressure Vessel

Compressed Gas Cylinder (Model 3005 only)

Pressure Regulator (Model 3005 only)

Test Gauge

Metering Valve

3 Way Control Valve

Preparation Board

FIG. 17.8 A commercially available pressure chamber, the Plant Water Status Console. A canister of gas is attached to the bottom of the hardwood base, making the unit self-contained. (Courtesy of Soilmoisture Equipment Corp., Santa Barbara, California.)

used widely, because they require patience and experience before meaningful data can be obtained. In addition, precise temperature control is needed.

However, care also is necessary to gather accurate readings with a pressure chamber. Samples must be protected against transpiration following excision. They must be measured immediately. In field experiments, the pressure chamber has to be protected against wind, so the exuded sap does not evaporate before a measurement can be recorded. When conditions are windy, the pressure chamber can be put in the open hatchback of a van. An operator can stand outside the van on the ground and use a magnifying glass to see the endpoint. The plant sample itself is out of the wind. If a large field is being sampled, "runners" carry the sample from the field to the pressure chamber, so a reading can be made within a matter of seconds after the stem has been cut. Sometimes samples are protected against water loss by putting them in a container with wet cheesecloth and then taking them to the pressure chamber. But the wet cheesecloth could provide water to the sample and result in an erroneous measurement.

Low pressurization rates must be used to avoid false endpoints (Tyree et al., 1978; Wenkert et al., 1978; Karlic and Richter, 1979; McCown and Wall, 1979; Turner and Long, 1980; Brown and Tanner, 1981; Leach et al., 1982). Brown and Tanner (1981) suggest that the pressurization rate should be 0.006 MPa s^{-1} (0.06 bar s^{-1}).

If the proper technique is used, measurements made with pressure chambers can agree with those made with thermocouple hygrometers (Faiz, 1983; Walker et al., 1983). Values for the flow of water through stems, obtained by applying pressure to plants in a pressure chamber, also agree with those obtained by applying vacuums to plants with vacuum pumps (Dryden and van Alfen, 1983).

The Scholander pressure chamber is not well suited to measurements of small plants such as grasses because a petiole must extend through the seal of the pressure chamber. Plants with tender tissues (e.g., new tillers on grasses) are easily damaged by the seal and cannot be used. "Telescoping" of inner leaves of grass tillers at high pressure is another problem. The inner leaves are pushed out of the seal by the high pressures in the chamber. Leaves of a substantial size must be sampled, and, if they exist in an experiment, the sampling results in rapid denudation of leaves. This is a problem in studies with limited plant material.

Large stems, like those of mature sunflower plants, cannot be measured, because commercially available pressure chambers do not have rubber grommets, which make the seal, wide enough to accommodate the large stems. There is an interest in exuding sap from plants such as sunflower in phytoremediation studies to determine if the pollutant has been taken up by the plant. If the stem is too big for the pressure chamber, the sap cannot be exuded.

Pressure chambers are heavy and cumbersome, not only because they require a heavy tank of high-pressure gas, but also because the equipment itself is heavy. The lightest-weight pressure chamber is sold by Plant Moisture Stress (Corvallis, Oregon), and its chassis weighs 6.1 kg. The gas supply limits the number of measurements that can be made in the field. The use of high-pressure gas can be dangerous for two reasons: 1) If not noncombustible, it is a fire hazard; 2) Plants can blow out of the chamber and hit a person in the eye. The commercially available pressure chambers (Figs. 17.7 and 17.8) use nitrogen (N_2) gas, which is noncombustible. Always wear glasses or safety glasses when using a pressure chamber, in case the sample blows out of the chamber and hits the eye.

Even though a measurement with a pressure chamber is faster than with a thermocouple psychrometer, it still takes about five minutes per sample. Pressure chambers and the constant supply of gas are expensive. Relatively unskilled workers can take measurements with a pressure chamber, but some training is required for reliable readings.

III. HYDRAULIC PRESS

The hydraulic press operates on the same principle as the pressure chamber, yet overcomes some of the pressure chamber's limitations (Campbell and Brewster, 1975) (Fig. 17.9). The press consists of a commercial 1.5 ton (1360 kg) hydraulic automobile jack modified to apply pressure through a thin rubber membrane to a leaf sample, which is observed through a 1.27-cm thick Plexiglas plate (Fig. 17.10) (Campbell and Brewster, 1975; Jones and Carabaly, 1980). The instrument applies pressure to the leaf and squeezes the leaf between the membrane and a Plexiglas plate. When the applied pressure equals the water potential, cell walls and intercellular spaces become saturated.

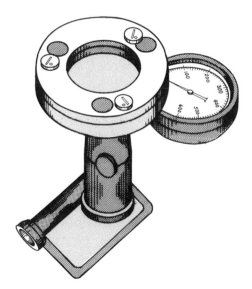

FIG. 17.9 A hydraulic press to measure water potential of plants. Hydraulic pressure beneath a flexible membrane is used to press a leaf or other plant tissue against a thick Plexiglass window. As pressure is applied, water will appear at the stem or cut edge of a leaf. Additional pressure will cause the leaf to change color and excrete water from the uncut surfaces. The pressure required to produce the color change often correlates with water potential measurements made using other techniques. (From a Campbell Scientific, Logan, Utah, brochure; the press is now sold by Decagon Devices, Inc. Reprinted by permission of Decagon Devices, Inc., Pullman, Washington.)

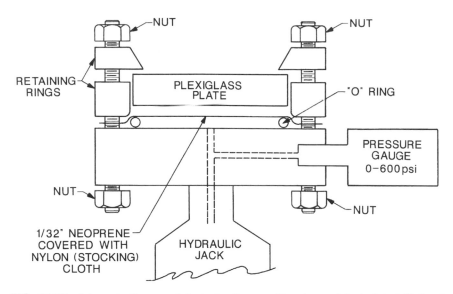

FIG. 17.10 Schematic diagram of the hydraulic press. The piston of the jack is drilled and then welded to the top of the jack. The bolts that hold the head together are 3/8 inch (0.95 cm) diameter, and two, or preferably three, are used. The metal parts of the head are aluminum. (From Campbell, G.S., and Brewster, S.F., Leaf water potential, matric potential and soil water content measured with a simple hydraulic press. Paper presented at the Western Regional Research Project W-67: Quantification of Water-Soil Plant Relations for Efficient Water Use. Honolulu, Hawaii, January, 1975. 11 pp. Reprinted by permission of Gaylon S. Campbell.)

On most leaves, there are three endpoints that occur as the pressure is increased (Campbell Scientific, Inc., undated):

1. A small amount of water is observed at the stem or cut edge. The significance of the first endpoint has not been established, but it correlates with nighttime water potential measured with a pressure chamber.
2. The leaf color changes (darker) and a larger amount of the water comes from the cut edge or stem. Usually, there is also water excreted from the uncut edges at this point. This endpoint correlates with the water potential measurements made with the pressure chamber in the daytime.
3. The leaf turns almost black, with a lot of water being excreted. This endpoint corresponds with osmotic water potential, because it is coincident with the first and second endpoints on leaves that are severely wilted. Heathcote et al. (1979) used the hydraulic press to measure osmotic potential.

The hydraulic press can be used with stems, twigs, needles, and soil. With soil there is only one endpoint, the first one, and it is when water first appears at any edge. A pressure chamber smaller than the Scholander pressure chamber has been made for use with needles of conifers, but it is not on the market (Roberts and Fourt, 1977).

According to Campbell Scientific Inc. (undated), the matric potential of the leaf can be measured with the hydraulic press. The leaf is frozen, thawed, and then placed in the press. When pressure is applied, there will be a pressure at which cell sap flows freely from the sample. This is the matric potential of the leaf. Campbell et al. (1979) used the hydraulic press to measure matric potential.

The hydraulic press has several advantages. A variety of soils and plants can be measured, including tender leaves and tillers. It weighs only 5 kg. It is rugged. No high-pressure gas is required. Measurements are fast (about twenty seconds per sample) and inexperienced workers can use it. It also is cheap.

The water potential of many plants has been measured with the hydraulic press (Rhodes and Matsuda, 1976; Jones and Carabaly, 1980; Bristow et al., 1981; Yegappan and Mainstone, 1981; Cox and Hughes, 1982; Radulovich et al., 1982; Palta, 1983; Rajendrudu et al., 1983; Markhart and Smit-Spinks, 1984; Hicks et al., 1986). In general, results show that measurements taken with the hydraulic press agree with those taken with other instruments. Hicks et al. (1986) found that measurements of leaf water potentials for sorghum [(*Sorghum bicolor* (L.) Moench] made with the hydraulic press and pressure chamber agreed well in the range of −0.5 to −3.5 MPa. Using the hydraulic press, Majerus (1987) found significant entry mean differences for flag-leaf water potential for eight sorghum parent lines, 15 F_1's, and 5 commercial checks ranging from susceptible to tolerant in reaction to water stress. They were grown in the field near Garden City, Kansas, for two years under dryland and irrigated conditions. The results indicated that sorghum could be screened for drought resistance using the hydraulic press.

Several papers, however, report that comparisons between the hydraulic press and the pressure chamber are erratic at low (dry) water potentials (Yegappan and Mainstone, 1981; Cox and Hughes, 1982; Radulovich et al., 1982; Palta, 1983). Although some papers report variability among workers in determining endpoints, Campbell and Brewster (1975) found no difference among operators when determining the relation between hydraulic-press measurements and pressure-chamber measurements. But it is important that each operator obtain his or her own correlation between the leaf press and pressure chamber. A calibration line between the hydraulic press and pressure chamber must be obtained for each plant before measurements can be made with the hydraulic press.

The main disadvantage of the hydraulic press appears to be that it does not have a sound theoretical basis (Shayo-Ngowi and Campbell, 1980). It is also difficult to get precise readings. But, because of its advantages, the instrument deserves study by theoreticians and plant physiologists. What, for example, is the effect of pressure on leaf cells? Why can a leaf in the hydraulic press turn completely black under pressure and then immediately spring back to its normal green color and apparent turgidity once the pressure is released? Markhart and Smit-Spinks (1984) suggested that the hydraulic press should be used only for crude estimates of water potential. This is probably good advice, until the physical meaning of the measurements is understood. Nevertheless, its value as an easy method to screen plants for drought resistance in the field should be recognized.

Because the hydraulic press is made in the United States, the gauge reads in lb/in^2 (0 to 600 lb/in^2). To convert lb/in^2 on the gauge into SI units (MPa), see Chapter 9, Section IV.

IV. PUMP-UP PRESSURE CHAMBER

In about 2000, Plant Moisture Stress (PMS) Instrument Company in Corvallis, Oregon, introduced a new type of pressure chamber (Fig. 17.11). It is different from the conventional gas chamber in that it does not require a source of compressed gas such as nitrogen, which can be dangerous to use, as noted in the preceding section. The pressure required to take water-potential readings is created by pumping the instrument as one would a bicycle pump. The relatively small chamber allows the user to achieve about 0.5 bar (7.25 psi) pressure per stroke (Fig. 17.12). The instrument is limited to 20 bars and is designed primarily for irrigation scheduling and monitoring, particularly for managing deficit irrigation. A picture of the instrument in use is shown by Goldhamer and Fereres (2001).

V. APPENDIX: BIOGRAPHY OF PER SCHOLANDER

Per Fredrik Scholander a physiologist, was born in Örebro, Sweden, on November 29, 1905, and he married in 1951 (American Men of Science, 1961). He got his M.D. degree in Oslo in 1932 and his Ph.D. in botany in 1934. He was an instructor of anatomy in Oslo between 1932 to 1934 and was a research fellow in comparative physiology between 1932 to 1939. He moved to the United States and became a naturalized citizen. He was a research associate in respiratory physiology at Swarthmore College in Swarthmore, Pennsylvania, from 1939 to 1943. He was a Rockefeller fellow

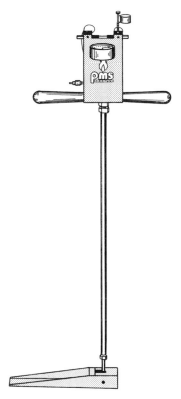

FIG. 17.11   Overall view of the pump-up pressure chamber, an alternative type of pressure chamber that does not use compressed gas. (From a PMS Instrument Company, Corvallis, Oregon, brochure. Reprinted by permission of PMS Instrument Company.)

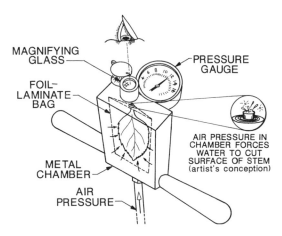

FIG. 17.12   A close-up of the top part of the pump-up pressure chamber. (From a PMS Instrument Company, Corvallis, Oregon, brochure. Reprinted by permission of PMS Instrument Company.)

from 1939 to 1941 and a research biologist from 1946 to 1949. He was a major for the U.S. Army Air Force Research from 1943 to 1946, and during this time was chief physiologist test officer, Air Force Base, Eglin Field (1943–1945), and an aviation physiologist at the aeromedical laboratory of Wright Field, Dayton, Ohio (1945–1946). From 1949 to 1951, he was a special research fellow in biochemistry at Harvard Medical School. He was a physiologist at the Oceanographic Institute in Woods Hole, Massachusetts, between 1952 to 1955. In 1955, he returned to Oslo, where he was a professor of physiology and director of the institute of zoophysiology until 1958. During this time (1955–1958) he also was an associate at the Oceanograpic Institute in Woods Hole. In 1958, he became a professor of physiology at the Scripps Institute of Oceanography in La Jolla, California, where he spent the rest of his career.

His honors included being an investigator in the Arctic Research Laboratory of the Office of Naval Research in Alaska and Panama from 1947–1949. He was a member of the polar research committee of the National Academy of Sciences and participated in arctic and tropical expeditions. He received the Legion of Merit in 1946. He was a member of the National Academy of Sciences, American Association for the Advancement of Science, Physiology Society, Society of Zoologists, Society of Plant Physiologists, Society of General Physiology, American Academy, Arctic Institute of North America, Norwegian Academy of Science, Norwegian Physiology Society, and Botanical Association of Norway. His major research areas were arctic botany, respiration of diving, cold adaptation, microtechniques, gas secretion, water and gas transport in plants, and gas in glaciers (American Men of Science, 1961).

According to the *Newsletter of the American Society of Plant Physiologists* (Vol. 7, No. 5, p. 4, October, 1980), Per Scholander died June 13, 1980, at the age of 74.

VI. APPENDIX: BIOGRAPHY OF JOHN BOYER

John Strickland Boyer, a biochemist and biophysicist, was born May 1, 1937, in Cranford, New Jersey (Marquis Who's Who, 2000). He married Jean R. Matsunami and they have two children. In 1961 he got his master's degree at the University of Wisconsin under the direction of Gerald C. Gerloff, a mineral nutritionist, and in 1964 he obtained his Ph.D. in botany at Duke University under the direction of Paul J. Kramer. The last book by Kramer was written jointly with Boyer (Kramer and Boyer, 1995). (Paul Kramer was born May 8, 1904, and died May 24, 1995.)

Boyer was visiting assistant professor of botany at Duke University from 1964 to 1965, and an assistant physiologist at the Connecticut Agricultural Experiment Station, 1965–1966. In 1966 he moved to the University of Illinois at Urbana and rose from assistant professor to professor of botany and agronomy. In 1978, he joined the USDA as a plant physiologist on the University of Illinois campus. Between 1984 and 1987 he was a professor at Texas A&M University. Since 1987 he has been du Pont Professor of marine biochemistry and biophysics at the University of Delaware.

He has won many recognitions. He is a member of the visitor committee, Carnegie Institute of Washington, Stanford University, and Harvard University. In 1983, he received the German Humboldt Senior Scientist award. He is a fellow of the Climate Laboratory (New Zealand), American Society of Agronomy, Crop Science Society of America, Australian National University, and the Japanese Society for the Promotion of Science. He is a member of the National Academy of Sciences and the American Society of Plant Physiologists. He was president of the American Society of Plant Physiologists in 1981–1982 and won the Shull award from the society in 1977 (Marquis Who's Who, 2000).

REFERENCES

American Men of Science. (1961). *A Biographical Directory*. 10th ed., p. 3598. The Physical and Biological Sciences. Jaques Cattell Press: Tempe, Arizona.

Boyer, J.S. (1967). Leaf water potentials measured with a pressure chamber. *Plant Physiol* 42; 133–137.

Bristow, K.L., van Zyl, W.H., and de Jager, J.M. (1981). Measurement of leaf water potential using the J14 press. *J Exp Bot* 32; 851–854.

Brown, P.W., and Tanner, C.B. (1981). Alfalfa water potential measurement: A comparison of the pressure chamber and leaf dew-point hygrometer. *Crop Sci* 21; 240–244.

Campbell Scientific, Inc. (Undated). Model J14 Leaf Press. Instruction Manual. Campbell Scientific Inc., P.O. Box 551, Logan, Utah.

Campbell, G.S., and Brewster, S.F. (1975). Leaf water potential, matric potential and soil water content measured with a simple hydraulic press. Paper presented at the Western Regional Research Project W-67: "Quantification of Water-Soil Plant Relations for Efficient Water Use." Honolulu, Hawaii, January, 1975. 11 pp. (Paper available from Dr. Gaylon S. Campbell, Decagon Devices, Inc., 950 NE Nelson Court, Pullman, WA 99163.)

Campbell, G.S., Papendick, R.I., Rabie, E., and Shayo-Ngowi, A.J. (1979). A comparison of osmotic potential, elastic modulus, and apoplastic water in leaves of dryland winter wheat. *Agronomy J* 71; 31–36.

Cochard, H., Forestier, S., and Améglio, T. (2001). A new validation of the Scholander pressure chamber technique based on stem diameter. *J Exp Bot* 52; 1361–1365.

Cox, J.R., and Hughes, H.G. (1982). Leaf water potentials of perennial grasses: Leaf press and pressure chamber evaluation. J. *Range Manage.* 35; 5–6.

Dryden, P., and van Alfen, N.K. (1983). Use of the pressure bomb for hydraulic conductance studies. *J Exp Bot* 34; 523–528.

Faiz, S.M.A. (1983). Use of pressure bomb in the determination of soil water potential. *Plant Soil* 73; 257–264.

Gandar, P.W., and Tanner, C.B. (1975). Comparison of methods for measuring leaf and tuber water potentials in potatoes. *Amer Potato J* 52; 387–397.

Goldhamer, D.A., and Fereres, E. (2001). Simplified tree water status measurements can aid almond irrigation. *California Agr* 55(3); 32–37.

Heathcote, D.G., Etherington, J.R., and Woodward, R.I. (1979). An instrument for non-destructive measurement of the pressure potential (turgor) of leaf cells. *J Exp Bot* 30; 811–816.

Hicks, S.K., Lascano, R.J., Wendt, C.W., and Onken, A.B. (1986). Use of a hydraulic press for estimation of leaf water potential in grain sorghum. *Agronomy J* 78; 749–751.

Jones, C.A., and Carabaly, A. (1980). Estimation of leaf water potential in tropical grasses with the Campbell-Brewster hydraulic press. *Trop Agr* 57; 305–307.

Karlic, H., and Richter, H. (1979). Storage of detached leaves and twigs without changes in water potential. *New Phytol* 83; 379–384.

Kramer, P.J. (1983). *Water Relations of Plants.* Academic Press: New York.

Kramer, P.J., and Boyer, J.S. (1995). *Water Relations of Plants and Soils.* Academic Press: San Diego.

Leach, J.E., Woodhead, T., and Day, W. (1982). Bias in pressure chamber measurements of leaf water potential. *Agr Meteorol* 27; 257–263.

Majerus, W.J. (1987). Comparison of Screening Technique for Traits Relating to Drought Tolerance in Grain Sorghum. M.S. Thesis. Kansas State University: Manhattan.

Markhart, A.H., III, and Smit-Spinks, B. (1984). Comparison of J-14 hydraulic press with the Scholander pressure bomb for measuring leaf water potential. *HortScience* 19; 52–54.

Marquis Who's Who. 2000. *Who's Who in America.* 54th ed. Marquis Who's Who: New Providence, New Jersey.

McCown, R.L., and Wall, B.H. (1979). Improvement of pressure chamber measurements of two legumes by constriction of stems. *Plant Soil* 51; 447–451.

Millar, B.D. (1982). Accuracy and usefulness of psychrometer and pressure chamber for evaluating water potentials of *Pinus radiata* needles. *Australian J Plant Physiol* 9; 499–507.

Oosterhuis, D.M., Savage, M.J., and Walker, S. (1983). Field use of in situ leaf psychrometers for monitoring water potential of a soybean crop. *Field Crops Res* 7; 237–248.

Palta, J.A. (1983). Evaluation of the hydraulic press for measurement of leaf water potentials in cassava. *J Agr Sci* 101; 407–410.

Radulovich. R.A., Phene, C.J., Davis, K.R., and Brownell, J.R. (1982). Comparison of water stress of cotton from measurements with the hydraulic press and the pressure chamber. *Agronomy J* 74; 383–385.

Rajendrudu, G., Singh, M., and Williams, J.H. (1983). Hydraulic press measurements of leaf water potential in groundnuts. *Exp Agr* 19; 287–291.

Rawlins, S.L. (1972). Theory of thermocouple psychrometers for measuring plant and soil water potential. In *Psychrometry in Water Relations Research* (Brown, R.W., and van Haveren, B.P., Eds.), pp. 43–50. Utah Agricultural Experiment Station, Utah State University: Logan.

Rhodes, P.R., and Matsuda, K. (1976). Water stress, rapid polyribosome reductions and growth. *Plant Physiol* 58; 631–635.

Roberts, J., and Fourt, D.F. (1977). A small pressure chamber for use with plant leaves of small size. *Plant Soil* 48; 545–546.

Scholander, P.F., Hammel, H.T., Hemmingsen, E.A., and Bradstreet, E.D. (1964). Hydrostatic pressure and osmotic potential in leaves of mangroves and some other plants. *Proc Nat Acad Sci* 52; 119–125.

Scholander, P.R., Hammel, H.T., Bradstreet, E.D., and Hemmingsen, E.A. (1965). Sap pressure in vascular plants. *Science* 148; 339–346.

Shay-Ngowi, A., and Campbell, G.S. (1980). Measurement of matric potential in plant tissue with a hydraulic press. *Agronomy J* 72; 567–568.

Turner, N.C., and Long, M.J. (1980). Errors arising from rapid water loss in the measurement of leaf water potential by the pressure chamber technique. *Australian J. Plant Physiol* 7; 527–537.

Tyree, M.T., MacGregor, M.E., Petrov, A., and Upenieks, M.I. (1978). A comparison of systematic errors between the Richards and Hammel methods of measuring tissue-water relation parameters. *Can J Bot* 56; 2153–2161.

Walker, S., Oosterhuis, D.M., and Savage, M.J. (1983). Field use of screen-caged thermocouple psychrometers in sample chambers. *Crop Sci* 23; 627–632.

Wenkert, W., Lemon, E.R., and Sinclair, T.R. (1978). Changes in water potential during pressure bomb measurement. *Agronomy J* 70; 353–355.

Yegappan, T.M., and Mainstone, B.J. (1981). Comparisons between press and pressure chamber techniques for measuring leaf water potential. *Exp Agr* 17; 75–84.

18

Stem Anatomy and Measurement of Osmotic Potential and Turgor Potential Using Pressure-Volume Curves

In this chapter we consider pressure-volume curves, which are used to determine osmotic potential and turgor potential. They are constructed from data obtained from the volume of sap exuded from the cut end of a stem protruding from a pressure chamber. Therefore, we need to understand stem anatomy before we turn to pressure-volume curves.

I. STEM ANATOMY

A. General Structure

The close association of the stem with the leaves makes the aerial part of the plant axis structurally more complex than the root (Esau, 1977,

p. 257). The term *shoot*, which refers to the stem and leaves as one system, serves to express this association. The stem, like the root, consists of three tissue systems: the dermal (epidermis), the fundamental or ground (pith and cortex), and the fascicular or vascular. The variations in the primary structure in stems of different species are based chiefly on differences in the relative distribution of the fundamental and vascular tissues (Esau, 1977, p. 257).

B. Dicotyledonous Stem

In dicotyledons, the vascular system of the internode commonly appears as a hollow cylinder delimiting an outer and an inner region of ground tissue, the cortex and the pith, respectively. Figure 18.1 shows bird's-foot trefoil, which has a typical dicotyledonous stem. The subdivisions of the vascular system, the vascular bundles, are separated from each other by more or less wide panels of ground parenchyma—the interfascicular parenchyma—that interconnects the pith and the cortex. This tissue is called interfascicular because it occurs between the bundles or fascicles (Esau, 1977, p. 257–258).

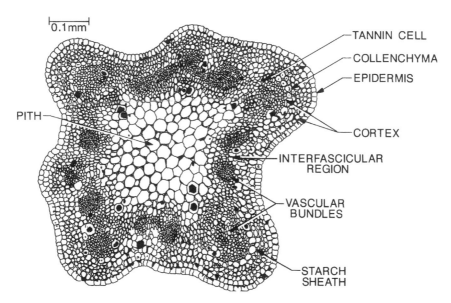

FIG. 18.1   Cross section of an herbaceous dicotyledon stem, *Lotus corniculatus* or birds-foot trefoil (Leguminosae family or legume family), in primary state of growth. The phloem is on the outside of each vascular bundle, and the xylem is on the inside. (From Esau, K., *Anatomy of Seed Plants*, 2nd ed., p. 258, ©1977, John Wiley & Sons, Inc.: New York. This material is used by permission of John Wiley & Sons, Inc.)

C. Monocotyledonous Stem

Stems of most monocotyledons have a complex arrangement of vascular tissues. The bundles may occur in more than one ring or may appear scattered throughout the cross section. The stems (culms) of the Poaceae (grass family), seen in cross section, have widely spaced vascular bundles not restricted to one circle (Esau, 1977, p. 313). The bundles are either in two circles (*Avena*, oat; *Hordeum*, barley; *Secale*, rye; *Triticum*, wheat; *Oryza*, rice) or scattered throughout the section (*Bambusa*, bamboo; *Saccharum*, sugar-cane; *Sorghum*, sorghum; *Zea*, corn). Figure 18.2 shows a cross section of a corn stem. The delimitation of the ground tissue into cortex and pith is less precise or does not exist when the vascular bundles do not form a ring in cross sections of internodes.

Monocotyledons other than Poaceae also have vascular bundles scattered or in rings near the periphery, as seen in stem transections. In *Tradescantia* (spiderwort) (Commelinaceae; spiderwort family), the central

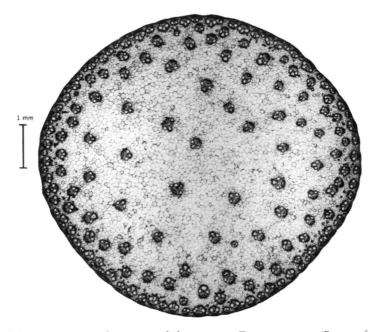

1 mm

FIG. 18.2 Cross section of a monocotyledonous stem, *Zea mays* or corn (Poaceae family or grass family). Note vascular bundles are distributed throughout the section, but are more numerous near the periphery. In each vascular bundle, the xylem is oriented toward the center of the stem and three large vessels ("monkey faces"), part of the metaxylem, are visible in each bundle. The phloem in each vascular bundle is oriented toward the outside of the stem. (From Esau, K., *Anatomy of Seed Plants*, 2nd ed., p. 206, ©1977, John Wiley & Sons, Inc.: New York. This material is used by permission of John Wiley & Sons, Inc.)

cylinder has scattered bundles. Compaction of vascular tissue characteristic of hydrophytes is common in aquatic monocotyledons. In *Potamogeton* (pondweed) (Potamogetonaceae; pondweed family), for example, a wide cortex consisting of aerenchyma encloses a compact vascular cylinder delimited by a small-celled endodermis. Variable amounts of pith tissue occur in different species of the genus (Esau, 1977, p. 314).

In general, we can say that dicotyledonous stems usually have a pith at the center of the stem surrounded by vascular bundles. Monocotyledonous stems often have scattered vascular bundles in a ground tissue, and no pith and cortex are delineated.

D. Stomata, Cortex, Pith, and Vascular Bundles in Primary Xylem

Stomata can be present on stems, but constitute a less prominent epidermal component in the stem than in the leaf (Esau, 1977, p. 259). The stem epidermis commonly consists of one layer of cells and has a cuticle and cutinized walls. It is a living tissue capable of mitotic activity, an important characteristic in view of the stresses to which the tissue is subjected during the primary and secondary increase in thickness of the stem. The epidermal cells respond to these stresses by enlargement and divisions (Esau, 1977, p. 259).

The cortex of stems contains parenchyma, usually with chloroplasts. Intercellular spaces are prominent, but sometimes are largely restricted to the median part of the cortex. In many aquatic angiosperms, the cortex develops as an aerenchyma with a system of large intercellular spaces (Esau, 1977, p. 259). The peripheral part of the cortex frequently contains collenchyma (Fig. 18-1). *Collenchyma* is a supporting tissue composed of more or less elongated living cells with unevenly thickened, nonlignified primary walls. It is in regions of primary growth in stems and leaves. In some plants, notably grasses, sclerenchyma rather than collenchyma develops as the primary supporting tissue in the outer region of the stem. *Sclerenchyma* is a tissue composed of sclerenchyma cells. A sclerenchyma cell is a cell variable in form and size and having more or less thick, often lignified, secondary walls. It is a supporting cell and may or may not be devoid of a protoplast at maturity.

As noted when we studied root anatomy (Chapter 14), the innermost layer of the cortex (endodermis) of roots of vascular plants has the casparian strip. Stems commonly lack a morphologically differentiated endodermis. In young stems, the innermost layer or layers may contain abundant starch and thus be recognized as a starch sheath (Fig. 18.1). Some dicotyledons, however, do develop casparian strips in the innermost cortical layer of the stem, and many lower vascular plants have a clearly differentiated stem endodermis (Esau, 1977, p. 259).

The pith of stems is commonly composed of parenchyma, which may contain chloroplasts. In many stems, the central part of the pith is destroyed during growth. Frequently, this destruction occurs only in the internodes, whereas the nodes retain their pith. The pith has prominent intercellular spaces, at least in the central part. The peripheral part may be distinct from the inner part in having compactly arranged small cells and greater longevity (Esau, 1977, p. 261).

The discrete individual strands of the primary vascular system of seed plants are commonly referred to as vascular bundles. The phloem and xylem show variations in their relative position in vascular bundles. The prevalent arrangement is *collateral*, in which the phloem occurs on one side (*abaxial*, or directed away from the axis) of the xylem (Figs. 18.1 and 18.2). That is, the phloem is closest to the outside of the stem, even in monocots with scattered vascular bundles (Fig. 18.2). The xylem in the corn plant shown in Fig. 18.2 makes "monkey faces" (two eyes and one large mouth) and is directed toward the center of the stem (away from the epidermis). In some dicotyledons (e.g., Cucurbitaceae, the squash family, and Solanaceae, the nightshade family, which includes potato), one part of the phloem occurs on the outer side and another on the inner side of the xylem. This arrangement is called *bicollateral*, and the two parts of the phloem are referred to as the *external* (abaxial) and the *internal* (adaxial) phloem (Esau, 1977, p. 261). *Adaxial* means directed toward the axis.

E. STRUCTURE OF SECONDARY XYLEM

In Chapter 14, Section IV, we considered secondary xylem when making calculations of Poiseuille-law flow through wood. Here we look at the structure of secondary xylem. A study of a block of wood reveals the presence of two distinct systems of cells (Fig. 18.3) (Esau, 1977, p. 101): the *axial* (longitudinal or vertical) and the *radial* (transverse or horizontal) or *ray* system. The axial system contains files of cells with their long axes oriented vertically in the stem or the root, that is, parallel to the main, or longitudinal, axis of these organs. The radial system is composed of files of cells oriented horizontally with regard to the axis of the stem or root.

Each of the two systems has its characteristic appearance in the three kinds of sections employed in the study of wood (Esau, 1977, p. 102). In the *transverse* section, that is, the section cut at right angles to the main

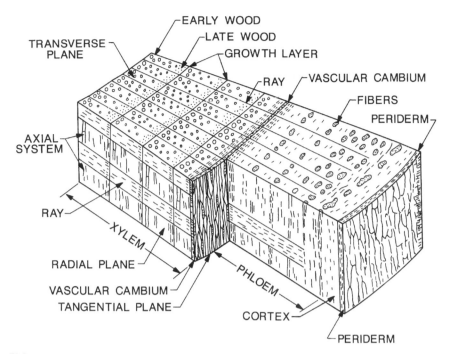

FIG. 18.3 Block diagram illustrating the basic features of secondary vascular tissue. (From Esau, K., *Anatomy of Seed Plants*, 2nd ed., p. 102, ©1977, John Wiley & Sons, Inc.: New York. This material is used by permission of John Wiley & Sons, Inc.)

axis of stem or root, the cells of the axial system are cut transversely and reveal their smallest dimensions. The rays, in contrast, are exposed in their longitudinal extent in a cross section. When stems or roots are cut lengthwise, two kinds of longitudinal sections are obtained: the *radial* (parallel to a radius) and the *tangential* (perpendicular to a radius) (Fig. 18.3).

With little or no magnification, the wood shows the layering resulting from the presence of more or less sharp boundaries between successive growth layers (Fig. 18.3). Each growth layer may be a product of one season's growth, but various environmental conditions may induce the formation of more than one growth layer in one season (Esau, 1977, p. 103). When conspicuous layering is present, each growth layer is divisible into early and late wood. The early wood is less dense than the late wood, because wider cells with thinner walls predominate in the early wood and narrower cells with thicker walls occur in the late wood.

Although woody stems are usually not used to make pressure-volume curves, we are interested in the structure of wood when we consider the rise of sap in plants. Its structure is a key part of the cohesion theory (see Chapter 19), which explains how water can ascend to the top of tall trees.

II. MEASUREMENT OF THE COMPONENTS OF THE WATER POTENTIAL

Differences in total water potential, osmotic potential, pressure potential, matric potential, and gravitational potential can develop in the water of part of a plant, for example, a leaf. (We say "potential," when we recognize that we mean "potential energy." "Potential" is shorter than "potential energy" and saves spaces in printing.) As we saw in Chapter 5 (Equation 5.3), when we were focusing on soil water, these five potentials for water at a particular point in a plant or soil are related by the equation:

$$\psi = \psi_s + \psi_p + \psi_m + \psi_g \tag{18.1}$$

in which the total water potential ψ for a particular unit mass of water (say a milligram) at a particular point is composed of four components, that is the potentials due to solutes, ψ_s, pressure, ψ_p, matrix, ψ_m, and gravity, ψ_g. The term ψ_m is associated with capillary or adsorption forces, which in a plant are forces such as those at the cell walls. Equation 18.1 can be compared, term by term, with the classical equation of plant physiology (Meyer et al., 1960, p. 56):

$$DPD = OP - TP, \tag{18.2}$$

where
    DPD = diffusion pressure deficit
     OP = osmotic pressure
     TP = turgor pressure

and terms corresponding to ψ_m, the matric potential, and ψ_g, the gravitational potential, are neglected. If ψ_m and ψ_g are ignored, the relation of ψ, ψ_s, and ψ_p to water content or cell volume may be described by means of a Höfler-type diagram (Höfler, 1920). In a Höfler-type diagram, the three potentials—water potential, osmotic potential, and turgor potential—are shown together. Figures 18.4 and 18.5 show two types of Höfler diagrams. A quantitative estimate of ψ is possible if the sum of ψ_s and ψ_p is known (Barrs, 1968, p. 236). In this chapter, we wish to discuss methods to measure ψ_s and ψ_p.

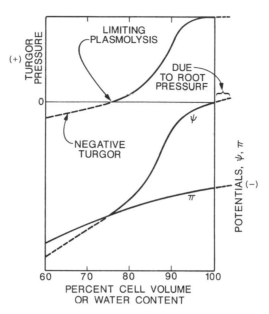

FIG. 18.4   A type of Höfler diagram; relationship between cell volume or water content and total water potential or its components for a single cell. (From Barrs, H.D., Determination of water deficits in plant tissues. In *Water Deficits and Plant Growth. Vol. 1. Development, Control, and Measurement* (T.T. Kozlowski, Ed.), pp. 235–386, ©1968, Academic Press: New York. Reprinted by permission of Academic Press.)

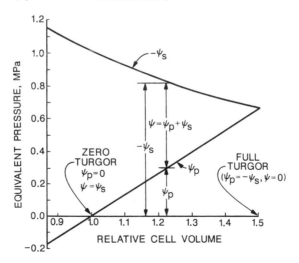

FIG. 18.5   A type of Höfler diagram; relationship between pressure potential (ψ_p), solute potential (ψ_s), and the resultant water potential (ψ) in an idealized (elastic) plant cell. (From Baker, D.A., Water relations, p. 297–318. In *Advanced Plant Physiology* (Wilkins, M.B., Ed.), ©1984, Pitman Publishing Limited: London. Reprinted by permission of Pearson Education Limited, Essex, United Kingdom, and Dennis A. Baker.)

III. OSMOTIC POTENTIAL (Ψ_S)

Two methods are commonly used to determine osmotic potential of plant leaves. In the first method, leaves are frozen, which breaks cell membranes and releases the solutes in the cell. Alternatively, the tissue can be crushed instead of frozen to break the cell membranes, but one has to devise a crushing device. With freezing, one is assured of breaking all the membranes after a tissue has been in a freezer overnight or in dry ice for a few minutes. Some people freeze the tissue and then squeeze out sap from the frozen tissue. With this method it is certain that the sap contains the solute component of the total water potential. The concentration of the solutes in the sap from crushed tissue or from the frozen and crushed tissue then can be measured with a thermocouple psychrometer or an osmometer (see this chapter, Section VIII, for a discussion of the osmometer).

Tissue can be handled in two ways in the frozen-tissue method. First, a piece of tissue can be put into a thermocouple psychrometer. Water potential, ψ, is determined. The chamber with the tissue is removed from the thermocouple, corked, and frozen. After freezing of the tissue, the chamber with the tissue is reattached to the thermocouple, and osmotic potential of the same tissue for which water potential was determined (same geometry, same cells), is measured. Second, water potential can be analyzed on a sample of a plant by using one instrument (e.g., a pressure chamber). The osmotic potential then is determined on another sample of the plant by freezing the sample, exuding sap from it, putting the sap on a piece of filter paper, and measuring its osmotic potential with a thermocouple psychrometer or osmometer (Clarke and Simpson, 1978). This second way has the disadvantage that, because water potential and osmotic potential are not determined on the same piece of tissue, osmotic-potential readings can be higher than water-potential readings (Singh et al., 1983). This is not possible and is due to experimental error or ignoring potentials other than the solute potential and turgor potential that contribute to the total water potential. Osmotic potential is lower or equal to the water potential, unless matric or gravitational potentials are significant.

IV. THEORY OF SCHOLANDER PRESSURE-VOLUME CURVES

The second method for determination of osmotic potential employs a pressure chamber to create a pressure-volume curve. We need to know the background of pressure-volume curves to understand them. The method now used is based on concepts developed by Scholander and colleagues (Scholander et al., 1964; 1965). We first need to define the term *hydrostatic pressure*, which Scholander et al. (1964, 1965) use. It is generally accepted that water is under tension (negative pressure) in the vessels in the xylem tissue. Scholander et al. (1964, 1965) refer to this pressure as "hydrostatic pressure." The *Handbook of Chemistry and Physics* (Weast, 1964, p. F-45) defines hydrostatic pressure as follows:

"Hydrostatic pressure at a distance h from the surface of a liquid of density d, is

$$P = hdg, \tag{18.3}$$

The total force on an area A due to hydrostatic pressure is

$$F = PA = Ahdg", \tag{18.4}$$

where
   F = force (dynes)
   P = hydrostatic pressure (dynes per cm^2)
   h = distance (cm)
   d = density (grams per cm^3)
   g = acceleration due to gravity (cm per s^2).

We sometimes denote d by the Greek letter rho, ρ. Nobel (1974, p. 38; 1983, p. 41) and Barrs (1968, p. 336) equate the hydrostatic pressure with turgor pressure. There is, however, no turgor pressure in mature xylem vessels, because cell membranes have disintegrated.

Scholander et al. (1964) show a leaf cell with vessel, as it exists outside and inside a pressure chamber, in a fresh state (Fig. 18.6) and in a wilted state (Fig. 18.7). They assume that negative hydrostatic pressure exists in a vessel connected to a living cell (Figs. 18.6 and 18.7, left side). They further assume that ambient air cannot enter the system because of surface tension. Water is extruded from the cut end of the stem when pressure is applied by using the pressure chamber (Figs. 18.6 and 18.7, right side). They assume that the membrane surrounding the cell (Figs. 18.6, 18.7) is semipermeable and that no solutes come out of the cell when pressure is applied. Therefore, they assume that the extruded liquid is plain water and that the rise in intracellular solute concentration (which occurs as water is pushed out of the cell) is proportional to the rise in the equilibrium pressure.

We now need to define *equilibrium pressure*. When the vessels (capillary tubes) are cut in a stem to put a leafy shoot in a pressure chamber, the water recedes from the cut (Figs. 18.6 and 18.7, left side). The reason for the recession is because atmospheric pressure is higher on the outside of the cut end than on the inside. If the same difference in pressure were

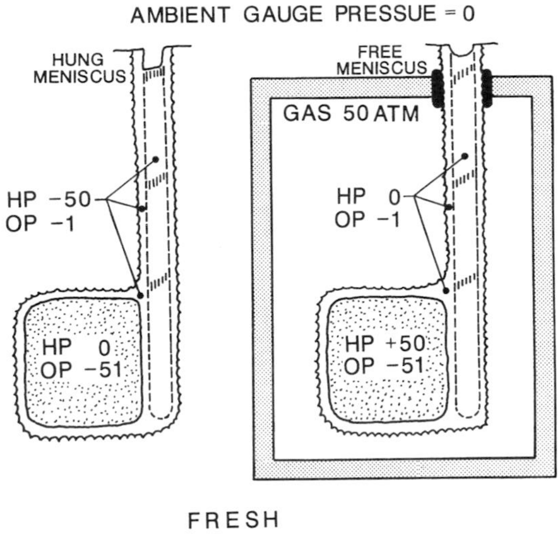

AMBIENT GAUGE PRESSUE = 0

HUNG MENISCUS

FREE MENISCUS

GAS 50 ATM

HP −50
OP −1

HP 0
OP −1

HP 0
OP −51

HP +50
OP −51

FRESH

FIG. 18.6 *Left*: Leaf cell with vessel. Air cannot enter the system because of surface tension (indicated as concave menisci). *HP* = hydrostatic pressure, *OP* = osmotic potential. *Right*: Balancing pressure on the same system, produced by compressed nitrogen. Notice the free meniscus at the cut end of the vessel. (From Scholander, P.F., Hammel, H.T., Hemmingsen, E.A., and Bradstreet, E.D., Hydrostatic pressure and osmotic potential in leaves of mangroves and some other plants. Proceedings National Academy of Sciences 52, 119–125, 1964. Reprinted by permission of Harold T. Hammel.)

AMBIENT GAUGE PRESSURE = 0

HUNG MENISCUS

FREE MENISCUS

GAS 100 ATM

HP −100
OP −1

HP 0
OP −1

HP 0
OP −101

HP +100
OP −101

WILTED

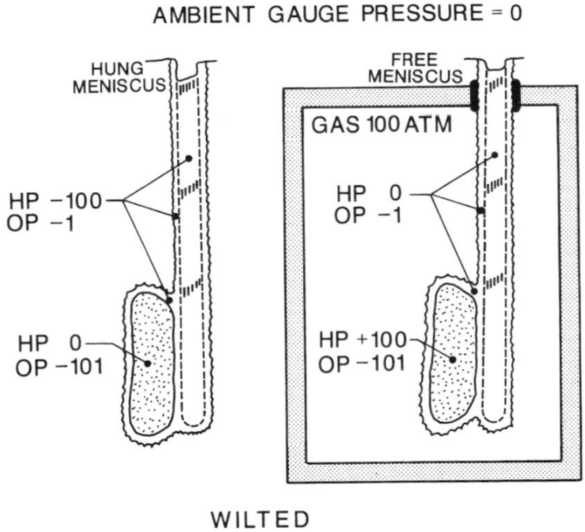

FIG. 18.7 Same system as in Fig. 18.6, but wilted, with about half of the intracellular water extruded. (From Scholander, P.F., Hammel, H.T., Hemmingsen, E.A., and Bradstreet, E.D., Hydrostatic pressure and osmotic potential in leaves of mangroves and some other plants. Proceedings National Academy of Sciences 52, 119–125, 1964. Reprinted by permission of Harold T. Hammel.)

reestablished, the meniscus would move back exactly to the cut. The cut shoot, therefore, is placed in a pressure chamber, leaving the vessels (capillaries) protruding. Gas pressure (usually nitrogen gas) is applied, and when the meniscus is back at the cut, we have the equilibrium pressure (Scholander et al., 1965, p. 340).

Using these assumptions, Scholander et al. (1965) provide the following analysis: If external gas pressure is applied in excess of the balancing pressure, pure water runs out, and at zero turgor the molal concentration in the cell should, therefore, be proportional to the sap pressure, according to the following equation:

$$S/(I\text{-}V) = KP \tag{18.5}$$

or

$$I - V = K_1 P^{-1} \tag{18.6}$$

where S stands for the intracellular solutes, I the cell volume, V the water that has run out, and P the equilibrium pressure. They do not define K or K_1. But they must be constants, and K_1 must equal S/K. If the inverse of pressure ($1/P$) is plotted against liquid removed (V), Scholander and colleagues say that a straight line results whenever the concentration is proportional to the pressure, and the intercept on the abscissa gives the volume of water that is being concentrated—that is, the intracellular water (I). The studies by Scholander and co-workers (1964, 1965) have been cited many times, as has a subsequent paper by Tyree and Hammel (1972).

We can refer to physical chemistry textbooks for the foundation of pressure-volume curves (see, for example, Moore, 1962, p. 135; Daniels and Alberty, 1966, p. 170). Let us follow the analysis of Daniels and Alberty (1966, pp. 170–172). When a solution is separated from the solvent by a semipermeable membrane, which is permeable by solvent but not by solute, the solvent flows through the membrane into the solution, where the chemical potential of the solvent is lower. This process is known as osmosis. This flow of solvent through the membrane can be prevented by applying a sufficiently high pressure to the solution. The osmotic pressure π is the pressure difference across the membrane required to prevent spontaneous flow in either direction across the membrane. Figure 5.1 shows a diagram of an osmometer, which can be used to measure osmotic pressure.

The phenomenon of osmotic pressure was described by Abbé Nollet in 1748. Pfeffer, a botanist, made the first direct measurments of it in 1877 (For a biography of Pfeffer, see the Appendix, Section IX.) Van't Hoff analyzed Pfeffer's data on the osmotic pressure of sugar solutions and found empirically that an equation analogous to the ideal gas law gave approximately the behavior of a dilute solution, namely (Daniels and Alberty, 1966, p. 170)

$$\pi \bar{V} = RT \tag{18.7}$$

where $\bar{V}$ is the volume of solution containing a mole of solute and R is the ideal gas constant and T is the absolute temperature. The origin of the pressure is different from that for a gas, however, and the equation of the form of the ideal gas equation is applicable only in the limit of low concentrations. See Fig. 18.8 for the ideal gas law.

Daniels and Alberty (1966) then proceed to derive the van't Hoff law using calculus, which we shall not do. (For a biography of van't Hoff, see the Appendix, Section X.) The van't Hoff law, which H.H. van't Hoff developed in 1885 (Moore, 1962, p. 135; see Hammel and Scholander, 1976, for references by van't Hoff) and which applies only to dilute solutions, is as follows (Daniels and Alberty, 1966, p. 171):

$$\pi = (cRT)/M \tag{18.8}$$

where c is the concentration of solute in grams per unit volume and M is the molecular weight of the solute. This is the approximate equation that

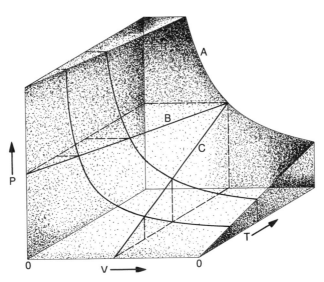

FIG. 18.8 Pressure-volume-absolute temperature relation for an ideal gas. (From Daniels, F., and Alberty, R.A., *Physical Chemistry*, 3rd ed., p. 8, ©1966, John Wiley & Sons, Inc.: New York. This material is used by permission of John Wiley & Sons, Inc.)

van't Hoff found empirically. Moore (1962, p. 135) writes the van't Hoff law as follows:

$$\pi = cRT \tag{18.9}$$

where $c = n/V$ and n = number of moles (grams/M) and V is volume. We can write Equation (18.9) as follows:

$$\pi = (nRT)/V. \tag{18.10}$$

Equation (18.10) is similar in form to Equation (18.6), developed by Scholander et al. (1965). That is, we have pressure inversely related to volume, if we are considering a dilute solution. However, we must note that Scholander et al. (1964, 1965) plot $1/P$ versus volume exuded (not volume left in the plant), and $1/P$ in their curves is inversely related to V. Note in Equation 18.10 and in Fig. 18.8 that P (or π), not $1/P$, is inversely related to V. Also, at constant temperature, P versus V is a rectangular hyperbola (Daniels and Alberty, 1966, p. 9) and the curve never touches the y or x axis, as it does in the curves developed by Scholander and colleagues (e.g., see Fig. 18.9 from Hammel and Scholander, 1976; note the right-hand-side ordinate is $1/P$).

Gardner and Rawlins (1965), who discussed the paper by Scholander et al. (1965), said that their procedure measured the difference in free

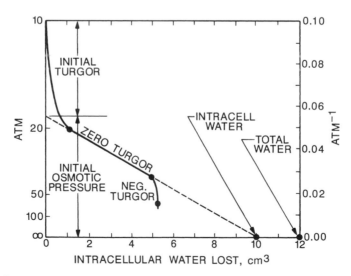

FIG. 18.9 Schematic presentation of a pressure-volume curve. (From Hammel, H.T. and Scholander, P.F., *Osmosis and Tensile Solvent*, Fig. 24, p. 36, ©1976, Springer-Verlag: Heidelberg, Germany. Reprinted by permission of Springer-Verlag and Harold T. Hammel.)

energy per unit volume between water in the plant and the same water outside of the plant. The pressure chamber operates on the same principle as the pressure-membrane apparatus used to measure the potential energy of water in soils (see Chapter 4, Fig. 4.5). Gardner and Rawlins said:

> When air pressure is applied to the sample chamber, the free energy of the water is raised. If this pressure increase is carried out isothermally, the free energy of the water would be raised by approximately $V\Delta P$, where V is the volume of water in the sample and ΔP is the pressure increase necessary to establish equilibrium between water in the system and that outside. It is common practice to express this energy difference in terms of energy per unit volume (the water potential), which, of course, is dimensionally the same as pressure. In the experiment of Scholander et al., the plant itself provides the membrane which is permeable to water but not to air.

V. HOW TO ANALYZE A PRESSURE-VOLUME CURVE

Now let us return to the actual measurement of osmotic potential using a pressure-volume curve. Pressure is applied incrementally to a plant sample. After each increase in pressure, the volume of exudate from the cut end of the plant (e.g., stem, petiole) is collected and measured, and a curve of the reciprocal of pressure versus cumulative volume exuded is plotted (Figs. 18.10, 18.11). From this pressure-volume curve, the osmotic potential at full turgor and the osmotic potential at zero turgor are indirectly determined by reading values on the ordinate (Fig. 18.10). Sometimes, instead of plotting pressure versus volume exuded, pressure versus water content (or relative water content) of the plant is plotted. In this case, the plot is called a water-release curve instead of a pressure-volume curve, and is similar to water-release curves developed for soils, in which pressure (or potential) is plotted versus soil water content. Turner (1981) (Fig. 18.10) illustrates how to determine osmotic potential from either a pressure-volume curve or a water-release curve.

Figure 18.11 can be used to learn how one gets ψ, ψ_s, and ψ_p from the $1/P$ values obtained with the pressure chamber. The top part of Fig. 18.11 has the data and the bottom part has the converted data. There are 11 data points that line up in the top and bottom part of the figure. The first 6 data points, starting in the upper left in the top part of the figure, are in the region of turgor potential. This is the region of turgor, because the relationship between $1/P$ and volume of sap exuded is curvilinear. (As we shall calculate, Point 6 is right at the break between the region of turgor and no

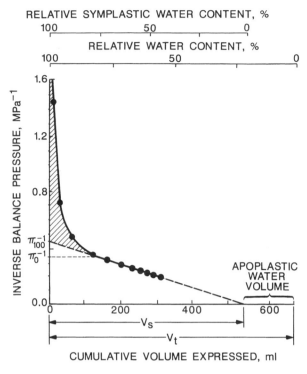

FIG. 18.10 A pressure-volume curve, i.e., the relationship between the inverse of the balance pressure and the cumulative volume of cell sap expressed, for wheat leaves. V_s is the volume of symplast water and V_t is the total volume of water in the leaf. π_{100}^{-1} and π_o^{-1} are the inverse of the osmotic potentials at full and zero turgor, respectively. The hatching indicates the inverse of the turgor pressure. (From Kluwer Academic Publishers journal *Plant and Soil*, Vol. 58, 1981, pp. 339–366, Turner, N.C. Techniques and experimental approaches for the measurement of plant water status, Fig. 7, p. 355, ©1981. With kind permission of Kluwer Academic Publishers and Neil C. Turner.)

turgor.) Points 7, 8, 9, 10, and 11 are in the region of zero turgor potential where a straight-line relation exists between $1/P$ and volume of sap expressed (compare the straight line in Fig. 18.10, top, with the straight line for zero turgor in Fig. 18.9). First we determine P (from $1/P$) and get the curve for ψ_w (this is the total water potential or ψ in Equation 18.1). The six values (in the region of turgor) are about –0.71, –1.11, –1.43, –1.82, –2.00, and –2.13 MPa. The water potential, ψ_w, is negative, so we need to add negative signs in front of these values. Then we determine $1/\psi_s$ by measuring down from the data point in the top part of the figure to the dashed line and then reading the value on the ordinate. We then take the

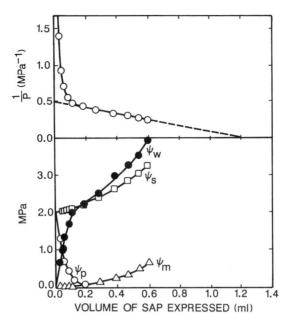

FIG. 18.11   The upper graph is a pressure volume curve for *Ilex opaca* (American holly). The ordinate is the reciprocal of the pressure, and the abscissa is the volume of water expressed, in milliliters. The dashed line is the calculated extension of the linear part of the curve. It intersects the ordinate at 0.0051, equal to an osmotic potential of −1.96 MPa, and the abscissa at 1.23 mL. The data in the lower graph were obtained by analysis of the pressure-volume curve. Curves are shown for turgor (ψ_p), osmotic (ψ_s), and matric (ψ_m) potentials, and total water potential (ψ_w). The turgor potential is positive, and all other potentials are negative. The abscissa of the lower graph represents water volume in milliliters, and the ordinate is given in megaPascals. (From Kramer, P.J., *Water Relations of Plants*, p. 53, ©1983, Academic Press: New York. Reprinted by permission of Academic Press.)

reciprocal of $1/\psi_s$ to get ψ_s. The six values for ψ_s in the region of turgor in Fig. 18.11 are about −1.98, −2.00, −2.02, −2.04, −2.08, and −2.13 MPa. We add negative signs because they are the solute potential. Subtracting ψ_s from ψ_w, we get the six values for turgor potential, ψ_p, which are 1.27, 0.89, 0.59, 0.22, 0.08, and 0.00 MPa (positive values). The last value is zero, showing that Point 6 is in the region of zero turgor.

In addition to water potential, osmotic potential, and turgor potential, the pressure-volume curves can be used to find modulus of elasticity (Melkonian et al., 1982; Sinclair and Venables, 1983). (In Chapter 21 we will determine modulus of elasticity of leaves, but not using pressure-volume curves.) Pressure-volume or water-release curves also can be obtained with dew-point hygrometers instead of pressure chambers (Richter, 1978; Wilson et al., 1979).

The osmotic potential of the sap exuded during a pressure-chamber measurement can be determined by placing the sap on filter paper and measuring its osmotic potential with a thermocouple psychrometer or an osmometer (Meyer and Ritchie, 1980). The sap in the dead cells that conduct water (vessel members or tracheids) has a much higher osmotic potential (less negative, i.e., it is very dilute) than that in the sap of living cells (Scholander et al., 1964; Ike et al., 1978).

There is disagreement as to which method (frozen-tissue; or pressure-volume curve, or water-release curve) provides the more reliable results. Brown and Tanner (1983) used alfalfa (*Medicago sativa* L.) to compare the two methods for determining osmotic potential. Osmotic potential of sap expressed from thawed tissue was 0.21 to 0.89 MPa lower than the osmotic potential obtained from water-release curves. They felt that the difference was due primarily to the production of solutes in thawed tissue by enzymatic hydrolysis and suggested that the water-release curve method was a better way to measure osmotic potential than the frozen-tissue method. In contrast, Rakhi et al. (1978) said that, because the pressure chamber dehydrated the tissues that they studied [*Carex physodes* M. Bieb. (sedge) and *Populus tremula* L. (European aspen)], the values for osmotic potential determined by freezing the tissue were more reliable than the ones derived from water-release curves. Walker et al. (1983) investigated wheat (*Triticum aestivum* L.) and found that the osmotic potential, as measured by the pressure-volume method, compared favorably with the osmotic potential, as measured on frozen tissue with psychrometers. Experiments with other plants are needed to find out if these two methods for determining osmotic potential give similar results.

VI. TURGOR POTENTIAL (Ψ_p)

Turgor potential normally is determined in one of two ways. First, turgor potential can be calculated, if the water potential and osmotic potential are measured with thermocouple psychrometers. The turgor potential is the difference between osmotic potential and water potential, assuming that matric potential (Barrs, 1968, p. 337) and gravitational potential are negligible. [Kirkham (1983) discusses situations in which gravity can be important.] Second, turgor potential can be estimated by using pressure-volume curves (Turner, 1981, Fig. 18.10; Melkonian et al., 1982; Sinclair and Venables, 1983). Both of these methods require excised samples.

Heathcote et al. (1979) describe an instrument, based on the use of small strain gauges, for the nondestructive measurement of turgor potential. Calibration showed that the instrument gave a voltage output that was linearly related to the pressure (turgor) potential of leaf cells, as determined with a pressure chamber. Turner and Sobrado (1983) tested Heathcote et al.'s (1979) instrument. They found no correlation between the output of the instrument and turgor potential of leaves, estimated from pressure-volume curves or by taking the difference between measured values of osmotic potential and water potential. They felt that variability in leaf thickness and the presence of large veins limited the instrument's usefulness for measurement of turgor potential.

For many years, people have tried to measure turgor potential (pressure) directly by using probes (Barrs, 1968, p. 336). The practical difficulties, however, have been enormous, because of the small size of most plant cells. Most of the work, until recent years, was done with large-celled algae (e.g., *Nitella, Chara*). Now a pressure probe has been developed to measure turgor potential of higher plants directly. Apparently, the first one was designed and constructed by Steudle and co-workers in Germany, and the fundamental structure is described in his papers, which are reviewed by Tomos and Leigh (1999). A couple of his early papers with co-workers are Steudle and Jeschke (1983) and Tyerman and Steudle (1984). In the United States, Nonami and Boyer (1984) were probably the first to publish on the pressure probe, and they reported the turgor potential of a higher plant, soybean (*Glycine max* L. Merr.). Boyer (1995) has described in detail the pressure probe. The equipment is expensive, complex, requires extensive training to learn how to use it, and takes precise control. The smallest vibration (such as an air movement or vibration of the floor) can cause pressure probe measurements to fail. A vibration-damping table is advised.

Tomos and Leigh (1999) review how the pressure probe has evolved from an instrument for measuring cell turgor into a device for sampling the contents of individual higher plant cells in situ in the living plant. In addition, the probe is being used to measure root pressure, xylem tension (negative pressure), hydraulic conductivity, the reflection coefficient of solutes, elasticity, solute concentrations, and enzyme activities at the resolution of single cells. They also review the controversy surrounding the interpretation of measurements of xylem tension obtained with the pressure probe. It is critical to know what these tensions (negative pressures) are, if we are to confirm or refute the cohesion theory for the rise of sap in plants (see Chapter 19). Tensions in the xylem measured with the pressure probe under transpiring conditions in both small plants and tall trees have been less negative than those obtained with the pressure chamber, and in many cases are positive, which would refute the cohesion theory, because it assumes negative pressures (tensions) in the xylem.

Apparently, the pressure probe can measure pressures over only a certain range, and the xylem has more negative pressures than the probe's limit, especially under dry conditions. Under well-watered conditions, the probe gives values that agree with other measurements, but under dry conditions the insertion of the probe appears to cause cavitation, so the true tension in the xylem cannot be measured. Tomos and Leigh (1999) conclude that the questions raised by the measurements with the pressure probe inserted into the xylem remain unresolved. But measurements with the pressure probe are the only direct determinations of pressure in the xylem that we can get. Despite the questions surrounding the pressure probe, its cost, and the difficulty in learning how to use it, the instrument gives us a basic understanding about water and solute relations at the cellular level.

VII. MEASUREMENT OF PLANT WATER CONTENT AND RELATIVE WATER CONTENT

Höfler-type curves plot plant potentials versus water content (Fig. 18.4) and pressure-volume curves plot the inverse balance pressure versus relative water content (Fig. 18.10). Therefore, we need to know how to measure both plant water content and relative water content.

Plant water status is usually described by two basic parameters: the content of water in the plant or the energy status of the water in the plant, expressed as the (total) water potential, ψ (Barrs, 1968, p. 236). We have studied the water potential and its components in this chapter, in Sections II through VI. Here we determine how to measure plant water content and relative water content.

Old techniques to measure plant water content expressed the water content on the following bases: dry weight, fresh weight, or leaf area. These are not acceptable because they are not stable. They change diurnally and seasonally, and the leaf area is reduced with drought.

Soil water content is expressed on a dry-weight basis as:

$$\text{soil water content } (\%) = [(\text{wet wt.} - \text{dry wt.})/\text{dry wt.}] \times 100. \quad (18.11)$$

The soil water content is almost always less than 100% when expressed on a dry-weight basis. Exceptions are highly organic soils, such as those that occur on forest floors. For example, the 0-to-5 cm, 5-to-10 cm, and 10-to-15 cm layers of forest soil from the Craigieburn Range in the South Island of New Zealand had water contents of 382.3%, 134.4%, and 89.1%, respectively (Kirkham and Clothier, 2000).

If we do express plant water content on a weight basis, then we choose the fresh-weight basis:

$$\text{plant water content } (\%) = [(\text{fresh wt.} - \text{dry wt.})/\text{fresh wt.}] \times 100. \quad (18.12)$$

If we put the water content on a dry-weight basis, the water content always would be greater than 100%, because of the high water content of plants.

Plant water content changes with age and condition of the plant. Therefore, some standard must be used in determining water content. The water content at full turgor has been used as a standard since the work of Stocker (1928, 1929) in Germany. He determined plant water content using the following equation:

$$\text{WD } (\%) = \text{WSD } (\%) = [(\text{turgid wt.} - \text{fresh wt.})/$$
$$(\text{turgid wt.} - \text{dry wt.})] \times 100 \quad (18.13)$$

where WD is the water deficit and WSD is the water saturation deficit; the two terms are equivalent. He determined the turgid weight by cutting off a whole leaf, or, when working with conifers, a small branch, and standing it in a little water in a closed container for 48 hours (Barrs, 1968, p. 243). The fresh weight is the weight at time of sampling. The dry weight is the weight after oven drying.

The method that Stocker developed was a reliable one to determine plant water content. However, some scientists ignored its importance and used the following equation (Barrs, 1968, p. 243):

$$\text{"WD"} = [(\text{turgid wt.} - \text{fresh wt.})/\text{dry wt.}] \times 100. \quad (18.14)$$

This equation neglects the importance of using the fully turgid weight as the basis for water content. Note that turgid weight does not appear in the denominator of Equation 18.14.

The following equation was used to avoid getting dry weight:

$$\text{RSD} = \text{DSH} = [(\text{turgid wt.} - \text{fresh wt.})/\text{turgid wt.}] \times 100, \quad (18.15)$$

where RSD = relative saturation deficit and DSH is the *déficit de saturation hydrique* (the French term for water saturation deficit). The DSH was confused in English with the WSD. However, the RSD or DSH method is not reliable. Dry weight needs to be determined. These many methods (WD, "WD," WSD, RSD, DSH) resulted in chaos in the literature and no standard technique existed.

In 1950, Weatherley (1950) standardized the technique. He used punched disks instead of standing whole leaves in water. He floated the disks for 24 hours (sometimes 48 hours) in closed Petri dishes that were exposed to diffuse daylight and a constant temperature. He then calculated relative turgidity, RT, as follows:

$$RT = [(\text{fresh wt.} - \text{dry wt.})/(\text{turgid wt.} - \text{dry wt.})] \times 100. \quad (18.16)$$

The relative turgidity is related to the water saturation deficit (or water deficit), as follows:

$$100 - RT = WSD. \quad (18.17)$$

Later Barrs and Weatherley (1962) revised Weatherley's 1950 method. They noted that there are two phases for water uptake: Phase I, which is in response to the initial water deficit and during which rapid uptake of water occurs, and Phase II in which a continued slow uptake of water continues due to growth of the tissue, even though it is excised (Fig. 18.12). The aim of the Barrs and Weatherley (1962) method to determine relative turgidity was to measure only Phase I. One needs to determine for each species when rapid uptake of water ceases (end of Phase I) and then float disks for this length of time. If an initial experiment to determine this time is not done, leaves can be floated for 3 to 6 hours, which is the normal time for Phase I for most plants. Four hours is the time most often chosen.

The Barrs and Weatherley (1962) technique is the standard method to measure relative turgidity, which is now called relative water content

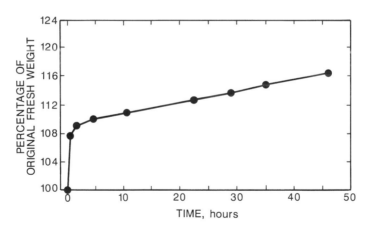

FIG. 18.12 Change over time in fresh weight of floating leaf disks. The leaves are from the castor-oil plant (*Ricinus communis* L.). Note the two phases in water uptake: a first phase in which water is taken up rapidly (0 to 5 hours) and a second phase in which water is taken up slowly (5 to 46 hours). (From Barrs, H.D., and Weatherley, P.E., A re-examination of the relative turgidity technique for estimating water deficits in leaves. *Australian Journal of Biological Sciences* 15; 413–428, ©1962. Permission to reprint granted by CSIRO Publishing, Collingwood, Victoria, Australia, the original publisher of the information.)

(RWC). The old term *relative turgidity* has been abandoned because early workers confused relative turgidity with turgor pressure (turgor potential) (Barrs, 1968, pp. 244–245). For example, Box and Lemon (1958) referred to "turgor pressure" instead of "relative turgidity" in discussing Weatherley's (1950) results with cotton. The turgor potential is an energy-based measurement. Relative turgidity (or relative water content) is a measure of plant water content.

The Barrs and Weatherley RWC technique (1962) consists of doing the following:

1. A punch is used to punch disks out of a leaf. The punch must be sharp to minimize cut-edge effects. When plant tissue is cut, cells are damaged, causing infiltration of water. This creates excessive uptake of water and gives spuriously low RWCs. These errors are hard to quantify and appear not always to be present (Barrs, 1968, p. 247). A sharp punch, along with a good sized leaf disk, avoids the cut-edge effect. Small disks are more prone to higher RWCs than are larger disks, and disks should not be smaller than 8 mm in diameter. A Number 8 cork borer, which has a diameter of 14 mm, is a good size to use.
2. The same diameter disks must be used in an experiment; 12 disks are recommended (Barrs, 1968, p. 251).
3. Mature or nearly mature leaves should be used rather than rapidly growing or senescent leaves. If mature leaves are used (e.g., third, fourth, or fifth leaf from the top of a plant), a time of 3 to 4 hours for floating should be sufficient and will coincide with Phase I (Barrs, 1968, p. 246).
4. Leaf disks must be placed in a closed Petri dish to maintain a constant humidity.
5. A constant temperature must be used. A thermometer should be placed on the laboratory bench to record the temperature. Results will differ depending on the temperature used (Fig. 18.13). Barrs and Weatherley (1962, p. 415) used 20°C in a constant temperature room.
6. The leaf disks must be exposed to diffuse light, such as that on a laboratory bench out of direct sunlight, to minimize growth and heating due to the sun.
7. After floating, the leaf disks are removed using tweezers from the Petri plate, blotted dry on a paper towel, and then put in a drying oven at 85°C (Barrs, 1968, p. 239).
8. All weights should be weighed on an analytical balance to the fourth decimal point and then rounded off to the third decimal point.

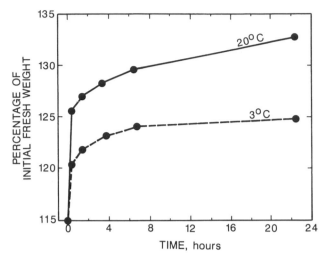

FIG. 18.13 Effects of floating on water at 20°C and 3°C on changes of fresh weight of leaf disks from the castor-oil plant (*Ricinus communis* L). (From Barrs, H.D., and Weatherley, P.E., A re-examination of the relative turgidity technique for estimating water deficits in leaves. *Australian Journal of Biological Sciences* 15; 413–428, ©1962. Permission to reprint granted by CSIRO Publishing, Collingwood, Victoria, Australia, the original publisher of the information.)

9. For conifer needles and grass leaves, which are not wide enough for punched disks, razors mounted on a block are used to cut constant-length segments. A length of 15 to 25 mm is normally used (Barrs, 1968, p. 250). Grass leaves can be floated, but conifer leaves should be placed upright in a beaker of water, because needles can become water-logged and sink.

The advantages of the Barrs and Weatherley method (1962) are:

1. It standardizes sampling.
2. It is sparing of tissue.
3. It allows some leaves to be sampled more than once.
4. It can be done in a shorter time (3 to 6 hours) than previous methods (24 to 48 hours).
5. Even though the measurements are tedious, the relative water content method is simple and requires only an analytical balance, Petri dishes, paper towels, a thermometer, tweezers, and a drying oven. No special skills other than carefulness and patience are needed to take the measurements.

Relative water content measurements are important for several reasons. First, they are used to construct pressure-volume curves (Fig. 18.10).

Second, relative water content can be used as a guide to irrigation. Ehrler and Nakayama (1984) in Arizona found that relative-water-content measurements were a good guide to schedule irrigations of guayule (a small shrub of northern Mexico and the southwestern part of the United States cultivated for the rubber obtained from its sap). Third, the measurements show variation of water status in different portions of a leaf, if the leaf is large enough for such sampling. Slavík (1963) measured water saturation deficit (see Equation 18.17) across a tobacco leaf and found that the WSD was least at the base of the leaf and most at the center edge and tip of the leaf. The high WSD at the tip of the leaf was associated with a low transpiration rate. The transpiration rate was highest at the base of the leaf.

Diaz-Perez et al. (1995) state that the relative-water-content method is a good measure of water status and is easier to measure than water potential. The value of relative-water-content measurements versus water-potential measurements has been debated (Kramer, 1988; Passioura, 1988; Schulze et al., 1988). Each has its place in measurement of plant-water status. However, only by measuring plant water potential can we determine the direction of movement of water in the plant. Water moves according to a potential-energy gradient (from high to low potential energy). The same holds true for water in the soil. It is important to know soil water content, but only measurements of soil water potential (or the total head) tell us the direction of movement of water (e.g., see Fig. 5.3).

In sum, by using the Barrs and Weatherley (1962) technique, the problems associated with measurements of plant water content (i.e., dry weight increases with time, continued increase in water content after attainment of full turgidity, and injection of water into the intercellular spaces at the cut edge) can be minimized. The method provides a standard technique that can be replicated by workers at any location. When it is used, the paper by Barrs and Weatherley (1962) always should be cited.

VIII. OSMOMETER

We have mentioned that osmotic potential can be determined with an osmometer. The osmometer measures osmolality, not osmotic pressure, so we must learn how we can relate osmolality to osmotic pressure (or its negative value, osmotic potential).

Osmolality expresses the total concentration of dissolved particles in a solution without regard for the particle size, density, configuration, or electrical charge (Wescor, 1989). All these items listed are *particle characters*. A *colligative property* depends on the number of solute and solvent particles present in a solution, not their character. Osmotic pressure is a colligative property, not a cardinal property, which we shall now discuss.

Consider a solvent and a solute (a solution). There are three cardinal properties of a solvent (e.g., water): freezing point, which is lowered by solutes; boiling point, which is raised by solutes; and vapor pressure, which is lowered by solutes. Measurement of solution concentration or osmolality can be made indirectly by comparing a colligative property of the solution (solute + solvent) with a corresponding cardinal property of the pure solvent (e.g., water).

The first instruments to measure osmolality were based on freezing point depression. The Wescor osmometer (5500 Series; Logan, Utah; Fig. 18.14) measures osmolality through measurement of vapor pressure by using thermocouple hygrometery (same principle that we saw for measuring water potential with thermocouple psychrometers in Chapter 16). The dew-point depression is determined using the Peltier effect. The relationship between vapor pressure depression and dew-point temperature depression is given by (Wescor, 1989) as:

$$\Delta T = \Delta e / S, \tag{18.18}$$

where ΔT is the dew-point temperature depression in degrees Celsius, Δe is the difference between saturated and chamber vapor pressure, and S is the slope of the vapor pressure-temperature function at ambient temperature (37°C in the osmometer). (The osmometer probably operates at 37°C, because it is used in medical clinics and the human body is normally at this temperature.) The Clausius-Clapeyron equation gives S as a function of temperature (T in degrees Kelvin), saturation vapor pressure (e_o), and the latent heat of vaporization (λ), as follows (Wescor, 1989)

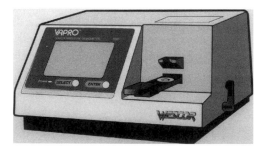

FIG. 18.14 The vapor pressure osmometer made by Wescor, Inc., Logan, Utah. (From a Wescor, Inc., Logan, Utah, brochure. Reprinted by permission of Wescor, Inc., Logan, Utah.)

$$S = (e_o\lambda)/(RT^2), \qquad (18.19)$$

where R is the universal gas constant. (For a biography of Clausius, see the Appendix, Section XI.)

Figure 18.15 shows the relation between vapor pressure and $1/T$ from a physical chemistry textbook (Daniels and Alberty, 1966, p. 127). Note the slopes for the different compounds are essentially linear.

The vapor pressure depression is a linear function of osmolality. A calibration line is obtained with two solutions that come with the equipment: 290 and 1,000 mmol. The salt in the calibrating solutions is not given because it does not matter, but the calibrating solutions are probably NaCl solutions.

The Wescor instrument gives us osmolality, but we would like to relate these values to osmotic pressure. Let us determine a relationship between osmolality and osmotic pressure. We know the van't Hoff law from Equation 18.8, $\pi = (cRT)/M$, or from Equation 18.10, $\pi = (nRT)/V$.

Example: What is the osmotic pressure, π, for 2 moles of NaCl ($58.5 \times 2 = 117.0$ grams)?

If we have NaCl, we have 2 ions from one molecule of salt.

$$\pi = (nRT)/V = (grams/volume)RT/M$$

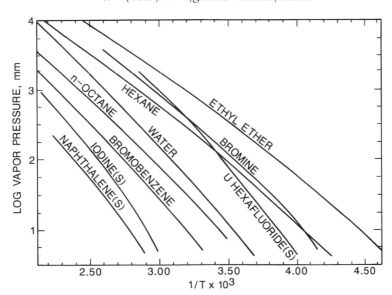

FIG. 18.15 Log of vapor pressure versus $1/T$ for various vaporization processes. T = temperature in degrees Kelvin. (From Daniels, F. and Alberty, R.A., *Physical Chemistry*, 3rd ed., p. 127, ©1966, John Wiley & Sons, Inc.: New York. This material is used by permission of John Wiley & Sons, Inc.)

$\pi = (117.0/1\ L)\ (0.0821\ L\text{-atm mol}^{-1}\ \text{deg}^{-1})(310K)/58.5 = 50.84\ \text{atm.}$
$50.84\ \text{atm} \times 2\ \text{ions} = 101.68\ \text{atm} = 103.02\ \text{bars.}$

The table published by Lang (1967) and reproduced by Barrs (1968, p. 288) gives 101.60 bars at 35°C. The 103.02 bars we calculated is close to the value given by Lang (1967).

If we want to convert osmolality to osmotic potential, and we do not know the number of ions (as we do with an NaCl or KCl solution), we use the following equation (Taiz and Zieger, 1992, p. 68):

$$\psi_s = -\ C_s RT, \qquad\qquad (18.20)$$

where

C_s = osmolality (in mol/kg or mol/L).

So if we measure with the osmometer 2 mmoles/1 kg = 0.002 molal

$$\psi_s = -0.002\ RT =$$
$$(-0.002\ \text{mol L}^{-1})$$
$$(0.0821\ L\text{-atm mol}^{-1}\ \text{deg}^{-1})\ (310°K) = -0.05\ \text{atm.}$$

Note that $(0.0821\ L\text{-atm mol}^{-1}\ \text{deg}^{-1})\ (310\ °K)$ = a constant = 25.45 or $(0.0832\ L\text{-bar mol}^{-1}\ \text{deg}^{-1})\ (310\ °K)$ = a constant = 25.8.

To summarize the method to use the osmometer:

1. With the osmometer, ΔT is measured to get Δe, and we know

$$\Delta T = \Delta e/S,$$

    where we get S from the Clausius-Clapeyron equation.
2. Then Δe is related to osmolality (linear function). Vapor pressure depression is a linear function of osmolality.
3. We make a calibration line with two solutions. Wescor provides 290 and 1,000 mmol/kg for the calibrating solutions.
4. We report the reading as osmolality (mmol/kg) or as osmotic pressure using

$$\psi_s = -C_s RT.$$

IX. APPENDIX: BIOGRAPHY OF WILHELM PFEFFER

Wilhelm Friedrich Philipp Pfeffer (1845–1920) was a German physiological botanist. He was born in Grebenstein, where his father owned a chemist's shop (pharmacy) (Frommhold, 1996), on March 9, 1845 (McIlrath, 1971).

There he learned fundamental knowledge and manual skills for his later profession. After earning a degree in botany and chemistry from Göttingen University in 1865, he spent the next several years studying botany and pharmacy at Marburg University. He continued his botanical studies at Berlin (1869–1870) and Würtzburg University (1870–1871). In 1871 he returned to lecture at Marburg, and in 1873 he was appointed lecturer at Bonn University. In 1877 he became a professor at Basel University and later at the universities of Tübingen (1878) and Leipzig (1887). In Leipzig, Pfeffer began his scientific research which lasted more than three decades (Frommhold, 1996). There he was director of the Botanical Institute and the Botanical Gardens. In 1884, Pfeffer married Henrika Volk (Frommhold, 1996), and in May of 1885 his son Otto was born.

His laboratories in the Botanical Institute in Leipzig were modern for the time. He had microscopes and a room with a constant temperature (the precursor of the growth chamber). He positioned measuring instruments so they were free from vibration. He installed a dark room and made devices by which alteration of light and dark conditions could be done automatically. He improved equipment such as the clinostat and the auxanometer. Pfeffer's institute attracted students and visiting scientists from around the world. One of his famous students was Carl Correns, who studied mutants in the Botanical Gardens. Pfeffer knew Wilhelm Ostwald, who also went to Leipzig in 1887.

Pfeffer made significant contributions in the following areas of plant physiology: respiration, photosynthesis, protein metabolism, and tropic and nastic movements. In 1881 he published the first part of his *Handbuch der Pflanzenphysiologie* (English translation by A.J. Ewart, *Physiology of Plants*, three volumes, 1906), which was an important text for many years.

During his life Pfeffer received numerous awards. His seventieth birthday (March 9, 1915) and his golden doctor's jubilee (February 10, 1915) were both celebrated on the same day in the first year of the First World War (1914–1918). The hardest blow in Pfeffer's life hit shortly before the end of the war. A few weeks before the armistice, he was informed that his son Otto was missing, and in the middle of 1919 he learned that Otto had been killed in France. He also suffered physically, because he, like many other Germans at the end of the war, did not have enough to eat. He destroyed all his scientific manuscripts because he felt that he could not complete his work. His research publications ended in 1916. However, he kept teaching, and in 1919–1920, his lectures often had to be offered twice, as the rooms were overcrowded with returned soldiers (Frommhold, 1996).

His last year of life was difficult. In addition to the loss of his son, he also faced the loss of his official residence upon enforced retirement in

1920 (Frommhold, 1996). On the day of his last physiology lecture on January 31, 1920, he died without having been seriously ill.

X. APPENDIX: BIOGRAPHY OF JACOBUS VAN'T HOFF

Jacobus Hendricus van't Hoff (1852–1911) was a Dutch physical chemist, who received the first Nobel Prize in chemistry (1901) for his work on chemical dynamics and osmotic pressure in solutions (Preece, 1971a). He was born in Rotterdam on August 30, 1852, and studied at the Polytechnic at Delft and at the University of Leiden. He then studied under Friedrich A. Kekulé von Stradonitz (1829–1896; German chemist) at Bonn, Charles A. Wurtz at Paris in the École de Médecine, and G.J. Mulder at Utrecht, where he obtained his doctorate in 1874. He was a lecturer in physics at the veterinary school in Utrecht (1876); professor of chemistry, mineralogy, and geology in Amsterdam University (1878); and professor at the Prussian Academy of Sciences in Berlin (1896), accepting an honorary professorship in the university so that he might lecture if he wished. He was elected a foreign member of the Royal Society in 1897 and awarded its Davy medal in 1893. He died in Berlin on March 1, 1911.

Van't Hoff's earliest important contribution was made in 1874. Starting with the results of the work of Johannes Wiclicenus (1835–1902; German chemist, who studied isomers), he showed that the four valencies of the carbon atom were probably directed in space toward the four corners of a regular tetrahedron. In this way optical activity, shown to be always associated with an asymmetric carbon atom, could be explained. An identical idea was put forward two months later (November 1874), independently, by Joseph Achille Le Bel (1847–1930; French chemist). Van't Hoff and Le Bel had been fellow students under Wurtz but had never exchanged a word about the carbon tetrahedron. The concept was attacked by Hermann Kolbe (1818–1884; German chemist), but its value was soon universally realized, and it laid the foundation stone of the science of stereochemistry (Preece, 1971a).

In 1877 van't Hoff published *Ansichten über die organischen Chemie*, which contains the beginnings of his studies in chemical thermodynamics. In *Études de dynamique chimique* (1884) he developed the principles of chemical kinetics, described a new method of determining the order of a reaction, and applied thermodynamics to chemical equilibriums. In 1886 he published the results of his study of dilute solutions and showed the analogy existing between them and gases, because they both obey equations of the type $pv = RT$. During the next nine years he developed this work in connection with the theory of electrolytic dissociation enunciated by Svante August Arrhenius (1859–1927; Swedish chemist). With Wilhelm Ostwald (1853–1932; German chemist who won the 1909 Nobel Prize in chemistry), he started the important *Zeitshrift für physikalische Chemie* in 1887, the first volume of which contained the famous paper by Arrhenius on electrolytic dissociation, along with the fundamental paper by van't Hoff (Preece, 1971a).

XI. APPENDIX: BIOGRAPHY OF RUDOLF CLAUSIUS

Rudolf Julius Emanuel Clausius (1822–1888) was a German physicist who made important contributions to molecular physics (Preece, 1971b). He was born in Köslin in Pomerania. In 1848 he got his degree at Halle, and in 1850 he was appointed professor of physics in the royal artillery and engineering school at Berlin and *Privatdocent* in the university. In 1855 he became an ordinary professor at the Zürich Polytechnic and professor at the University of Zürich. Clausius moved to Würzburg in Germany in 1867 as professor of physics, and two years later he was appointed to the same chair at Bonn, a position that he held until his death.

The work of Clausius, who was a mathematical rather than an experimental physicist, was concerned with many of the most abstruse problems of molecular physics. He made thermodynamics a science; he enunciated the second law, in a paper contributed to the Berlin Academy in 1850, in the well-known form, "Heat cannot of itself pass from a colder to a hotter body." He applied his results to an exhaustive development of the theory of the steam engine.

The kinetic theory of gases owes much to his researches. He raised it to the level of a theory, and he carried out many numerical determinations in connection with it, such as determining the mean free path of a molecule. Clausius also made an important advance in the theory of electrolysis, suggesting that molecules in electrolytes are continually interchanging atoms. This view found little favor until 1887, when it was taken up by S.A. Arrhenius, who made it the basis of the theory of electrolytic dissociation.

REFERENCES

Baker, D.A. (1984). Water relations. In *Advanced Plant Physiology* (Wilkins, M.B., Ed.), pp. 297–318. Pitman: London.

Barrs, H.D. (1968). Determination of water deficits in plant tissues. In *Water Deficits and Plant Growth. Vol. 1. Development, Control, and Measurement* (Kozlowski, T.T., Ed.), pp. 235–368. Academic Press: New York.

Barrs, H.D., and Weatherley, P.E. (1962). A re-examination of the relative turgidity technique for estimating water deficits in leaves. *Australian J Biol Sci* 15; 413–428.

Box, J.E., and Lemon, E.R. (1958). Preliminary field investigations of electrical resistance-moisture stress relations in cotton and grain sorghum plants. *Soil Sci Soc Amer Proc* 22; 193–196.

Boyer, J.S. (1995). *Measuring the Water Status of Plants and Soils.* Academic Press: San Diego.

Brown, P.W., and Tanner, C.B. (1983). Alfalfa osmotic potential: A comparison of the water-release curve and frozen-tissue methods. *Agronomy J* 75; 91–93.

Clarke, J.M., and Simpson, G.M. (1978). Leaf osmotic potential as an indicator of crop water deficit and irrigation need in rapeseed (*Brassica napus* L.). *Agr Water Manage* 1; 351–356.

Daniels, F., and Alberty, R.A. (1966). *Physical Chemistry.* 3rd ed. John Wiley and Sons: New York.

Diaz-Perez, J.C., Shackel, K.A., and Sutter, E.G. (1995). Relative water content and water potential of tissue-cultured apple shoots under water deficits. *J Exp Bot* 46; 111–118.

Ehrler, W.L., and Nakayama, F.S. (1984). Water stress status in guayule as measured by relative leaf water content. *Crop Sci* 24; 61–66.

Esau, K. (1977). *Anatomy of Seed Plants.* 2nd ed. Wiley: New York.

Frommhold, I. (1996). The 150th birthday of Wilhelm Pfeffer. *JSPP Newsletter*, August issue, pp. 10–15. Japanese Society of Plant Physiologists: Kyoto, Japan.

Gardner, W.R., and Rawlins, S.L. (1965). Sap pressure in plants. *Science* 149; 920.

Hammel, H.T., and Scholander, P.F. (1976). *Osmosis and Tensile Solvent.* Springer-Verlag: Berlin.

Heathcote, D.G., Etherington, J.R., and Woodward, F.I. (1979). An instrument for non-destructive measurement of the pressure potential (turgor) of leaf cells. *J Exp Bot* 30; 811–816.

Höfler, K. (1920). Ein Schema für die osmotische Leistung der Pflanzenzelle [A scheme for the osmotic capacity of plant cells]. *Berichte der Deutschen Botanischen Gesellschaft* [Report of the German Botanical Society] 38; 288–298.

Ike, I.F., Thurtell, G.W., and Stevenson, K.R. (1978). Evaluation of the pressure chamber technique for measurement of leaf water potential in cassava (*Manihot* species). *Can J Bot* 56; 1638–1641.

Kirkham, M.B. (1983). Physical model of water in a split-root system. *Plant Soil* 75; 153–168.

Kirkham, M.B., and Clothier, B.E. (2000). Infiltration into a New Zealand native forest soil. In *A Spectrum of Achievements in Agronomy* (Rosenzweig, C., Ed.). ASA Special Pub. No. 62. American Society of Agronomy, Crop Science Society of America, and Soil Science Society of America: Madison, Wisconsin.

Kramer, P.J. (1983). *Water Relations of Plants.* Academic Press: New York.

Kramer, P.J. (1988). Changing concepts regarding plant water status. *Plant Cell Environ* 11; 565–568.

Lang, A.R.G. (1967). Osmotic coefficients and water potentials of sodium chloride solutions from 0 to 40°C. *Australian J Chem* 20; 2017–2023.

McIlrath, W.J. (1971). Pfeffer, Wilhelm (Friedrich Philipp). *Encyclopaedia Britannica* 17; 791.

Melkonian, J.J., Wolfe, J., and Steponkus, P.L. (1982). Determination of the volumetric modulus of elasticity of wheat leaves by pressure-volume relations and the effect of drought conditioning. *Crop Sci* 22; 116–123.

Meyer, B.S., Anderson, D.B., and Böhning, R.H. (1960). *Introduction to Plant Physiology.* D. Van Nostrand: Princeton, New Jersey.

Meyer, W.S., and Ritchie, J.T. (1980). Resistance to water flow in the sorghum plant. *Plant Physiol* 65; 33–39.

Moore, W.J. (1962). *Physical Chemistry.* 3rd ed. Prentice-Hall: Englewood Cliffs, New Jersey.

Nobel, P.S. (1974). *Introduction to Biophysical Plant Physiology.* W.H. Freeman: San Franciso, California.

Nobel, P.S. (1983). *Biophysical Plant Physiology and Ecology.* W.H. Freeman: San Francisco, California.

Nonami, H., and Boyer, J.S. (1984). Regulation of growth of soybean at low water potentials. *Plant Physiol* 75 (supplement); 174.

Passioura, J.B. (1988). Response to Dr P.J. Kramer's article, "Changing concepts regarding plant water relations," Volume 11, Number 7, pp. 565–568. *Plant Cell Environ* 11; 569–571.

Preece, W.E. (General Ed.) (1971a). Van't Hoff, Jacobus Hendricus. Encyclopaedia Britannica 22; 887.

Preece, W.E. (General Ed.) (1971b). Clausius, Rudolf Julius Emanuel. Encyclopaedia Britannica 5; 888.

Rakhi, M.O., Zavadskaya, I.G., and Bobrovskaya, N.I.. (1978). Errors in determining components of the leaf water potential with the aid of a pressure chamber. *Soviet Plant Physiol* 25 (No. 4, Part 2); 689–696.

Richter, H. (1978). Water rleations of single drying leaves: Evaluation with a dewpoint hygrometer. *J Exp Bot* 29; 277–280.

Scholander, P.F., Hammel, H.T., Hemmingsen, E.A., and Bradstreet, E.D. (1964). Hydrostatic pressure and osmotic potential in leaves of mangroves and some other plants. *Proc Nat Acad Sci* 52; 119–125.

Scholander, P.F., Hammel, H.T., Bradstreet, E.D., and Hemmingsen, E.A. (1965). Sap pressure in vascular plants. *Science* 148; 339–346.

Schulze, E.-D., Steudle, E., Gollan, T., and Schurr, U. (1988). Response to Dr P.J. Kramer's article, "Changing concepts regarding plant water relations," Volume 11, Number 7, pp. 565–568. *Plant Cell Environ* 11; 573–576.

Sinclair, R., and Venables, W.N. (1983). An alternative method for analysing pressure-volume curves produced with the pressure chamber. *Plant Cell Environ* 6; 211–217.

Singh, P., Kanemasu, E.T., and Singh, P. (1983). Yield and water relations of pearl millet genotypes under irrigated and nonirrigated conditions. *Agronomy J* 75; 886–890.

Slavík, B. (1963). The distribution of transpiration rate, water saturation deficit, stomata number and size, photosynthetic and respiration rate in the area of the tobacco leaf blade. *Biol Plant* 5; 143–153.

Steudle, E., and W.D. Jeschke, W.D. (1983). Water transport in barley roots. Measurements of root pressure and hydraulic conductivity of roots in parallel with turgor and hydraulic conductivity of root cells. *Planta* 158; 237–248.

Stocker, O. (1928). Das Wasserhaushalt ägyptischer Wüsten-und Salzpflanzen. *Botanische Abhandlungen* 13.

Stocker, O. (1929). Das Wasserdefizit von Gefässpflanzen in verschiedenen Klimazonen. *Planta* 7; 382–387.

Taiz, L., and Zeiger, E. (1991). *Plant Physiology.* The Benjamin/Cummings: Redwood City, California.

Tomos, A. D., and Leigh, R.A. (1999). The pressure probe: A versatile tool in plant cell physiology. *Annu Rev Plant Physiol Plant Mol Biol* 50; 447–472.

Turner, N.C. (1981). Techniques and experimental approaches for the measurement of plant water status. *Plant Soil* 58; 339–366.

Turner, N.C. and Sobrado, M.A. (1983). Evaluation of a non-destructive method for measuring turgor pressure in *Helianthus. J Exp Bot* 34; 1562–1568.

Tyerman, S.D., and Steudle, E. (1984). Determination of solute permeability in *Chara* internodes by a turgor minimum method. *Plant Physiol* 74; 464–468.

Tyree, M.T., and Hammel, H.T. (1972). The measurement of the turgor pressure and the water relations of plants by the pressure-bomb technique. *J Exp Bot* 23; 267–282.

Walker, S., Oosterhuis, D.M., and Savage, M.J. (1983). Field use of screen-caged thermocouple psychrometers in sample chambers. *Crop Sci* 23; 627–632.

Weast, R.C. (Ed.) (1964). *Handbook of Chemistry and Physics*. 45th ed. Chemical Rubber Co: Cleveland, Ohio.

Weatherley, P.E. (1950). Studies in the water relations of the cotton plant. I. The field measurement of water deficits in leaves. *New Phytol* 49; 81–97.

Wescor, Inc. (1989). *Instruction/Service Manual M2448-4 for the 5500 Vapor Pressure Osmometer*. Wescor, Inc.: Logan, Utah. Pages not numbered sequentially. (See Section 8, p. 8-1 to 8-4, for the theory of operation.)

Wilson, J.R., Fisher, M.J., Schulze, E.-D., Dolby, G.R., and Ludlow, M.M. (1979). Comparison between pressure-volume and dewpoint-hygrometer techniques for determining the water relations characteristics of grass and legume leaves. *Oecologia* 41; 77–88.

19

The Ascent of Water in Plants

The problem of the rise of water in tall plants is as old as the science of plant physiology. In this chapter we consider the cohesion theory, which is the best formulation to explain how water can get to the top of tall trees and vines.

I. THE PROBLEM

Let us consider why it is hard for water to get to the top of trees. A suction pump can lift water only to the barometric height, which is the height that is supported by atmospheric pressure (1.0 atm) or 1033 cm (10.33 m; 33.89 feet) (Salisbury and Ross, 1978, p. 49). If a hose or pipe is sealed at one end and filled with water, and then placed in an upright position with the open end down and in water, atmospheric pressure will support the water column to 10.33 meters, theoretically. At this height the pressure equals the vapor pressure of water at its temperature. Above this height of 1033 cm, water turns to vapor. When the pressure is reduced in a column of water so that vapor forms or air bubbles appear (the air coming out of solution), the column is said to cavitate (Salisbury and Ross, 1978, p. 49). My father, Don Kirkham, and his students tried to see how far they could climb the outside back stairs of the Agronomy Building at Iowa State University with a hose, closed end in hand and with the hose's bottom in a water bucket on the ground. The column of water in the hose collapsed before they climbed 10.33 meters. This was probably because of impurities on the hose wall.

II. HOW WATER GETS TO THE TOP OF TALL BUILDINGS AND ANIMALS

How does water get to the top of tall trees? The tallest tree in the world is 111.6 m, and in 1895 a tree 127 m tall was felled in British Columbia, Canada (Salisbury and Ross, 1978, p. 49). Let us first consider how water gets to the top of skyscrapers. Modern buildings in cities use electrical pumping systems to get water to high floors. But before electrical pumps were available, wooden tanks that hold water were used, and still are used, to raise water. People who live in a tall building and are not getting a good strong shower are probably too close to the holding tank (Weber, 1989). In those buildings in which the plumbing requires the help of gravity to create sufficient water pressure, a tank needs to be elevated at least 25 feet (762 cm) above a building's highest standpipe. One gets 1 lb/in² (0.06896 bars) of pressure for every 2.3 (70 cm) feet in height. In New York City, the skyline is dotted with more than 10,000 of these tanks, and they vary in size from 5000 to 50,000 gallons (19,000 to 190,000 L) and run from 12 to 20 feet (3.658 to 6.096 m) high. They have been in use since 1890. Tanks last for 60 years (Weber, 1989). To make a wooden tank, lumber is cut from yellow cedar from British Columbia or from California redwood. The people who replace the wooden tanks are highly trained and have a difficult and dangerous job getting the planks to great heights (National Public Radio, 2002).

Physicists who question how water can get to the top of trees point out that animals have pumps (hearts) that plants do not have. So let us consider how fluids get to the top of a giraffe, probably the tallest animal. An upright giraffe ought to suffer massive edema in its feet; moreover, when it lowers its head to drink, the blood should rush down into it and be unable to flow up again (Pedley, 1987). But pressure measurements in the giraffe reveal why neither of these things happens.

A counter-gravitational gradient of venous pressure (P_v) exists in the giraffe's neck. Measurements of the gravitational (or hydrostatic) gradient of pressure with height, in an upright animal 3.5 m tall, show that blood pressure in an artery in the head is as much as 110 mm Hg (about 1.5 m H_2O or about 15 kPa) lower than the level of the heart, which is about 200 mm Hg above atmospheric pressure, double the human value. This high arterial pressure near the giraffe's heart provides normal blood pressure and perfusion to the brain (Hargens et al., 1987).

Two features of the peripheral circulation that inhibit edema in a standing giraffe are: 1) a high resistance to flow in the thick-walled arterioles, which keeps venous pressure, and hence capillary pressure, well below arterial pressure [an arteriole is a small branch of an artery leading into capillaries (Hickman, 1961, p. 511)]; and 2) very tight skin in the lower legs (an "anti-gravity" suit"), which allows tissue pressure to be much higher than in man (in man, tissue pressure is about 0).

Even so, there is a net filtration pressure of more than 80 mm Hg, and quietly standing giraffes will be susceptible to some edema. In the ambulant giraffe, however, the "muscle pump" comes into play, as in man, squeezing blood up out of the lower veins as the skeletal muscles contract, and sucking it in again through the capillaries as they relax, backflow in the veins being prevented by valves. These pressures move fluid upward against gravity. The giraffe's jaw muscles (chewing actions) do the same to pump blood up the neck. [A vein is about twice the cross section of its corresponding artery. Veins, especially in the lower parts of the body, are provided with valves to prevent the backflow of blood (Hickman, 1961, p. 629).] Dinosaurs were even taller than giraffes, and they may have had several hearts to raise fluids to their heads (Dr. Octave Levenspiel, Sigma Xi Lecture, "A Chemical Engineer Visits Dinosaurland," Kansas State University, April 8, 2002).

III. COHESION THEORY

In plants, no standing tanks, pumps (hearts), or valves have been observed. If one looks through books on plant anatomy, one sees no such structures (Esau, 1965; 1977). So, again we ask, how does water get to the top of tall trees?

At present, the *cohesion theory*, or sap-tension theory, is the theory generally accepted as the one that explains most satisfactorily the way that water ascends in plants. [The dictionary defines *sap* as "the juices of a plant, especially the watery solution which circulates through the vascular tissue" (*Webster's Collegiate Dictionary*, 1939).] Here we will use interchangeably the terms "sap" and "water in the tracheary cells of the xylem tissue." We recognize that the fluid in the tracheary cells is not pure water, but a dilute aqueous solution (Nobel, 1974, p. 393). Even in mangroves, which grow in salt water, the sap in the xylem tissue is very nearly salt free and changes the melting point of water <0.1°C (Hammel and Scholander, 1976, p. 32). A 0.1°C depression of the freezing point of water would be brought about by a 0.027 molal solution of NaCl at 25°C, which is about −1 bar (Lang, 1967). The <0.1°C depression in the freezing point, as observed by Hammel and Scholander (1976), would mean that the sap has an osmotic potential of >−1.0 bar. Scholander et al. (1965) show that the osmotic potential (ψ_π) in the sap in mangrove is −0.3 atm (Fig. 19.1).

THE ASCENT OF SAP

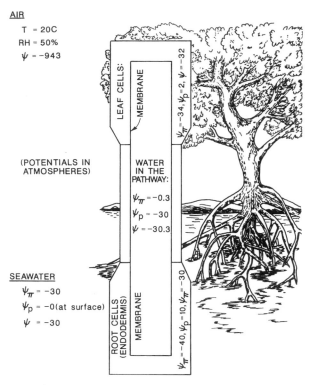

FIG. 19.1 Water relations of a mangrove growing with its roots immersed in sea water. The diagram indicates the essential parts of the mangrove tree from the standpoint of water relations, particularly the membranes of the endodermis and of the leaf cells. It is important to note that if leaf membranes should suddenly cease to be differentially permeable, salt and subsequently water would move from leaf cells into the pathway due to the high tension of water there, and thus the leaf cells would collapse. Data from Scholander et al. (1965). (From Salisbury, F.B., and Ross, C.W., *Plant Physiology*, 2nd ed., p. 61, ©1978. Wadsworth Publishing Company, Inc: Belmont, California. Reprinted with permission of Brooks/Cole, a division of Thomson Learning: www.thomsonrights.com. Fax 800 730-2215.)

Let us review xylem tissue (see Chapter 14, Section I, Part G). The xylem tissue is made up of four types of cells (Table 14.1): vessel members (also called xylem elements), the conducting cells that occur only in angiosperms (the flowering plants), the most highly evolved plants; tracheids, the conducting cells that occur in angiosperms and gymnosperms (e.g., the conifers); fibers, which give structural support; and parenchyma, which store carbohydrates and assist in lateral movement of water and solutes into and out of the conducting cells. At maturity, vessel members, tracheids, and fibers are dead. Only the parenchyma cells are living.

Also, at maturity, the end walls of vessel members disintegrate, and, consequently, a long tube, called a xylem vessel, is formed (Esau, 1977, pp. 101–124).

The cohesion theory of the ascent of sap was foreshadowed by Stephen Hales (1677–1761; English clergyman, physiologist, chemist, and inventor, famous for his pioneering studies in animal and plant physiology), Julius von Sachs (1832–1897; German botanist and outstanding plant physiologist), and Eduard Strasburger (1844–1912; German botanist and one of first to realize the importance of the nucleus and chromosomes in heredity). They all concluded that transpiration produces the pull causing the ascent of sap (Kramer, 1983, p. 282). The first successful attempt to measure cohesion in water experimentally seems to have been made by Berthelot in 1850 (Greenidge, 1957). He obtained values to 50 atm by a method similar to that used later by Dixon and Joly (1895). Boehm (1893) demonstrated that transpiring branches could raise mercury above barometric pressure, but the demonstration by Askenasy (1895) and Dixon and Joly (1895), showing that water has considerable tensile strength, was necessary to make the cohesion theory acceptable (Kramer, 1983, p. 282).

The cohesion theory, as first set forth by Dixon and Joly (1895) and Dixon (1895, 1897, 1914), assumes that diffusion of water from the noncollapsible xylem elements in contact with the leaf cells creates a state of tension within the water columns in the xylem vessels. This tension is possible because of the cohesion of water molecules and their adhesion to the hydrophilic walls of the xylem elements. Tension in the water columns is assumed to lift water from the roots to the leaves, in addition to reducing the potential energy of the water in the root xylem tissue until water diffuses from the soil into the root during absorption of the water. The cohesion theory assumes continuity of water columns, laterally and vertically, in the conducting elements of the xylem tissue. These water columns ultimately are placed under tensile strain. But widespread rupture is believed not to occur in the water columns under tensile strain owing to the purported cohesive properties of water when entrapped in small capillaries. Figure 19.2 from Salisbury and Ross (1978, p. 58) outlines the cohesion theory. (For a biography of Dixon, see the Appendix, Section IX, and for that of Joly, see the Appendix, Section X.)

IV. LIMITATIONS OF THE COHESION THEORY

Even though most plant physiologists feel that the cohesion theory is probably the correct explanation for the rise of water in plants, the theory has

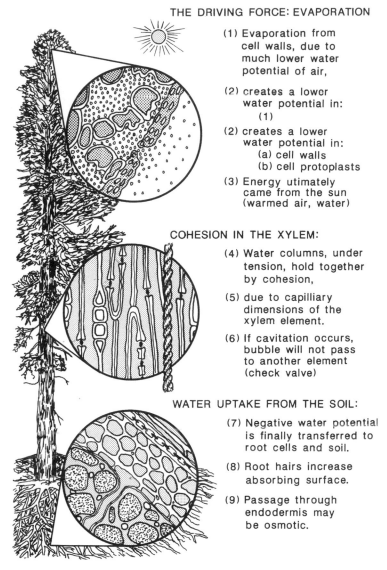

THE DRIVING FORCE: EVAPORATION

(1) Evaporation from cell walls, due to much lower water potential of air,

(2) creates a lower water potential in:
   (1)

(2) creates a lower water potential in:
   (a) cell walls
   (b) cell protoplasts

(3) Energy utimately came from the sun (warmed air, water)

COHESION IN THE XYLEM:

(4) Water columns, under tension, hold together by cohesion,

(5) due to capilliary dimensions of the xylem element.

(6) If cavitation occurs, bubble will not pass to another element (check valve)

WATER UPTAKE FROM THE SOIL:

(7) Negative water potential is finally transferred to root cells and soil.

(8) Root hairs increase absorbing surface.

(9) Passage through endodermis may be osmotic.

FIG. 19.2 The cohesion theory of the ascent of sap summarized. (From Salisbury, F.B., and Ross, C.W., *Plant Physiology*, 2nd ed., p. 58, ©1978. Wadsworth Publishing Company, Inc: Belmont, California. Reprinted with permission of Brooks/Cole, a division of Thomson Learning: www.thomsonrights.com. Fax 800 730-2215.)

limitations. The main difficulty is that it postulates a system of potentially great instability and vulnerability, although it is clear that the water-conducting system in plants must be both stable and invulnerable. Objections to the theory include three major points (Kramer, 1983, p. 283; Salisbury and Ross, 1978, p. 58–60):

1. The tensile strength of water is inadequate under the great tensions necessary to pull water to the top of plants, especially tall plants.
2. There is insufficient evidence for the existence of continuous water columns (that is, water columns under tension are not stable and they cavitate, or form cavities, hollows, or bubbles).
3. It seems impossible to have tensive channels in the presence of free air bubbles, which can occur when trees in cold climates freeze and then thaw.

Let us consider each point. First, is the tensile strength of water adequate to pull water to the top of plants? Tensile strength is defined as the "resistance to lengthwise stress, measured by the greatest load in weight per unit area pulling in the direction of length that a given substance can bear without tearing apart" (*Webster's New World Dictionary of the American Language,* 1959).

Nobel (1970, pp. 35–36, 40; 1974, pp. 46–47, 52–53) calculates the tensile strength of water. Let us do the calculations that Nobel does. We must consider the structure of ice (Nobel, 1974, p. 46). Ice is a coordinated crystalline structure in which essentially all the water molecules are joined by hydrogen bonds (see Fig. 3.2). When heat is added so that the ice melts, some of these intermolecular hydrogen bonds are broken. The heat of fusion of ice at 0°C is 80 cal/gm or 1.44 kcal/mole. (Remember: 18 gm/mole for water; 80 cal/gm × 18 gm/mole = 1,400 cal/mole = 1.44 kcal/mole.) The total rupture of the intermolecular hydrogen bonds involving each of its hydrogens would require 9.6 kcal/mole of water. The 9.6 kcal/mole is a given value. Nobel (1974, p. 46) gives references for the value. He cites work by Eisenberg and Kauzmann (1969; e.g., see p. 145, 269) and Pauling (1964; e.g., see p. 456) for references on the hydrogen bond energy. Nobel (1974, p. 46) points out that the actual magnitude of the hydrogen bond energy assigned to ice depends somewhat on the particular operational definition used in the measurement of the various bonding energies. Therefore, the quoted values vary somewhat.

The heat of fusion thus indicates that (100) (1.44)/(9.6), or at most 15%, of the hydrogen bonds are broken when ice melts. Some energy is needed to overcome van der Waal's attractions, so that less than 15% of the hydrogen bonds are actually broken upon melting (Nobel, 1974, p. 46).

Conversely, over 85% of the hydrogen bonds remain intact for liquid water at 0°C. Because 1.00 cal is needed to heat 1 gram of water 1°C, (1.00 cal/gm°)(25°)(18 gm/mole)(0.001 kcal/cal), or 0.45 kcal/mole is required to heat water from 0 to 25°C. If all of this energy were used to break hydrogen bonds, over 80% of the bonds would still remain intact at 25°C. (Note: 0.45/9.6 = 0.047, which is less than 5%; so 85% − less than 5% = greater than 80%.) The extensive amount of intermolecular hydrogen bonds present in the liquid state contributes to the unique and biologically important properties of water, including its high tensile strength, which is of interest to us now.

If 80% of the hydrogen bonds are intact in water at 25°C (Nobel, 1974, p. 52), then the energy will be (0.80)(9.6) or 7.7 kcal/mole, which is (7.7 kcal/mole)/(18 gm/mole) or 0.43 kcal/gm of water. For a density of 1.00 gm/cm^3, and replacing kcal by 4.184×10^{10} ergs, we calculate that the energy of the hydrogen bonds is 1.8×10^{10} ergs/cm^3. (Remember: 1 joule = 10^7 ergs; 1 cal = 4.184 joule; therefore, 1 cal = 4.184×10^7 ergs; 1 kcal = 4.184×10^{10} ergs.) The tension that is applied to a water column acts against this attractive energy of the hydrogen bonds.

When the fracture is just about to occur at each hydrogen bond, the maximum possible tensile strength is developed. Thus, the maximum tensile strength would represent an input of 1.8×10^{10} ergs/cm^3. Since an erg = dyne-cm and a bar = 10^6 dyne/cm^2, the maximum tensile strength of water corresponds to 18,000 bars.

Nobel's (1974) theoretical considerations, therefore, show that the calculated value for the tensile strength of water is large (18,000 bars) and would permit rise of water in plants even under great tensions. Tensions in higher (more evolved) plants probably never exceed 100 atm. Lower plants such as fungi apparently can grow in soil with a tension (or absolute value of matric potential) of |400| bars (Harris, 1981, p. 26). What values of the tension of water have been measured experimentally? Dixon and Joly (1895) estimated that water entrapped in glass tubes of small diameter could withstand tensions exceeding 200 atm without fracture. Ursprung (1929) calculated that tensions on the order of 300 atm were reached in annulus cells of discharging fern sporangia. Briggs (1950) employed a centrifugal method to obtain values of about 220 atm for the tensile strength of water. (See Fig. 19.3 for a method of measuring the cohesive properties of water using a bent centrifuge tube.)

In contrast to the foregoing rather large values, a number of other investigators have demonstrated that water may have a relatively low tensile strength. Loomis et al. (1960) suggested that Ursprung's values of

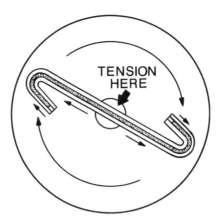

FIG. 19.3 Method of measuring the cohesive properties of water using a centrifuged Z-tube. Small arrows indicate direction of centrifugal force and principle of balancing due to the Z-tube. These tubes are centrifuged causing tension on the water at the center of the tube. The tension present when the water column breaks can be calculated. (From Salisbury, F.B., and Ross, C.W., *Plant Physiology,* 2nd ed., p. 59, ©1978. Wadsworth Publishing Company, Inc: Belmont, California. Reprinted with permission of Brooks/Cole, a division of Thomson Learning: www.thomsonrights.com. Fax 800 730-2215.)

the tensile strength of water were open to question because of a confusion of adsorption forces with cohesion. Scholander et al. (1955), through centrifugation in glass tubes, observed tensive values from 10 to 20 atm without producing cavitation of water. When the experiments were repeated using plant material, they observed much lower values (1–3 atm). Also, they were unable to fit hydrostatic pressures in transpiring grape vines into a pattern that followed the cohesion theory. Measured pressure did not indicate cohesion tension at any time, and hydrostatic pressures in transpiring tall vines were higher at the top rather than lower, as they should have been if the transpiring stream were under tension. Measurements taken on Douglas fir trees, however, did follow the pattern that one would expect if water were rising in the plants according to the cohesion theory (Scholander et al., 1965) (Fig. 19.4). That is, the hydrostatic pressure at the top of the trees was more negative than at the bottom of the trees. Greenidge (1957) discusses different techniques used to measure the tensile strength of water that yield values ranging from 0.05 to 10 atm.

It appears that, experimentally, water can withstand negative pressures (tensions) only up to about 300 bars without breaking (Nobel, 1974, p. 52). The observed tensile strength depends on the wall material, the diameter of the xylem vessel, and any solutes present in the water. Local

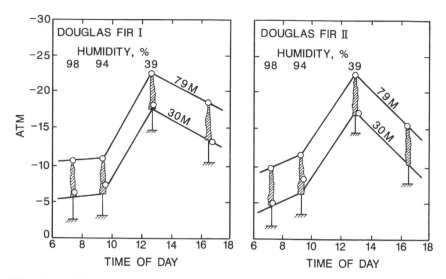

FIG. 19.4 Differences in hydrostatic pressure in upper and lower parts of crowns of Douglas fir trees at various times of day as measured on excised twigs in a pressure chamber. (Reprinted with permission from Scholander, P.F., Hammel, H.T., Bradstreet, E.D., and Hemmingsen, E.A., Sap pressure in vascular plants. *Science* 148; 339–346, ©1965, American Association for the Advancement of Science.)

imperfections in the semicrystalline structure of water, such as those caused by H^+ and OH^-, which are always present, even in pure water, reduce the observed tensile strength from the maximum value predicted based on hydrogen bond strengths. Nevertheless, the measured tensile strength for water provided by the intermolecular hydrogen bonds (up to 300 bars) is nearly 10% of that for copper or aluminum, and is sufficiently high to meet the demands encountered for water movement in plants (Nobel, 1974, p. 53). The tensile strength of copper is 4140 to 4830 bars; for aluminum, it is 2070 to 2760 bars, (Weast, 1964, p. F-15).

Let us now consider the second problem with the cohesion theory. Are water columns in the xylem tissue stable under tension? Much has been written about the instability of water columns under tension and the ease with which they break by cavitation in glass capillary tubing (Kramer, 1969, p. 275). It has been suggested that if they break as easily in the xylem of trees, they would soon become inoperative because of shocks such as those caused by swaying in the wind.

Considerations of nucleation prompted Silver (1942) to infer that the tensile strength of water is negligible. There is evidence of widespread fracture of stretched water columns and a high percentage of gas-filled, nonfunctional elements under field conditions (Preston, 1938; Greenidge, 1957; Scholander, 1958). However, it seems probable that the nature of the walls of the dead xylem tissue, which is filled with imbibed water, makes the water columns in the stems of plants more stable than those in glass tubes. If cavitation caused by air entry should occur in the conducting tubes of the xylem tissue, the matric potential component attributed to the hydrophilic nature of the surfaces involved could be expected to maintain surface films of water capable of transporting water up the stem (Gardner, 1965). The cell walls in the conducting cells of the xylem tissue are probably charged and exhibit double-layer characteristics. Thus, even if a column breaks, there is a thin layer of adsorbed water, with a concave curvature, which ensures that entrapped air eventually will be dissolved (W.R. Gardner, personal communication, February 29, 2000).

Greenidge (1955, 1957) did experiments in which a dye was injected into the xylem of trees after severing all vessels of the stem by two or more opposing saw cuts. Dye not only moved readily to the top of such trees, but it completely stained the wood immediately below and above the saw cuts, indicating capillary movement under low tension and showing no evidence of rupture of stressed water columns. In other experiments, the dye moved to the top of tree trunks from which all leafy branches had been removed and all vessels severed at one or more points.

It is true that the water columns in large xylem vessels often break and smaller vessels remain water filled. If vessels become filled with air (gas bubbles), the bubbles usually cannot spread beyond the vessel members or tracheids in which they developed (Kramer, 1983, p. 284). Thus, the entire conducting system is not suddenly blocked by expanding bubbles. In a study of moisture relations in tall lianas, Scholander et al. (1957) found that allowing vessels of a cut vine to become plugged with air caused a lowered hydrostatic pressure in the plant, but did not reduce the rate of water uptake, indicating that water movement was shifted to the numerous tracheids of the stem. Again they found no direct evidence of cohesion tension. The work by Scholander (1958) and Scholander et al. (1957) indicated that there is a large "safety factor" (Kramer, 1983, p. 284) in the xylem. Although partial blockage increases resistance to flow of water, the volume of flow is not necessarily reduced.

The ability to hear the water columns break is supporting evidence that the columns are under tension, and, when they cavitate, the sound can be picked up acoustically. Milburn and Johnson (1966) developed an acoustic detector, and subsequent experimenters have monitored cavitation using the technique (e.g., see Tyree et al., 1986; Jackson and Grace, 1996).

The method has been used to monitor water stress in crop plants and to tell when to irrigate (Senft, 1986).

Let us now consider the third problem. Microscopic observations have shown that air blockade occurs when some trees in cold climates are frozen (Johnson, 1977). Inability to restore the water columns in the spring may well be the factor that excludes certain trees and especially vines with large vessels from these regions (Salisbury and Ross, 1978, p. 60) (Fig. 19.5). But how do trees grow in such regions?

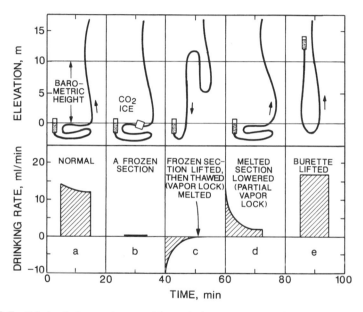

FIG. 19.5 Scholander's experiments with tropical rattan vines (*Calamus* sp.). (a) The vine is cut off under water and a burette is attached, allowing measurement of the rate of water uptake. If the burette is stoppered, water continues to be taken up until a vacuum is created in the burette, and the water boils. (b) To freeze the water in the vine, the burette first had to be taken off so that air entered all the xylem elements, vapor-locking the system. Then after freezing, the vapor-locked portion (about 2 m) was cut off under water and the burette attached again. There was still no water uptake, indicating that freezing had indeed blocked the system. (c) If the vapor-locked portion was hoisted above the barometric height and allowed to thaw, some water ran out, but there was no uptake, indicating that the system was now vapor-locked. (d) If the vapor-locked portion was lowered to the ground, there was a rapid initial uptake as vapor condensed to water, breaking the vapor lock, but then uptake was slower than originally because some air had been excluded from freezing. (e) When the burette was elevated 11 m, the rate of water uptake returned to the original level, indicating that the vapor lock had now been completely eliminated. Data from Scholander et al. (1961). (Reprinted with permission from Scholander, P.F., Hemmingsen, E., and Garey, W., Cohesive lift of sap in rattan vine. *Science* 134; 1835–1838, ©1961, American Association for the Advancement of Science.)

Imagine a northern tree thawing in the spring. As the ice melts, the tracheids become filled with liquid containing the many bubbles of air that had been forced out by freezing. As melting continues and transpiration begins, tension begins to develop in the xylem tissue. Because of the small dimensions of the tracheids involved, the pressure difference across the curved air-water interface bounding the bubbles would be considerable, resulting in much higher pressure in an air bubble than would exist in the water. Any bubbles that form should dissolve fairly readily, restoring the integrity of the water column (Gardner, 1965). Studies of wood in the spring indicate that about 10% of the tracheids are filled with vapor, but the remaining 90% appear ample to handle sap movement (Salisbury and Ross, 1978, p. 60; Kramer, 1983) (Fig. 19.6). Gymnosperms with their tracheids are especially well adapted to cold climates. Trees, and especially vines with large, long vessels, are practically absent from cold climates, but are abundant in the tropics.

Dividing cambial cells in the spring also produce new water-filled conducting cells in the xylem tissue. In some ring-porous trees (trees with large vessel members) virtually all the water moves in these newly formed tubes (Salisbury and Ross, 1978, p. 60).

V. ALTERNATIVE THEORY TO THE COHESION THEORY

For many decades, the cohesion theory was accepted and essentially no experiments related to it were performed between about 1960 and 1995. With the advent of the pressure probe, measurements made with it contradicted the cohesion theory. The measurements with the pressure probe showed (Canny, 1995a):

1. The necessary high tensions in the xylem are not present (i.e., the operating tension in the xylem is around 2 bars and not 20 or more bars).
2. The necessary gradient of tension with height is not present.
3. The measurements of tension with the pressure chamber, believed to verify the cohesion theory, conflict with those made with the xylem-pressure probe.

Canny (1995a), therefore, put forward a theory, called the *compensating-pressure theory* to account for the rise of water in plants. He noted that the xylem has ray cells throughout it (see Fig. 18.3). These are living parenchyma cells. He said that the compensating pressure is provided by the tissue pressure of xylem parenchyma and ray cells, pressing onto the closed fluid spaces of the tracheary elements and squeezing them.

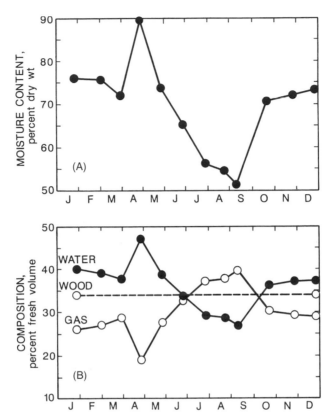

FIG. 19.6 (A) Seasonal changes in water content of yellow birch trunks calculated from disks cut from the base, middle, and top of the trunks. (B) Seasonal changes in gas and water content of yellow birch tree trunks calculated as percentage of total volume. (From Kramer, P.J., *Water Relations of Plants*, p. 285, ©1983, Academic Press: New York. Reprinted by permission of Academic Press.)

The driving force is provided, as in the cohesion theory, by evaporation and the tensions generated in curved menisci in the wet cell walls of the leaf. The force is transmitted, as in the cohesion theory, by tension in the water in the tracheary elements. But this tension is maintained by compression from tissue pressure around the tracheary elements. The gravitational gradient of tension up a tall tree is then compensated by increasing tissue pressure of the xylem parenchyma with height, and the need for a tension gradient to sustain the standing columns disappears. He discusses his theory in other papers (Canny, 1995b; 1997; 1998; 2001).

Canny's theory was challenged by Comstock (1999), who pointed out the following:

1. Canny's attempt to alter xylem pressure by the application of tissue pressure violates basic tenets about plant-water relations. He introduces new concepts of "wall pressure" and "tissue pressure," which combine to make "turgor," but these are not components of the total water potential, which is made up of the osmotic potential and turgor potential (ignoring gravitational and matric potential energies). So Canny's theory does not fit into the classic concepts of plant-water relations, as has been taught for several decades.
2. Tissue pressure is likely to be ubiquitous but small. Extreme reinforcement would be needed to sustain the tissue pressures postulated in his model, not just one or two cell layers with thickened walls.
3. Canny postulates a pump-and-valve system, which is essential to the working of his model, but no viable mechanism has been identified. Comstock (1999) states that the xylem in Canny's model is a contained volume, which can be pressurized by surrounding tissues, and flow characteristics are set by an active water pump in the roots and a one-way, regulating valve in the leaves. This assemblage of valves and pumps ensures that the flow rate through the xylem is independent of the action of tissue pressure on the middle of the pathway. It also ensures that the tissue pressure cannot squeeze water out of the pipes, but merely pressurizes them.

For Canny, the pump is the endodermis (at the inner boundary of the root cortex) and hypodermis (at the outer boundary of the root cortex), so he is not envisioning pumps and valves such as those that occur in animals (hearts and vein valves). The closest structure in a plant that looks like a valve is the pit membrane, which is part of the intercellular layer and primary cell wall that limits a pit cavity in a cell wall. But pit membranes do not act in the same ways as valves do in veins. The pit membrane, as shown in Fig. 19.7 (Esau, 1965, p. 40), is a middle lamella (not a living membrane), which is a layer of intercellular material, chiefly pectic substances, cementing together the primary walls of contiguous cells (Esau, 1977, p. 516). The torus in the center of the pit can move to one side to plug the pit, behaving like an inanimate valve.

Canny's theory apparently considers that solutes in the parenchyma cells of the tissue around the tracheids and vessel members cause an imbibing of water and create a pressure on the tracheary elements. This pressure on the tracheary elements keeps the water in them from cavitating. Canny (1995, p. 351) notes the close association of xylem and phloem. He states that the phloem, as the most powerful generator of tissue pressure,

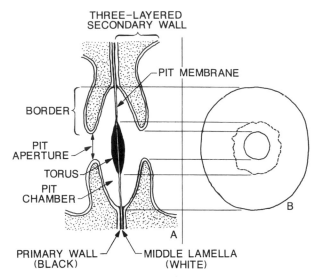

FIG. 19.7   Bordered pit-pair of *Pinus* in sectional (**A**) and face (**B**) views. The pit membrane consists of two primary walls and the intercellular lamella, but is thinner than the same triple structure in the unpitted part of the wall. The torus is formed by thickening of the primary wall. In **B**, outline of the torus is uneven. (From Esau, K., *Plant Anatomy*, 2nd ed., p. 40, ©1965, John Wiley & Sons, Inc: New York. This material is used by permission of John Wiley & Sons, Inc.)

probably lies next to the xylem to provide protection against cavitation. In stems, xylem and phloem are together in vascular bundles (see Chapter 18, Part I, Sections A, B, and C). However, in roots, xylem and phloem are in separate bundles (see Chapter 14, Part I, Sections C and D). Comstock (1999) pointed out that, if the water potential gradient were such that the parenchyma cells have a lower water potential than the tracheary cells, water would move from the tracheary cells to the parenchyma.

Zimmermann et al. (1994, 1995) said that the cohesion theory should be reappraised, because direct measurements of the xylem pressure in single vessels of tall trees, using the xylem pressure probe, indicate that the xylem tension in leaves is often much smaller than that predicted for transpiration-driven water ascent through continuous water columns. Canny's theory (1995a) has support from the work by Kargol et al. (1995), who said that water is transported along the xylem vessels by "graviosmotic mechanisms." They also showed the importance of a root pump in getting water up a plant.

Others defend the cohesion theory (Holbrook et al., 1995; Pockman et al., 1995; Steudle, 1995; Tyree, 1997; Stiller and Sperry, 1999; Wei et al.,

1999). Most of the negation of the cohesion theory comes from measurements made with the xylem pressure probe, and the probe may be measuring inaccurate values. Apparently, the probe is incapable of measuring pressures more negative than about −0.6 MPa, either because of an imperfect seal between the probe and the xylem wall or the creation of micro-fissures in the xylem cell wall when the probe is inserted (Tomos and Leigh, 1999). In both cases, cavitation via "air-seeding" is proposed to occur at pressures less negative than those normally sustained by the xylem. Wei et al. (1999) reported that direct measurements of xylem pressure support the cohesion-tension theory. They used a cell pressure probe filled with silicone oil instead of with water. If the pressure probe can be used to monitor pressure (or tension) in the tracheary elements, then perhaps it can be inserted into a ray cell to see what pressure exists there, which is pushing against the tracheary elements.

VI. NEW TECHNIQUES TO CONFIRM THE COHESION THEORY

Experiments to study tensive values of water in plants have been done with plants that have been punched with manometers (Scholander et al., 1955), cut (Scholander et al., 1957), sawed (Greenidge, 1955), punctured with a pressure probe (Tomos and Leigh, 1999), frozen (Cochard et al., 2000), or otherwise disturbed. If it were possible to study plants under natural conditions, when they were intact, one might come to a better understanding as to what tension water is under in plants, and if tensions are built up, if they are sufficient to account for the rise of water.

New equipment is being developed that can be used to measure non-destructively the characteristics of water transport in the soil-plant-atmosphere continuum, such as nuclear magnetic resonance imaging (Scheenen et al., 2000) or sap flow gauges (Green and Clothier, 1988). The difficulty is in getting the equipment to the top of giant trees. Tall platforms have been constructed to access the top of forest canopies (12 m high) (Ellsworth, 1999). (See the cover of *Plant, Cell Environment*, May, 1999, issue for a photo of a platform.) For example, sap flow gauges could be put at the top of a tall tree to see if the flux was upwards, as predicted by the cohesion theory, or downwards. Reverse flow occurs in plants if the water potential gradients allow it (Kirkham, 1983; Emerman, 1996; Song et al., 2000; Huang, 1999). Maybe Scholander's measurements were correct (1955, 1957) and the tension does not increase with height. (It has been said that Scholander got his measurements of tension in the top of tall vines and trees by shooting branches down with a gun and then putting them in his pressure chamber.) It may be that tall trees and vines absorb water from the air during rainfall and that water does not need to rise from the roots. Measurements of the direction of sap flow in the tops of intact tall plants are needed.

VII. CONTROVERY ABOUT THE COHESION THEORY

In spite of difficulties in demonstrating, in some experiments, appreciable values of tension in water columns of plants, most plant physiologists continue to assume that high tension values are readily obtainable and that the cohesion theory is correct (Kramer, 1983; Baker, 1984). Feelings get heated when scientists are either defending or refuting the cohesion theory, and this has been the case for decades. When a physicist published a book questioning the validity of the cohesion theory (Bose, 1923), plant physiologists who reviewed the book used strong language to show that he was wrong. For example, MacDougal and Overton (1927) said, "Every page of Bose's book on the ascent of sap . . . is utterly lacking in scientific significance. Such books appearing on the lists of scientific publications constitute a menace and danger to sound science." Other reviewers were critical of Bose's work (Anonymous, 1929; Shull, 1923). The Bose questioning the cohesion theory was Sir Jagadis Chunder Bose, who was the teacher of Satyendra Nath Bose (Ghosh, 1992). S.N. Bose was the Bose of the Bose-Einstein condensation (BEC), a purely quantum phenomenon whereby a macroscopic number of identical atoms occupy the same single-particle state (Wyatt, 1998). Even though outstanding physicists have challenged the cohesion theory, it has been vigorously defended by plant physiologists for many years, but more experiments are needed to accept fully its assumptions.

VIII. POTENTIALS IN THE SOIL-PLANT-ATMOSPHERE CONTINUUM

No matter how water gets to the top of tall trees, the gradient in water potential from the soil to the top of the tree is calculated to be large. Nobel (1974, p. 402; 1983, p. 507; 1991, p. 521) shows representative values for the water potential, ψ, and its components (ψ_m, matric potential; ψ_s, solute potential; ψ_g, gravitational potential; ψ_p, turgor potential) in the soil-plant-atmosphere continuum. Let us choose three values of the water potential and its components that he gives: one for the soil, one for the plant, and one for the atmosphere just over the plant (the reference level is at the soil surface):

Soil: 0.1 m below ground and 10 mm from the root:
ψ, −0.3 MPa
ψ_m, −0.2 MPa
ψ_s, −0.1 MPa
ψ_g, 0.0 MPa

Vacuole of leaf mesophyll cell at 10 m:
ψ, −0.8 MPa
ψ_p, 0.2 MPa
ψ_s, −1.1 MPa
ψ_g, 0.1 MPa

Air just across boundary layer of the leaf at 50% relative humidity:
ψ, −95.1 MPa
ψ_p, 0.0 MPa
ψ_s, 0.0 MPa
ψ_g, 0.1 MPa
Potential of the air at 50% RH: −95.2 MPa

Therefore, we see that the water potential changes from −0.3 MPa in the soil to −95.1 MPa in the air just outside the leaf—a change of 94.8 MPa.

The representative values will, in actuality, be affected by three factors. First, the water potential goes through a diurnal cycle (Nobel, 1983, pp. 516–517) (Fig. 19.8). The water potential is typically at its highest value just before dawn, when the plant has had a chance to rehydrate during the night, and it is usually lowest right after midday. Because the water potential is always changing throughout a day, it is important to measure it at the same time each day during an experiment, unless one wishes to document the diurnal changes. Figure 19.8 is for a general situation. The format of Fig. 19.8 was originally drawn by Gardner and Nieman (1964), who presented actual data for a pepper (Fig. 19.9).

Second, marked changes in the hydrostatic pressure in the xylem can cause plants to have measurable diurnal fluctions in their diameters (Nobel, 1974, p. 404). When the transpiration rate is high, the large tension within the xylem vessel members is transmitted to the water in the cell walls of the xylem vessels, then to water in adjacent cells, and eventually all the way across the stem. The decrease in hydrostatic pressure in a trunk can, therefore, cause a whole tree to contract during the day. At night, the hydrostatic pressure in the xylem may become positive, and the tree diameter then increases, generally by about 1%. Such changes in tree diameter, and therefore volume, represent net release of water during the day and storage at night.

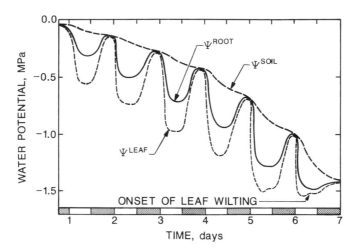

FIG. 19.8 Schematic representation of daily changes in the water potentials in the soil, root, and leaf of a plant in an initially wet soil that dries out over a one-week period. ψ^{soil} is the water potential in the bulk soil, ψ^{root} is that in the root xylem, and ψ^{leaf} is the value in a leaf mesophyll cell. (From Nobel, P.S. *Biophysical Plant Physiology and Ecology*, p. 517, ©1983 by W.H. Freeman and Company: San Francisco. Used with permission.)

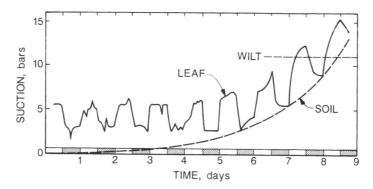

FIG. 19.9 Diurnal fluctuation of the suction (diffusion pressure deficit) of a pepper leaf (solid line) and average soil suction in root zone (dashed line). Solid bars along the abscissa indicate twelve-hour dark periods. The horizonal dashed line indicates leaf suction at which wilting symptoms appear. The plant was grown in a 3-gallon (11.4-L) jar containing clay loam soil. (Reprinted with permission from Gardner, W.R., and Nieman, R.H., Lower limit of water availability to plants. *Science* 143; 1460–1462, ©1964, American Association for the Advancement of Science.)

Strain gauges are used to monitor the change in stem diameter. In an experiment done at Iowa State University, strain gauges were cemented to poplar trees (Iowa State University, 1984). When light intensity increased at sunrise, the stem contracted in response to increased evaporation. The stem reached its smallest size around 3:00 P.M. As evaporative demand dropped in the evening, the stem expanded. Two hours after sunset the stem was back to the size of the previous night plus the growth of the day. As the tree dried out, the stem contracted, and eventually it did not recover at night and the stem did not expand. The measurements with strain gauges can be used to determine when to irrigate; however, they cannot be used on small stems, like those of soybean plants, because they do not stay attached.

Third, the water potential in the vacuole of the leaf will depend on osmotic adjustment. Osmotic adjustment is the lowering of the osmotic potential of a plant when the osmotic potential of the root medium decreases (Bernstein, 1961, 1963). At the cell level, osmotic adjustment is defined as the net accumulation of solutes in a cell in response to a fall in the water potential of the cell's environment. As a consequence of this net accumulation, the osmotic potential of the cell is lowered, which, in turn, attracts water into the cell and tends to maintain turgor potential (Blum et al., 1996). Osmotic adjustment is under genetic control (Zhang et al., 1999). Therefore, the representative value for osmotic potential given by Nobel (1974, 1983, 1991) (−1.1 MPa) can be changed by osmotic adjustment. Under arid conditions or in halophytes, the osmotic potential could fall to as low as −5.0 MPa.

IX. APPENDIX: BIOGRAPHY OF HENRY DIXON

Henry Horatio Dixon (1869–1953), Irish botanist, was born in Dublin, Ireland, in 1869, the son of George and Rebecca (Yeates) Dixon. He got his Sc.D. at Trinity College, Dublin, and also was educated at the University of Bonn in Germany (Marquis Who's Who, 1968). He married Dorothea Mary Franks in 1907, and they had three sons. Between 1892–1904, he rose from assistant to professor of botany at Dublin University and was university professor of botany from 1904 to 1950. He was professor of plant biology at Trinity College, Dublin, from 1922; director of the botanical gardens at Trinity College from 1906–1951; and its keeper of the herbarium from 1910–1951. He was a trustee of the Imperial Library of Ireland. He became a commander of the Irish Lights from 1924. He was a visiting professor at the University of California in 1927. He was honorary chairman of the 6th International Botanical Congress held in Amsterdam, The Netherlands, in 1935; and he was honorary president of the International Botanical Congress held in Stockholm, Sweden, in 1950.

He was recipient of the Boyle Medal in 1917. In 1908 he became a fellow of the Royal Society and was its Croonian lecturer. He was a member of the International Institute of Agriculture and was its chairman for the Committee on Biochemistry in 1927. He was a member of the Royal Dublin Society and was its president from 1945 to 1949. He was a corresponding member of the American Society of Plant Physiologists. He was a member of the British Association for the Advancement of Science and was the president of its Botanical Section in 1922. He was author of several books, including *Transpiration and the Ascent of Sap in Plants* (1914), *Practical Plant Biology* (1922), and *The Transpiration Stream* (1924). He is best known for his research on plant transpiration. He died December 20, 1953 (Marquis Who's Who, 1968).

X. APPENDIX: BIOGRAPHY OF JOHN JOLY

John Joly (1857–1933), a British physicist and geologist, was born in 1857 (Calef, 1971). He was the son of J.P. and Julia (de Lusi) Joly. He was educated at Trinity College, Dublin, Ireland, where he got a B.A., M.A., and D.Sc. (Marquis Who's Who). He got an LL.D. at the University of Michigan; an Sc.D. (honorary) at Cambridge University, England; and an Sc.D. at the National University of Ireland. At Trinity College, he was a demonstrator in civil engineering from 1882–1891, a demonstrator of experimental physics in 1893, and from 1897 a professor of geology and mineralogy. He was Warden, Alexandra College for Higher Education of Women.

He was senior commander of the Irish Lights and was science adviser to Dr. Steevens' Hospital, Dublin. He was a member of the British educational mission to the United States in 1918. He became a fellow of the Royal Society in 1892 and he got its Royal Medal in 1910. He was a fellow of the Geological Society and got its Murchison Medal in 1923. He was a member and president of the Royal Dublin Society and got its Boyle Medal in 1911. He was a member of the British Association and was president of the Geological Section in 1908. He was an honorary member of the Academy of Science of Russia.

He was author of several books, including *On the Specific Heats of Gases at Constant Volume, On a Method of Photography in Natural*

Colours (1896), *Radio-activity and Geology, The Local Application of Radium in Therapeutics* (1914), *The Birth-Time of the World and Other Scientific Essays* (1915), *Synchronous Signalling in Navigation, Reminiscences and Anticipations* (1920), *Radioactivity and the Surface History of the Earth* (1924), and *The Surface History of the Earth* (1925; 2nd ed., 1930). He was editor (with others) of the *Philosophical Magazine* from 1901. He devised the diffusion phytometer, meldometer, and steam calorimeter. (The meldometer, described in an article by Joly in *Nature* in 1885, was an apparatus that was an adjunct to the mineralogical microscope, and it allowed the approximate determination of the melting point of minerals.) He developed a uniform radiation method for use in cancer treatment and was a color-photography pioneer (Marquis Who's Who, 1968).

Joly's work on crust formation of the earth is his best known, in which he presented a theory on continental origins based on the process of convection (Calef, 1971). He proposed that heat was generated in the interior of the earth by decay of radioactive elements. Because the heat could not escape sufficiently rapidly by other means, it started convection currents which carried hot material toward the surface where it cooled and sank, thus setting up a convective cell circulation. He suggested that the earth's crust was dragged sidewise at the top of the cell, which caused buckling and folding and thus mountain making, and the crust collapsed above the sinking portion of the convection cell. The greatest difficulty with this theory was a lack of observational data indicating any convection currents or cells (Calef, 1971). Joly died December 8, 1933.

REFERENCES

Anonymous. (1929). News and Views. *Nature* 124; 103.
Askenasy, E. (1895). Ueber [sic] das Saftsteigen. *Botanisches Centralblatt* 62; 237–238.
Baker, D.A. (1984). Water relations. In *Advanced Plant Physiology* (Wilkins, M.B., Ed.), pp. 297–318. Pitman: London.
Bernstein, L. (1961). Osmotic adjustment of plants to saline media. I. Steady state. *Amer J Bot* 48; 909–918.
Bernstein, L. (1963). Osmotic adjustment of plants to saline media. II. Dynamic phase. *Amer J Bot* 50; 360–370.
Blum, A., Munns, R., Passioura, J.B., and Turner, N.C. (1996). Genetically engineered plants resistant to soil drying and salt stress: How to interpret osmotic relations? *Plant Physiol* 110; 1051.
Boehm, J. (1893). Capillarität und Saftsteigen. *Berichte der Deutschen Botanischen Gesellshaft* 11; 203–212.
Bose, J.C. (1923). *The Physiology of the Ascent of Sap*. Longmans, Green: London.
Briggs, L.J. (1950). Limiting negative pressure of water. *J Appl Physics* 21; 721–722.
Calef, W.C. (1971). Continent. *Encyclopaedia Britannica* 6; 417–419.

Canny, M.J. (1995a). A new theory for the ascent of sap—Cohesion supported by tissue pressure. *Ann Bot* 75; 343–357.

Canny, M.J. (1995b). Apoplastic water and solute movement: New rules for an old space. *Annu Rev Plant Physiol Plant Mol Biol* 46; 215–236 (see p. 229 and following).

Canny, M.J. (1997). Vessel contents of leaves after excising—a test of Scholander's assumption. *Amer J Bot* 84; 1217–1222.

Canny, M.J. (1998). Applications of the compensating pressure theory of water transport. *Amer J Bot* 85; 897–909.

Canny, M.J. (2001). Contributions to the debate on water transport. *Amer J Bot* 88; 43–46.

Comstock, J.P. (1999). Why Canny's theory doesn't hold water. *Amer J Bot* 86; 1077–1081.

Cochard, H., Bodet, C., Améglio, T., and Cruiziat, P. (2000). Cryo-scanning electron microscopy observations of vessel content during transpiration in walnut petioles. Facts or artifacts? *Plant Physiol* 124; 1191–1202.

Dixon, H.H. (1895). On the ascent of sap. *Phil Trans Roy Soc London* B186; 563–576.

Dixon, H.H. (1897). The tensile strength of cell-walls. *Ann Bot* 11; 585–588.

Dixon, H.H. (1914). *Transpiration and the Ascent of Sap in Plants*. Macmillan and Co: London.

Dixon, H.H., and Joly, J. (1895). The path of the transpiration-current. *Ann. Bot* 9; 403–420.

Eisenberg, D., and Kauzmann, W. (1969). *The Structure and Properties of Water*. Oxford University Press: Oxford.

Ellsworth, D.S. (1999). CO_2 enrichment in a maturing pine forest: Are CO_2 exchange and water status in the canopy affected? *Plant, Cell Environ* 22; 461–472.

Emerman, S.H. (1996). Towards a theory of hydraulic lift in trees and shrubs. In *Proceedings of the 16th Annual American Geophysical Union Hydrology Days. April 15–18, 1996* (Morel-Seytoux, H.J., Ed.), pp. 147–157. Colorado State University, Fort Collins. Hydrology Days Publications: Atherton, California.

Esau, K. (1965). *Plant Anatomy*. 2nd ed. Wiley: New York.

Esau, K. (1977). *Anatomy of Seed Plants*. 2nd ed. Wiley: New York.

Gardner, W.R. (1965). Dynamic aspects of soil-water availability to plants. *Annu Rev Plant Physiol* 16; 323–342.

Gardner, W.R., and Nieman, R.H. (1964). Lower limit of water availability to plants. *Science* 143; 1460–1462.

Ghosh, A. 1992. Vignette: Looking toward Calcutta. *Nature* 257; 1775.

Green, S.R., and Clothier, B.E. (1988). Water use of kiwifruit vines and apple trees by the heat-pulse technique. *J Exp Bot* 39; 115–123.

Greenidge, K.N.H. (1955). Studies in physiology of forest trees. III. The effect of drastic interruption of conducting tissues in moisture movement. *Amer J Bot* 42; 582–587.

Greenidge, K.N.H. (1957). Ascent of sap. *Annu Rev Plant Physiol* 8; 237–256.

Hammel, H.T., and Scholander, P.F. (1976). *Osmosis and Tensile Solvent*. Springer-Verlag: Berlin.

Hargens, A.R., Millard, R.W., Pettersson, K., and Johansen, K. (1987). Gravitational haemodynamics and oedema prevention in the giraffe. *Nature* 329; 59–60.

Harris, R.F. (1981). Effect of water potential on microbial growth and activity. In *Water Potential Relations in Soil Microbiology* (Parr, J.F., Gardner, W.R., and Elliott, L.F., Eds.), pp. 23–95. SSSA Special Pub. No. 9. Soil Science Society of America: Madison, Wisconsin.

Hickman, C.P. (1961). *Integrated Principles of Zoology*. 2nd ed. The C.V. Mosby: St. Louis.

Holbrook, N.M., Burns, M.J., and Field, C.B. (1995). Negative xylem pressures in plants: A test of the balancing pressure technique. *Science* 270; 1193–1994.

Huang, B. (1999). Water relations and root activities of *Buchloe dactyloides* and *Zoysia japonica* in response to localized soil drying. *Plant Soil* 208; 179–186.

Iowa State University. (1984). And you thought watching trees was boring. *The Iowa Stater* 10(7); 8.

Jackson, G.E., and Grace, J. (1996). Field measurements of xylem cavitation: Are acoustic emissions useful? *J Exp Bot* 47; 1643–1650.

Johnson, R.P.C. (1977). Can cell walls bending round xylem vessels control water flow? *Planta* 136; 187–194.

Kargol, M., Kosztołowicz, T., and Przestalski, S. (1995). About the biophysical mechanisms of the long-distance water translocation in plants. *Int Agrophys* 9; 243–255.

Kirkham, M.B. (1983). Physical model of water in a split-root system. *Plant Soil* 75; 153–168.

Kramer, P.J. (1969). *Plant and Soil Water Relationships: A Modern Synthesis*. McGraw-Hill: New York.

Kramer, P.J. (1983). *Water Relations of Plants*. Academic Press: New York.

Lang, A.R.G. (1967). Osmotic coefficients and water potentials of sodium chloride solutions from 0 to 40°C. *Australian J Chem* 20; 2017–2023.

Loomis, W.E., Santamaria, R., and Gage, R.W. (1960). Cohesion of water in plants. *Plant Physiol* 35; 300–306.

MacDougal, D.T., and Overton, J.B. (1927). Sap flow and pressure in trees. *Science* 65; 189–190.

Marquis Who's Who. (1968). *World Who's Who in Science: From Antiquity to the Present*. 1st ed. Marquis-Who's Who: Chicago.

Milburn, J.A., and Johnson, R.P.C. (1966). The conduction of sap. II. Detection of vibrations produced by sap cavitation in *Ricinus* xylem. *Planta* 69; 43–52.

National Public Radio. (2002). The Water Tower Man. In *All Things Considered*. Transcript of February 13, 2002, p. 7–11. National Public Radio: Washington, DC. (Transcript produced by Burrelle's Information Services, Box 7, Livingston, New Jersey 07039.)

Nobel, P.S. (1970). *Plant Cell Physiology*. W.H. Freeman: San Francisco.

Nobel, P.S. (1974). *Introduction to Biophysical Plant Physiology*. W.H. Freeman: San Francisco.

Nobel, P.S. (1983). *Biophysical Plant Physiology and Ecology*. W.H. Freeman: San Francisco.

Nobel, P.S. (1991). *Physicochemical and Environmental Plant Physiology*. Academic Press: San Diego.

Pauling, L. (1964). *College Chemistry*. 3rd ed. W.H. Freeman: San Francisco.

Pedley, T.J. (1987). Haemodynamics. How giraffes prevent oedema. *Nature* 329; 13–14.

Pockman, W.T., Sperry, J.S., and O'Leary, J.W. (1995). Sustained and significant negative water pressure in xylem. *Nature* 378; 715–716.

Preston, R.D. (1938). The contents of the vessels of *Fraxinus americana* L., with respect to the ascent of sap. *Ann Bot* (new series) 2; 1–21.

Salisbury, F.B., and Ross, C.W. (1978). *Plant Physiology*. 2nd ed. Wadsworth: Belmont, California.

Scheenen, T.W.J., van Dusschoten, D., de Jager, P.A., and Van As, H. (2000). Quantification of water transport in plants with NMR imaging. *J Exp Bot* 51; 1751–1759.

Scholander, P.F. (1958). The rise of sap in lianas. In *The Physiology of Forest Trees. A Symposium Held at the Harvard Forest, April, 1957* (Thimann, K.V., Ed.), pp. 3–17. Ronald Press: New York.

Scholander, P.F., Love, W.E., and Kanwisher, J.W. (1955). The rise of sap in tall grapevines. *Plant Physiol* 30; 93–104.

Scholander, P.F., Ruud, B., and Leivestad, H. (1957). The rise of sap in a tropical liana. *Plant Physiol* 32; 1–6.

Scholander, P.F., Hemmingsen, E., and Garey, W. (1961). Cohesive lift of sap in rattan vine. *Science* 134; 1835–1838.

Scholander, P.F., Hammel, H.T., Bradstreet, E.D., and Hemmingsen, E.A. (1965). Sap pressure in vascular plants. *Science* 148; 339–346.

Senft, D. (1986). Microphones monitor plants' water needs. *Agr Res* 34(3); 13.

Shull, C.A. (1923). Book reviews. Physiology of ascent of sap. *Bot Gaz* 76; 316–317.

Silver, R.S. (1942). Tensile strength of water and liquid structure theory. *Nature* 150; 605.

Song, Y., Kirkham, M.B., Ham, J.M., and Kluitenberg, G.J. (2000). Root-zone hydraulic lift evaluated with the dual-probe heat-pulse technique. *Australian J Soil Res* 38; 927– 935.

Steudle, E. (1995). Trees under tension. *Nature* 378; 663–664.

Stiller, V., and Sperry, J.S. (1999). Canny's compensating pressure theory fails a test. *Amer J Bot* 86; 1082–1086.

Tomos, A. D., and Leigh, R.A. (1999). The pressure probe: A versatile tool in plant cell physi ology. *Annu Rev Plant Physiol Plant Mol Biol* 50; 447–472.

Tyree, M.T. (1997). The cohesion-tension theory of sap ascent: Current controversies. *J Exp Bot* 48; 1753–1765.

Tyree, M.T., Fiscus, E.L., Wullschleger, S.D., and Dixon, M.A. (1986). Detection of xylem cav itation in corn under field conditions. *Plant Physiol* 82; 597–599.

Ursprung, A. (1929). The osmotic quantities of the plant cell. *Proc Int Congr Plant Sci* 2; 1081–1094.

Weast, R.C. (Ed.) (1964). *Handbook of Chemistry and Physics.* Chemical Rubber Co: Cleveland, Ohio.

Weber, B. (1989). Keeping up the pressure. *New York Times Magazine*, September 3, 1989; 62.

Webster's Collegiate Dictionary (1939). 5th ed. G. & C. Merriam Co: Springfield, Massachusetts.

Webster's New World Dictionary of the American Language (1959). College ed. World Publishing: Cleveland, Ohio.

Wei, C., Steudle, E., and Tyree, M.T. (1999). Water ascent in plants: Do ongoing controversies have a sound basis? *Trends Plant Sci* 4; 372–375.

Wyatt, A.F.G. (1998). Evidence for a Bose-Einstein condensate in liquid ^{4}He from quantum evaporation. *Nature* 391; 56–59.

Zhang, Jingxian, Hguyen, H.T., and Blum, A. (1999). Genetic analysis of osmotic adjustment in crop plants. *J Exp Bot* 50; 291–302.

Zimmermann, U., Meinzer, F.C., Benkert, R., Zhu, J.J., Schneider, H., Goldstein, G., Kuchenbrod, E., and Haase, A. (1994). Xylem water transport: Is the available evidence consistent with the cohesion theory? *Plant Cell Environ* 17; 1169–1181.

Zimmermann, U., Meinzer, F., and Bentrup, F.-W. (1995). How does water ascend in tall tress and other vascular plants? *Ann Bot* 76; 545–551.

20

Electrical Analogues for Water Movement through the Soil-Plant-Atmosphere Continuum

Electrical analogues have long been used to study the movement of water in soil. The analogy between the flow of electricity through conducting media and the flow of water through porous media (i.e, soil) was pointed out by Slichter (1899), a mathematician at the University of Wisconsin. The analogy later was expanded to include movement of water through the entire soil-plant-atmosphere continuum. We now consider why the analogy works and its application.

I. THE ANALOGY

The analogy can be seen when we compare Ohm's law with Darcy's law. Ohm's law states (Kirkham and Powers, 1972, p. 183):

$$I = -\sigma A (V_2 - V_1)/L = V/R, \qquad (20.1)$$

where

I = quantity of electricity flowing per unit time (coulombs of electricity per unit time) (coulombs per second or amperes)

σ = specific electrical conductivity or electrical conductivity (Siemens per cm)

L = length of element through which current flows (cm)

A = cross-sectional area of the element (cm^2)

V_2 and V_1 = voltages (volts)

$(V_2 - V_1)/L$ = potential gradient

R = resistance (ohms)

For water flow in porous media (Darcy's law), we can write (Kirkham and Powers, 1972, p. 183):

$$Q = -KA(\phi_2 - \phi_1)/L, \qquad (20.2)$$

where

Q = cm^3 of water flowing per unit time (cm^3 per second)

K = hydraulic conductivity (cm per second)

L = length (cm)

A = cross-sectional area (cm^2)

ϕ_2 and ϕ_1 = hydraulic heads (cm)

$(\phi_2 - \phi_1)/L$ = hydraulic gradient

We see immediately the close analogy between Ohm's law, Equation 20.1, and Darcy's law, Equation 20.2. (For a biography of Ohm, see the Appendix, Section VIII.)

II. MEASUREMENT OF RESISTANCE WITH THE WHEATSTONE BRIDGE

To measure R (resistance in ohms; the Greek letter capital omega, Ω, is used to symbolize resistance in ohms), we use a Wheatstone bridge, which is an instrument for measuring the value of an unknown resistance by comparing it with a standard. This method, devised in 1833 by S. Hunter Christie, was brought to public attention by the English physicist, Sir Charles Wheatstone (1802–1875) and has remained associated with his name (Hausman and Slack, 1948, p. 388). (For a biography of Wheatstone, see the Appendix, Section IX.) The Wheatstone bridge is the most convenient, and at the same time accurate, way of measuring resistances of widely different values (Ingersoll et al., 1953). It works on the principle of a divided circuit, which is illustrated in Fig. 20.1. The current from the battery divides between the two branches abc and adc. Because the potential drop is the same along the two branches, corresponding intermediate points b and d may be found which are at the same potential. Under these circumstances, no current will flow through the galvanometer, G, connected between b and d. The bridge is then said to be balanced, and

$$R_1/R_2 = R_3/X. \qquad (20.3)$$

Thus, any one of the four resistances may be obtained in terms of the three others.

A Wheatstone bridge often looks like a black box with knobs on the top (Fig. 20.2), but there are also "slide-wire" Wheatstone bridges. In the slide-wire form of the bridge, one of the branches (e.g., abc in Fig. 20.1), consists of a wire of uniform cross section. The point b is located by a sliding contact. The unknown resistance X is placed in one arm of the other branch, the remaining arm containing the known resistance R_3 usually in the form of a resistance box. Because only the ratio of the resistances R_1 and R_2 is required, this ratio may be replaced by the ratio of the lengths of the two arms of the slide-wire (Ingersoll et al., 1953, p. 135).

III. LAW OF RESISTANCE

We know R, by measuring it with the Wheatstone bridge, but now we need to know conductivity. To determine conductivity, we use the

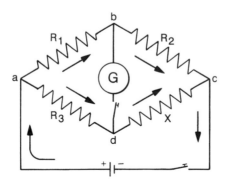

FIG. 20.1   Wheatstone bridge circuit. (From Ingersoll, L.R., Martin, M.J., and Rouse, T.A., *A Laboratory Manual of Experiments in Physics*, p. 134, ©1953 McGraw-Hill Book Co.: New York. This material is reproduced with permission of The McGraw-Hill Companies.)

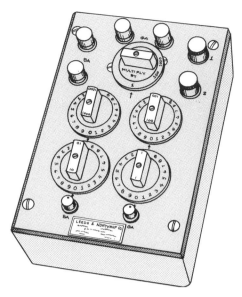

FIG. 20.2 A Wheatstone bridge. (From a brochure of Leeds and Northrup Co., North Wales, Pennsylvania. Courtesy of Honeywell International, Inc.)

law of resistance. The law of resistance is true by experimentation, and states:

$$R = \rho L/A, \tag{20.4}$$

where

   R = resistance (ohms)
   L = length (cm)
   A = area (cm^2)
   ρ = resistivity (ohm-cm)

Figure 20.3 illustrates how we can apply the law of resistance. If we have a cube of material that is 1 cm on a side for a total cross-sectional area of 1 cm^2 and a length of 1 cm through which the electricity flows, and we have a 1 volt potential difference in our circuit and we have 1 amp of electricity flowing (I) (1 amp = 1 coulomb/s), we have 1 ohm of resistance, because by Ohm's law R (ohms) = V (volts)/I (amperes). We know R, L, A, and we can determine resistivity from the law of resistance. From resistivity, we determine conductivity, as follows:

$$\sigma = \text{conductivity} = 1/\text{resistivity} = 1/\rho.$$

The units of conductivity = 1/(ohm-cm) or mho/cm. (These are not SI units; we will change these to SI units in the next section.)

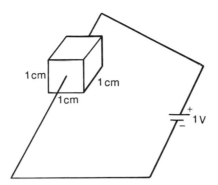

FIG. 20.3 Illustration showing how to use the law of resistance to get resistivity. (From a sketch by Don Kirkham.)

Resistivity (ρ) varies with materials. Resistivity of materials can be found in an early edition of the *Handbook of Chemistry and Physics* (Hodgman, 1959, p. 2598). The resistivity of rocks and soils is high. For example, the resistivity of granite varies between 10^7 to 10^9 ohm-cm, and the resistivity of sand varies between 10^5 to 10^6 ohm-cm. (The temperature at which the resistivity of rocks and soils was determined is not stated in the handbook.) The resistivity of metals is small (Hodgman, 1959, pp. 2587–2593). For example, the resistivity of aluminum at 20°C is 2.828×10^{-6} ohm-cm. The resistivity of copper at 20°C is 1.77×10^{-6} ohm-cm. The resistivity of gold at 20°C is 2.44×10^{-6} ohm-cm. The resistivity of silver at 18°C is 1.629×10^{-6} ohm-cm.

IV. UNITS OF ELECTRICAL CONDUCTIVITY

Electrical conductivity is used to measure the salinity of a soil. Old units of electrical conductivity were mmho/cm; the USDA Handbook No. 60, edited by L.A. Richards (1954), still in use for standard measurements of saline soils, uses the old unit of mmho/cm. We need to know how to convert mmho/cm into SI units. The SI unit for conductance is the Siemens. (For biographies of members of the Siemen family, see the Appendix, Section X.) Conductance is $1/R$, and its non-SI unit is the mho, which is ohm spelled backward.

$$1 \text{ Siemen} = 1/R = 1/1 \text{ ohm} = 1 \text{ mho}.$$

The SI unit for electrical conductivity is the deciSiemen/m.

$$1 \text{ deciSiemen/m} = 1 \text{ dS/m} = 1 \text{ mmho/cm}.$$

Example: Assume that we have saltwater in a container that is 24 cm long and 5 cm wide. The saltwater stands to a height of 9 mm in the container. We measure a resistance of 400 ohms with a Wheatstone bridge. What is the conductance? What is the electrical conductivity? (Hint: From the law of resistance, get resistivity and take its reciprocal.)

Conductance = 1/400 ohms = 0.0025 mhos.
Area = 0.9 cm × 5 cm = 4.5 cm^2.
400 ohms = (ρ 24 cm)/4.5 cm^2.
ρ = 75 ohm-cm. This is the resistivity.
1/75 ohm-cm = 0.013 mho/cm = 13 mmho/cm = 13 dS/m. This is the electrical conductivity.

V. EXAMPLE OF AN ELECTRICAL ANALOGUE APPLIED TO SOIL WITH WORMHOLES

The same container cited in the preceding example (24 cm long and 5 cm wide) was used to determine, in an electrical-analogue study, the water and air conductance in soil with earthworms (Kirkham, 1982). The objective was to quantify the relationship between conductance and wormholes of different sizes oriented in the horizontal and vertical directions, which simulated wormholes oriented horizontally to the soil surface or perpendicularly to the soil surface. Copper pipes of different diameters, placed horizontally and vertically in the center of the electrolyte (tap water) in the container, simulated the wormholes. The results showed that wormholes, when their diameter and/or length is increased, cause an increase in soil conductance. Large increases (e.g., 100%) in conductance did not occur for holes in the vertical direction (and flow perpendicular to the holes) until a hole had a diameter that was greater than 70% of the length of the unit volume. Similarly, large increases in conductance did not occur in the horizontal direction unless the wormhole length was an appreciable amount of the soil length associated with the hole.

The experiment simulated the concentration of oxygen in the moving air, or nutrients in the moving soil water, in the wormholes (Fig. 20.4). The increased concentration of oxygen in air, or nutrients in water, may be one reason why roots concentrate in wormholes. The increased concentration of oxygen or nutrients in the hole will occur even when the wormholes are not directly connected to the soil surface, as was the case in this experiment (Kirkham, 1982). The experiment showed that electrical-analogue studies can provide information that is not easily measured in the field.

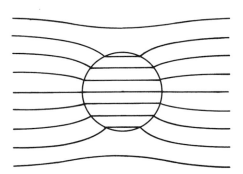

FIG. 20.4 Electrical analogue of oxygen-flow concentration or nutrient-flow concentration (flow lines are close together) in an isolated vertical wormhole with flow perpendicular to the hole axis. The figure is for a conductivity in the "wormhole" equal to five times that in the soil. (From Smythe, W.R., *Static and Dynamic Electricity*, 2nd ed., p. 68, ©1950 McGraw-Hill Book Co: New York. This material is reproduced with permission of The McGraw-Hill Companies.)

VI. VAN DEN HONERT'S EQUATION

The analogy for water movement through the entire soil-plant-atmosphere continuum and the flow of electricity has been in the literature for decades. According to van den Honert (1948), "It was Gradmann's [1928] idea to apply an analogue of Ohm's law to this water transport as a whole." However, if one looks at the Gradmann paper, one sees no place where Ohm is mentioned. So Gradmann must have suggested a linear flow law, like Ohm's law, without mentioning Ohm specifically. We remember that Ohm's law is one of the linear flow laws that is so important in transport (Table 7.1). Kramer (1983, p. 190) cites Huber (1924) as the originator of the idea, but also lists Gradmann (1928) as one who developed it. Despite the uncertainty about who originated the idea that water flow through the soil-plant-atmosphere system is similar to the flow of electricity, the paper published by van den Honert (1948) (in English) is the most cited paper on the topic.

Let us first look at the more simple form of Ohm's law in Equation 20.1, $V = IR$. A current I of electricity exists in a conductor whenever electric charge q is being transferred from one point to another in that conductor. If charge is transferred at the uniform rate of 1 coulomb per second, then the constant current existing in the conductor is 1 ampere (Schaum, 1961, pp. 146–147). The potential difference V between two points in a conductor is measured by the work W required to transfer unit charge from one point to the other. The volt is the potential difference between two points in a conductor when 1 joule of work is required to transfer 1 coulomb of charge from one point to the other. The resistance R of a conductor is the property that depends on its dimensions, material, and temperature, and that determines the current produced in it by a given potential difference. The *ohm* is the resistance of a conductor in which there is a current of 1 ampere when the potential difference between its ends is 1 volt. Ohm's law states that the value of the steady electrical current I in a metallic conductor at a constant temperature is equal to the potential difference V between the ends of the conductor divided by the resistance R of the conductor (Fig. 20.5), or

$$I \text{ (current)} = V \text{ (potential difference)}/R \text{ (resistance)} \qquad (20.5)$$

$$I \text{ (amperes)} = V \text{ (volts)}/R \text{ (ohms)}$$

Ohm's law may be applied to any part of a circuit or to the entire circuit. Thus the potential difference, or voltage drop, across any part of a conductor is equal to the current I in the conductor multiplied by the resistance R of that part, or $V = IR$. When Ohm's law is applied to the soil-plant-atmosphere continuum, the following analogies are made:

V is the potential difference between any two parts in the system. The potential in each part of the system is the (total) water potential (ψ_w), which is measured, for example, with a thermocouple hygrometer or pressure chamber and is usually expressed using the unit of MPa.
I is the flow of water (transpiration rate). This is what Nobel (e.g., 1974, p. 142) calls J_v or volume flow measured in units such as m s^{-1}.
R is the resistance. Its units depend upon how V (or ψ_w) and I have been defined.

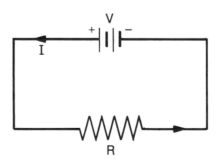

FIG. 20.5 Diagram illustrating Ohm's law. (Adapted from Schaum, D., *Theory and Problems of College Physics*, p. 149, ©1961, Schaum Publishing Co: New York. This material is reproduced with permission of The McGraw-Hill Companies.)

Van den Honert (1948) uses the Ohm's law analogy to develop an equation similar to the following, which Baker (1984, p. 310) modified using modern terminology (ψ in place of the old terminology of diffusion pressure deficit, which van den Honert used):

$$J_v = (\Delta\psi)/r = (\psi_{soil} - \psi_{root})/r_1 = (\psi_{root} - \psi_{stem})/r_2 = (\psi_{stem} - \psi_{leaf})/r_3 =$$
$$(\psi_{leaf} - \psi_{air})/r_4 = (\psi_{soil} - \psi_{air})/(r_1 + r_2 + r_3 + r_4), \tag{20.6}$$

where J_v is the steady rate of water flow, ψ is the water potential at different parts in the system (the subscript designates the location), and $r_1, r_2, r_3,$ and r_4 are the resistances between the soil and root, the root and stem, the stem and leaf, and the leaf and air, respectively. This equation has been reproduced in many textbooks (e.g., see Kramer, 1983, p. 190).

VII. PROOF OF VAN DEN HONERT'S EQUATION

It is not obvious why the string of equations in Equation 20.6 should equal each other. In fact, some have questioned the "equal" signs in Equation 20.6 and suggested that they should be "plus" signs. So we shall now prove Baker's (van den Honert's) equation (Equation 20.6), or we shall prove that

$$J_v = (\psi_{soil} - \psi_{air})/(r_1 + r_2 + r_3 + r_4).$$

We divide up the string of equations into individual equations.

$$J_v = (\psi_{soil} - \psi_{root})/r_1 \tag{20.7}$$

$$J_v = (\psi_{root} - \psi_{stem})/r_2 \tag{20.8}$$

$$J_v = (\psi_{stem} - \psi_{leaf})/r_3 \tag{20.9}$$

$$J_v = (\psi_{leaf} - \psi_{air})/r_4 \tag{20.10}$$

We now multiply the first equation (Equation 20.7) through by r_1; Equation 20.8 through by r_2; Equation 20.9 through by r_3; and Equation 20.10 through by r_4.

$$J_v r_1 = (\psi_{soil} - \psi_{root}) \tag{20.11}$$

$$J_v r_2 = (\psi_{root} - \psi_{stem}) \tag{20.12}$$

$$J_v r_3 = (\psi_{stem} - \psi_{leaf}) \tag{20.13}$$

$$J_v r_4 = (\psi_{leaf} - \psi_{air}) \tag{20.14}$$

In Equations 20.11 through 20.14, we add up the left sides and add up the right sides and then equate the resultant left and right sides to get one equation (Equation 20.15):

$$J_v r_1 + J_v r_2 + J_v r_3 + J_v r_4 = (\psi_{soil} - \psi_{root}) + (\psi_{root} - \psi_{stem}) +$$
$$(\psi_{stem} - \psi_{leaf}) + (\psi_{leaf} - \psi_{air}) \tag{20.15}$$

We now cancel units in Equation 20.15 and factor the left-hand side.

$$J_v(r_1 + r_2 + r_3 + r_4) = \psi_{soil} - \psi_{air} \tag{20.16}$$

We divide each side of Equation 20.16 by $(r_1 + r_2 + r_3 + r_4)$:

$$J_v(r_1 + r_2 + r_3 + r_4)/(r_1 + r_2 + r_3 + r_4) = (\psi_{soil} - \psi_{air})/(r_1 + r_2 + r_3 + r_4) \tag{20.17}$$

We simplify the left-hand side of Equation (20.17) and get the equation as shown by Baker (1984):

$$J_v = (\psi_{soil} - \psi_{air})/(r_1 + r_2 + r_3 + r_4) \quad Q.E.D.$$

(Q.E.D. is used in mathematics, and it is Latin for *"quod erat demonstrandum"* or "which was to be proved.")

VIII. APPENDIX: BIOGRAPHY OF GEORG OHM

Georg Simon Ohm (1789–1854) was an ingenious German investigator who, although removed from the influence of personal contact with the renowned physicists of his time and working independently and alone, discovered the great law bearing his name (Cajori, 1929, p. 234). He was born in Erlangen on March 16, 1787, and was educated at the university there (Preece, 1971a). He then taught school at Gottstadt, Neufchâtel, and Bamberg. In 1817, he became teacher of mathematics and physics in the Jesuits' college in Cologne, and taught there for nine years with great success. A pupil of that time, who later attained fame as a mathematician, was Lejene Dirichlet (1805–1859).

Ohm wanted to do research, but the want of leisure and books, as well as the lack of suitable apparatus, made progress difficult. The mechanical skill that he had acquired as a boy from his father, a locksmith, enabled him to construct much apparatus for himself (Cajori, 1929, p. 235).

Ohm's first experiments were on the relative conductivity of metals. In these tests he was troubled by variations in his batteries ("Wogen der

Kraft" or "surge in power"). He adopted thermo-electric elements as the sources of current that were free from this trouble. He published the experimental results that were the basis for his famous law in 1826. The following year he published a book entitled *Die galvanische Kette, mathematisch bearbeitet* (Mathematical Work on the Galvanic Chain) published in Berlin, 1827. It contained a theoretical deduction of his law, and became far more widely known than his article of 1826, giving the experimental deduction. There was unfavorable reception of his conclusions. In the Berlin *Jahrbücher für wissenschaftliche Kritik*, Ohm's theory was "named a web of naked fancies, which can never find the semblance of support from even the most superficial observation of facts; he who looks on the world with the eye of reverence must turn aside from this book as the result of an incurable delusion, whose sole effort is to detract from the dignity of nature" (Cajori, 1929, p. 238).

Because Ohm's great ambition was to secure a university professorship, we can understand how this criticism affected him. To write his book of 1827, he had secured leave of absence and had gone to Berlin, where the library facilities were better than at Cologne. Not only did he fail to secure promotion by the publication of his book, but he incurred the ill will of a school official, who was a supporter of Hegelianism and, therefore, opposed to experimental research. In consequence, Ohm resigned his position in Cologne (Cajori, 1929, p. 238).

For six years Ohm lived in Berlin, giving three mathematical lessons a week in the *Kriegsschule* for a small salary. In 1833, he obtained an appointment at the polytechnic in Nürnberg. Gradually his electric researches called forth respect and admiration, particularly from foreigners, including Gustav Fechner (1801–1887) in Germany, Wheatstone in England (see Section IX), Heinrich Lenz (1804–1865) in Russia, and Joseph Henry (1797–1878) in America. In 1841, the Royal Society of London awarded him its Copley Medal, and, in 1842, it made him a foreign member. Ohm's experience reminds us of the biblical saying, "A prophet is not without honor, save in his own country" (Matthew 13:57).

In 1849, at the age of 62, the ambition of Ohm's youth was finally attained. He was appointed extraordinary professor at the University of Munich and, in 1852, ordinary professor. His writings were numerous. In addition to a number of papers on mathematical subjects, Ohm wrote a textbook, *Grundzuge der Physik* (Main Features of Physics) (1854). He died in Munich on July 7, 1854 (Preece, 1971a).

IX. APPENDIX: BIOGRAPHY OF CHARLES WHEATSTONE

Sir Charles Wheatstone (1802–1875) was an English physicist whose name is associated with the Wheatstone bridge for measuring electrical resistance. He was born near Gloucester in February 1802. He became a manufacturer of musical instruments, but in 1834 accepted the chair of experimental physics at King's College, London. About this time Wheatstone measured (with a revolving mirror) the great speed of electric discharge in conductors. Applying this speed for sending messages, he and William Fothergill Cooke (1806–1879, an English inventor) patented an early form of electric telegraph in 1837. Wheatstone's inventions included a cryptographic machine, the concertina (a small musical instrument of the accordion type, with bellows and keys), and a form of stereoscope. He wrote papers on the transmission of sound in solids and on the physiology of vision, binocular vision, and color. Wheatstone showed that the electrical sparks from different metals give different spectra. He played a prominent part in the early development of electric generators and of telegraphy with submarine cables (Preece, 1971b).

Wheatstone, a great admirer of Ohm, perceived the necessity of more accurate means of measuring resistances. The measurement of resistance had been brought to perfection chiefly by those interested in the development of the telegraph. Wheatstone invented the rheostat, but this had been superseded by the resistance box. The earlier methods of measuring resistance had the defect of depending on the constancy of the batteries used (Cajori, 1929, p. 239). Wheatstone overcame the trouble by adopting a method suggested in 1833 by Samuel Hunter Christie (1784–1865; British mathematician) (Marquis Who's Who, 1968). A footnote in a book by James Clerk Maxwell (1892, p. 495) states, "Sir Charles Wheatstone, in his paper on 'New Instruments and Processes,' *Phil. Trans.*, 1843, brought this arrangement [Wheatstone's bridge] into public notice, with due acknowledgment of the original inventor, Mr. S. Hunter Christie, who had described it in his paper on 'Induced Currents,' *Phil. Trans.*, 1833, under the name of a Differential Arrangement."

Wheatstone was an experimentalist of extraordinary skill, but disliked speaking in public. In fulfillment of his duties at King's College he delivered a course of eight lectures on sound, but his habitual (though unreasonable) distrust of his own powers of speech proved to be an invincible obstacle, and he soon discontinued his lectures. Nevertheless, he retained the professorship for many years. In private, people were charmed by his able and lucid exposition, but in public, including at the Royal Society, his attempt to repeat the same information invariably proved unsatisfactory (Cajori, 1929, p. 239). For this reason some of his more important investigations were brought before the Royal Society by Faraday.

Wheatstone's *Scientific Papers* were collected and published by the Physical Society of London in 1879. Wheatstone retired to private life, living on the income from his inventions, particularly that of the telegraph. He died in Paris on October 19, 1875 (Preece, 1971b).

X. APPENDIX: BIOGRAPHIES OF MEMBERS OF THE SIEMENS FAMILY

There were four important men in the Siemens family: Werner, William, Friedrich, and Alexander.

Werner von Siemens (1816–1892) was the chief founder of the electrical firm with his name. He was born on December 13, 1816, at Lenthe, Hanover, Germany. Between 1838 and 1848 he held a commission in the artillery, was entrusted with many specialized undertakings, and, in particular, became acquainted with the recently developed electric telegraph. In 1847 he founded, together with skilled mechanic J.G. Halske, the firm of Siemens and Halske for the manufacture of telegraphic apparatus. This firm, under Siemens's guidance, became one of the most important electrical companies in the world, with branches in different countries. The branches in England and Russia were particularly important. It carried out large telegraphic projects and expanded into other electrical fields, as new applications of electricity were developed (Weston, 1971).

Many of Werner von Siemens's inventions related to telegraphic apparatus. He used gutta-percha, a rubberlike substance from trees in Malaysia, as an insulator for telegraphic cable in 1847. This form of insulation was later widely used for electric light cables. The Siemens armature, which he invented in 1856 for use in telegraphy, was used in large generators and has evolved into the modern armature. One of the most important of Siemens's discoveries was that of the dynamo-electric principle, which governs the self-excitation of the dynamo. He died at Charlottenburg, Berlin, on December 6, 1892 (Weston, 1971).

Sir William Siemens (Karl Wilhelm; 1823–1883) was Werner's brother, and is known for his work in electricity and in the application of heat. In both fields he combined the functions of innovator, manufacturer, and successful businessman. He was born at Lenthe, Hanover, on April 4, 1823. After attending the University of Göttingen, he entered, as a pupil, the manufacturing concern of Count Stolberg at Magdeburg. At the age of 19 he first visited England in the hope of introducing an electroplating process invented by himself and Werner, which he succeeded in selling. He returned to Germany, but in 1844 was again in England, this time with another invention, the "chronometric," or differential, governor. Finding that British patent law afforded the inventor a protection then lacking in Germany, he henceforth made England his home.

The next few years were spent in trying to develop his inventions, of which at this time his water meter was commercially the most successful. His activities made him a respected figure in scientific circles. His paper "On the Conservation of Heat Into Mechanical Effect," read to the Institution of Civil Engineers in 1853, gained him the Telford Medal, and in 1862 he was elected a member of the Royal Society. William's chief work in the field of heat was concerned with regenerative heating and consequent improvements in steelmaking processes.

In the field of electricity, William became an acknowledged authority and leader. From 1848 onward, he represented the firm of Siemens and Halske in London, and when in 1865 the separate firm of Siemens Brothers was established he became a partner and director. At first, the chief business was the erection of overland telegraph lines and the laying of submarine telegraph cables. William was, however, in constant close liaison with all the ideas and projects of his brother Werner in Berlin and, when the latter discovered the dynamo-electric principle, William introduced it to England by reading a paper about it to the Royal Society in 1867. Gradually, in the late 1870s and 1880s, the electric-light side of the business grew. One of the last projects with which William was associated was the Portrush electric railway in the north of Ireland, opened in 1883, which utilized water turbines driving a Siemens dynamo. William Siemens was knighted in 1883, and he died in London the same year on November 19 (Weston, 1971).

Friedrich Siemens (1826–1904) was the brother of Werner and William. He was born in Mentzendorff, Germany, on December 8, 1826 (Marquis Who's Who, 1968). Friedrich, along with William, first tried to apply the regenerative condenser to the steam engine, using the heat from the regenerator to preheat the boiler feed water. When this did not succeed, other applications were sought and the idea occurred of applying the principle to furnaces, using the heat regained from the flue gases to heat the air supply to the furnace. This was patented by Friedrich in 1856 and met with great success for use both in glassmaking and in steel manufacture. Later the use of gas instead of solid fuel greatly extended the use of the regenerative furnace (Weston, 1971). He died May 26, 1904.

Alexander Siemens (1847–1928), William's nephew, was born in Hanover, Germany, on January 22, 1847. In 1867 he went to England, where he worked first in the workshops of Siemens Brothers at Woolwich, and then in the erection of the Indo-European telegraph line in Persia (1868) and in the laying of the Black Sea cable (1868). In 1878 he became a naturalized British subject. The following year he took over the management of the electric-light department of Siemens Brothers, and was responsible for the installation of electric light at Godalming, Surrey, the first English town to be so lighted.

Like many other members of the family, Alexander patented several inventions. After the death of Sir William he became a director of the company, a position he retained until 1918. He took an active part in public activities associated with his profession, was a member of several important committees, and was twice president of the Institution of Electrical Engineers. He died at Milford-on-Sea, Hampshire, on February 16, 1928 (Weston, 1971).

Siemens is still an important name in business today, and the company is often noted in the *Wall Street Journal*.

REFERENCES

Baker, D.A. (1984). Water relations. In *Advanced Plant Physiology* (Wilkins, M.B., Ed.), pp. 297–318. Pitman: London.

Cajori, F. (1929). *A History of Physics*. Macmillan: New York.

Clerk Maxwell, J. (1892). *A Treatise on Electricity and Magnetism*. Vol. II. Reprinted 1904. Oxford University Press: London. (Reprinted photographically in Great Britain in 1937 by Lowe and Brydone, Printers, London, from sheets of the Third edition.)

Gradmann, H. (1928). Untersuchungen über die Wasserverhältnisse des Bodens als Grundlage des Pflanzenwachstums. *Jahrbucher für Wissenschaftliche Botanik* 69; 1–100.

Hausmann, E., and E.P. Slack. (1948). *Physics*. D. Van Nostrand: New York.

Hodgman, C.D. (1959). *Handbook of Chemistry and Physics*. 40th ed. Chemical Rubber Publishing: Cleveland, Ohio.

Huber, B. (1924). Die Beurteilung des Wasserhaushaltes der Pflanze. Ein Beitrag zur vergleichenden Physiologie. *Jahrbucher für Wissenschaftliche Botanik* 64; 1–120.

Ingersoll, L.R., Martin, M.J., and Rouse, T.A. (1953). *A Laboratory Manual of Experiments in Physics*. McGraw-Hill: New York.

Kirkham, D., and Powers, W.L. (1972). *Advanced Soil Physics*. Wiley: New York.

Kirkham, M.B. (1982). Water and air conductance in soil with earthworms: An electrical-analogue study. *Pedobiologia* 23; 367–371.

Kramer, P.J. (1983). *Water Relations of Plants*. Academic Press: New York.

Marquis Who's Who. (1968). *Who's Who in Science From Antiquity to the Present*. 1st ed. Marquis-Who's Who: Chicago.

Nobel, P.S. (1974). *Introduction to Biophysical Plant Physiology*. W.H. Freeman: San Francisco.

Preece, W.E. (General Ed.) (1971a). Ohm, Georg Simon. Encyclopaedia Britannica 16; 896–897.

Preece, W.E. (General Ed.) (1971b). Wheatstone, Sir Charles. Encyclopaedia Britannica 23; 473.

Richards, L.A. (Ed.). (1954). Diagnosis and Improvement of Saline and Alkali Soils. *Agr Handbook* No. 60. United States Department of Agriculture: Washington, DC.

Schaum, D. (1961). *Schaum's Outline of Theory and Problems of College Physics.* 6th ed. Schaum: New York.

Slichter, C.S. (1899). Theoretical investigations of the motion of ground-water. *US Geol Survey Annu Rep* 19; 295–384.

Smythe, W.R. (1950). *Static and Dynamic Electricity.* 2nd ed. McGraw-Hill: New York.

van den Honert, T.H. (1948). Water transport in plants as a catenary process. *Discussions Faraday Soc* 3; 146–153.

Weston, M.K. (1971). Siemens. *Encyclopaedia Britannica* 20; 484–485.

21

Leaf Anatomy
and Leaf Elasticity

In this chapter we learn how to measure leaf elasticity and calculate moduli of elasticity for the leaves. But before we study elasticity, we need to look at leaf anatomy to understand the type of organ for which we are making the calculations.

I. LEAF ANATOMY

Plants are usually classified according to their water relations, as follows: xerophytes, mesophytes, and hydrophytes (Esau, 1977, p. 351). The xerophytes are adapted to dry habitats. Mesophytes require abundant available soil water and a relatively humid atmosphere. Hydrophytes (or hygrophytes) depend on a large supply of moisture or grow partly or completely submerged in water. The structural features typical of plants of the different habitats are referred to as xeromorphic, mesomorphic, or hydromorphic, respectively. The characteristics that distinguish plants of the various habitats are most striking in leaves. Here we consider dicotyledonous and monocotyledonous leaves and focus mainly on mesophytes, and then we look at special adapations of xerophytes.

A. Dicotyledonous Leaves

Figure 21.1 shows a dicotyledonous leaf. It is a leaf of the shrub, lilac (*Syringa vulgaris* L.) (Torres and Costello, 1963, p. 124). It is composed of an upper (adaxial) and lower (abaxial) epidermis. One stoma is evident in the lower epidermis. The thin, colorless layer deposited on the walls of the upper epidermal cells is called the cuticle and is composed of a waxy material called cutin (Torres and Costello, 1963, p. 43). The mesophyll is divided into an upper palisade mesophyll and a lower spongy mesophyll. One or two layers of columnar, compact cells lie beneath the upper epidermis. These cells make up the palisade mesophyll (also called the palisade parenchyma). Between the palisade tissue and the lower epidermis, there is a layer of large, irregular, loosely packed cells with many intercellular spaces between them. This tissue is the spongy mesophyll (also called the spongy parenchyma). The stoma in Fig. 21.1 is located near the intercellular spaces, which allows easy transport of carbon dioxide to the mesophyll. Chloroplasts, oval-shaped bodies, are present in the mesophyll cells. Dispersed throughout the mesophyll are the veins of the leaf. A vein is a strand of vascular tissue in a flat organ such as a leaf (Esau, 1977, p. 531). The largest central vein is known as the midrib. The vein is in the center of Fig. 21.1, and the conducting tissue of the vein consists of xylem and phloem. The xylem tissue is closest to the adaxial surface because, when the vascular tissue bends over from the stem into the leaf, the xylem, which is closer to the center of the plant in each vascular bundle than the phloem (see Chapter 18, Section I, for stem anatomy), comes out on top of the vascular bundle. Thus, the xylem is on top of the phloem in the leaf vascular bundle. In the figure, the cells in the xylem are shown with thick

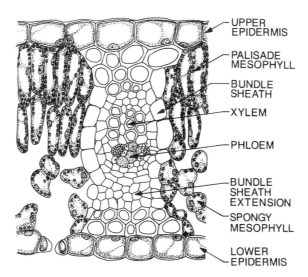

FIG. 21.1 A transverse section of a lilac leaf, a dicotyledonous leaf. (From Torres, A.M., and Costello, W.L., *A Laboratory Manual for General Botany*, p. 124, ©1963, Wm. C. Brown Book Co.: Dubuque, Iowa. This material is reproduced with permission of The McGraw-Hill Companies.)

walls and are empty because the vessel members are dead at maturity. A bundle sheath surrounds the vascular tissue. The bundle sheath extensions link the bundle sheath to both the upper and lower epidermis. The bundle sheath extensions have thick cell walls because they are made up of sclerenchyma.

In the dicotyledons, the supporting tissue in leaves may be collenchyma or sclerenchyma. The vascular bundles themselves also contribute to the support of the blades. The collenchyma occurs along the larger veins, on one or both sides. Sclerenchyma occurs in the form of bundle sheaths and bundle-sheath extensions composed of fibrous cells, and as sclereids in the mesophyll.

Many herbaceous dicotyledons have leaves with a relatively undifferentiated mesophyll (Esau, 1977, pp. 355–357). The palisade tissue is absent or weakly developed, the intercellular volume is large, and the leaf is often thin. The epidermis bears a thin cuticle, and the stomata are more or less raised. Examples of leaves with relatively undifferentiated mesophyll are those of *Pisum sativum* (pea) and *Lactuca sativa* (lettuce). A thin, loosely organized mesophyll with a single row of palisade cells is found in *Raphanus sativus* (radish), *Solanum tuberosum* (potato), and *Lycopersicon esculentum* (tomato). Leaves of the species *Gossypium* (cotton) have long palisade cells that occupy approximately one-third to one-half of the blade thickness.

Various shrubby and woody species furnish examples of leaves with well-differentiated palisade parenchyma on the adaxial side of the leaf (e.g., *Vitis*, grape; *Syringa*, lilac, Fig. 21.1; *Ligustrum*, privet; and *Pyrus*, pear) (Esau, 1977, pp. 356–357).

B. Monocotylendous Leaves

The leaves of the monocotyledons vary in form and structure, and some resemble those of the dicotyledons (Esau, 1977, p. 359). Monocotyledonous leaves may have petioles and blades, for example *Canna* (common name is also canna) and *Hosta* (plantain-lily). But the majority are differentiated into blade and sheath, and the blade is relatively narrow. The venation is typically parallel. In contrast, dicotyledonous leaves normally show a reticulate pattern of venation (Bowes, 2000, p. 10).

The anatomic structure of monocotyledonous leaves ranges from hydromorphic to extreme xeromorphic. Hydrophytes in the monocotyledons show the same basic features as those in the dicotyledons, and both have an abundance of aerenchyma. Aerenchyma is parenchyma tissue containing particularly large intercellular spaces of schizogenous, lysigenous, or rexigenous origin. *Schizogenous* is a term applied to an intercellular space originating by separation of cell walls along the middle lamella (Esau, 1977, p. 524), *lysigenous* is a term applied to an intercellular space originating by a dissolution of cells (Esau, 1977, p. 514), and *rexigenous* is a term applied to an intercellular space originating by the rupture of cells (Esau, 1977, p. 524).

Numerous monocotyledonous leaves develop large amounts of sclerenchyma, which in some species serves as an important source of commercial hard leaf fibers. The fibers are associated with the vascular bundles or appear as independent strands (Esau, 1977, p. 360).

The grass leaf typically consists of a more or less narrow blade and a sheath enclosing the stem. Vascular bundles of different sizes alternate rather regularly with one another, as typified by the wheat leaf (Fig. 21.2). The median bundle may be the largest (Esau, 1977, p. 360). The mesophyll of grasses shows, as a rule, no distinct differentiation into palisade and spongy mesophyll (parenchyma), although sometimes the cell rows beneath both epidermal layers are more regularly arranged than in the rest of the mesophyll. In some grasses, the mesophyll cells surround the vascular bundles in an orderly manner, each cell oriented with its longer diameter at right angles to the bundle so that in transverse sections the mesophyll cells appear to radiate from the bundles (Esau, 1977, p. 360).

The epidermis of grasses contains a variety of cells. The narrow guard cells of the stomata are associated with subsidiary cells (see Chapter 22,

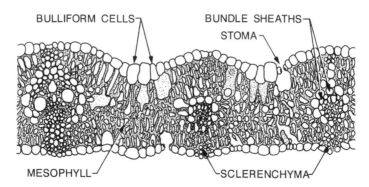

FIG. 21.2 A transverse section of a wheat leaf, a monocotyledonous leaf. The adaxial epidermis bears bulliform cells in grooved parts of the blade. Subepidermal cells are elongated like palisade cells. There is an inner thick-walled and an outer thin-walled bundle sheath. Sclerenchyma in the ribs are connected with the bundle sheath. (From Esau, K., *Plant Anatomy*, 2nd ed., p. 700, ©1965, John Wiley & Sons, Inc.; New York. This material is used by permission of John Wiley & Sons, Inc.)

Section II, for stomatal anatomy). Silica cells, cork cells, and trichomes may be present. Enlarged epidermal cells, referred to as *bulliform cells* (Fig. 21.2), are cells participating in folding movements of grass leaves. In a number of xeric grasses, enlarged epidermal cells line adaxial grooves between the vein ribs and are continuous with similarly enlarged mesophyll cells, called the *hinge cells*. During excessive loss of water, the bulliform cells, or the hinge cells, or both, become flaccid and enable the leaf to fold or to roll. But the shrinkage of the various large, thin-walled cells is only one factor causing folding, because leaves without such cells also respond to loss of moisture by rolling. Differential shrinkage of other tissues, distribution of sclerenchyma, and cohesive forces among tissues also contribute to rolling and folding of leaves (Esau, 1977, p. 362).

Grass leaves have strongly developed sclerenchyma. Commonly, fibers extend from the large vascular bundles to the epidermis. The leaf margins may have fibers, as do leaves of wheat (not shown in Fig. 21.2 because the figure shows only the center of the leaf; but see Fig. 19.8C in Esau, 1977, p. 363, for the sclerenchyma at the edge of a wheat leaf).

C. Grass Leaf Structure and Type of Photosynthesis

The bundle sheaths of grasses show variations that are significant taxonomically and as indicators of the type of photosynthesis characteristic of the species (Esau, 1977, p. 362). After the discovery of the C_4 or Hatch-Slack (1966) pathway of photosynthesis in sugar cane [see Laetsch (1974) for a review of the history of the discovery of the C_4 photosynthetic pathway], comparative grass leaf anatomy became the object of intensive investigation in relation to photosynthesis (Esau, 1977, p. 364).

The most common photosynthetic cycle is the C_3 or Calvin-Benson pathway. In C_3 plants, carbon dioxide from the atmosphere is fixed as phosphoglyceric acid, a 3-carbon compound. In C_4 plants, an additional mechanism is involved, in which atmospheric carbon dioxide is fixed as oxaloacetic acid, a 4-carbon molecule. The leaves of C_3 and C_4 plants differ in morphology as well as in the chemical mechanisms of carbon-dioxide fixation (Mellor and Jensen, 1986). In C_3 plants, chloroplasts are found in mesophyll cells throughout the leaf cross sections. Bundle sheath cells that surround the vascular bundles in C_3 plants are lacking or essentially lacking in chloroplasts.

In leaves of C_4 plants, such as corn leaves, however, the chloroplasts that fix carbon dioxide by the C_3 mechanism are highly concentrated in the bundle sheath cells. In a C_4 leaf, the chloroplasts that fix carbon dioxide by the C_4 mechanism are located in the relatively large mesophyll cells that make up the body of the leaf. The 4-carbon malic and aspartic acids (formed from the initial 4-carbon product, oxaloacetic acid) produced by the chloroplasts in these mesophyll cells are transported to the C_3 chloroplasts in the bundle sheaths. Enzymes there split off carbon dioxide from the malic and aspartic acids, and this carbon dioxide is taken up by the C_3 mechanism to form phosphoglyceric acid, the compound that is metabolized to produce the various carbohydrates, proteins, and other compounds that make up the major components of the plant. The 3-carbon molecule left after the splitting of carbon dioxide from malic and aspartic acid is pyruvic acid. The pyruvic acid is returned to the C_4 chloroplasts, where it is activated by transfer of a high-energy phosphate group from adenosine triphosphate to form phosphoenol-pyruvate. The phosphoenol-pyruvate in turn reacts with incoming carbon dioxide from the atmosphere to form oxaloacetic acid, with the loss of phosphate. This completes the C_4 cycle. Thus, in C_4 plants there are two connected carbon-dioxide fixing cycles. The C_4 cycle feeds the C_3 cycle (Mellor and Jensen, 1986).

The C_4 cycle is characteristic of plants that require relatively high temperatures for growth. In the angiosperms, this cycle has been recorded in representatives of some ten families (Amaranthaceae, Aizoaceae, Chenopodiaceae, Asteraceae, Cyperaceae, Euphorbiaceae, Poaceae, Nyctaginaceae, Portulacaceae, and Zygophyllaceae) (Esau, 1977, p. 364). About half of the species of the Poaceae are included among the C_4 plants. The C_4 plants are of tropical origin and occur widely in xerophytic environments. Because so few angiosperms are specialized for the C_4 photosynthetic cycle, the C_4 condition is considered to be of more recent origin than the C_3 condition. Woolhouse (1978) shows a map with the percentage distribution of C_4 grasses in the flora of North America. In far northern regions of Canada and Alaska, 0% of the flora has the C_4 photosynthetic pathway; in the southern United States, high percentages occur (e.g., 80% in southern Florida).

D. Xerophytic Adaptations

Plants overcome adverse conditions of a particular environment in different ways (Esau, 1977, p. 351). In a habitat deficient in water, for example, some plants develop features protecting the aerial parts from excessive loss of water; others form underground water storage organs, or develop roots reaching great depths [e.g., the deep roots of sunflower; see Rachidi et al. (1993), who found depletion of water by sunflower roots at the 2.7 m depth]; and still others control the problem by having a short life span restricted to the time when water supply is most abundant. Availability of water is an especially important factor affecting the form and structure of plant leaves. Xeromorphic characteristics of leaves include:

1. Thick cuticle (wax)
2. Small intercellular spaces
3. A large proportion of mechanical tissue (sclereids, fibers)
4. Relatively small cells
5. Multiple epidermes
6. Several layers of palisade cells between the epidermis and the spongy parenchyma
7. Sunken stomata
8. Presence of hairs in stomatal pits (crypts)
9. Presence of water storage cells
10. Spines
11. Lignified cells

Fahn and Cutler (1992) survey morphological and anatomical adaptations enabling plants to grow in arid and semi-arid regions.

II. INTERNAL WATER RELATIONS

Plants have little storage capacity for water compared with the amounts that pass through them each day (Baver et al., 1972, p. 394). They must regulate their water status to survive. To understand this regulatory process, we must discuss the internal water balance of plants, including elasticity. Here we look at leaf elasticity from a physical point of view. We follow the analysis of Gardner and Ehlig (1965), which also has been partially reproduced in Baver et al. (1972, pp. 394–398).

As we saw in Equation 18.1, under equilibrium conditions the state of the water in plant leaf cells may be written in terms of the various components of the potential energy, as follows

$$\psi = \psi_s + \psi_p + \psi_m + \psi_g, \tag{21.1}$$

where ψ is the total water potential, ψ_s is the osmotic (solute) potential component, ψ_p is the pressure potential component (turgor pressure), ψ_m is the component due to adsorption forces such as those in the cell wall, and ψ_g is the component due to gravity. We usually ignore gravity, so Equation 21.1 becomes

$$\psi = \psi_s + \psi_p + \psi_m, \qquad\qquad (21.1a)$$

The partition of energy between the osmotic and adsorption components is somewhat arbitrary, because some of the water in the leaf tissue may be subject to both osmotic effects and adsorption forces, particularly at low leaf-water content. In the vacuole, the osmotic component is important.

If the cell solution were to behave ideally, the osmotic pressure would be directly proportional to the solute concentration. There would exist, then, a simple relation between osmotic potential and cell water content:

$$\psi_s = \psi^{\circ}_s / \theta, \qquad\qquad (21.2)$$

where ψ°_s is the osmotic pressure at full turgor and θ is the relative water content of the cell. θ is the ratio of the water content of the cell to that water content it has when in equilibrium with free water at the same temperature and pressure. If the amount of bound water is appreciable, then this amount should be subtracted from θ. Some investigators have found appreciable amounts of bound water (as much as 30%; Slavík, 1963). However, we shall consider the amount of bound water to be small, and we shall not subtract it from θ.

The osmotic and pressure components of the potential are not independent. Because of the elastic nature of the cell wall, changes in turgor pressure cause changes in cell volume, due to changes in cell water content. An increase in the turgor pressure results in an expansion of the cell walls. This is accomplished by the uptake of water. Unless this uptake also is accompanied by a proportional uptake of solutes, the solute concentration decreases with a consequent increase in the osmotic potential. Solute transport across the membranes can and does occur, but at a rate that is generally slower than the rate of water movement, so that the immediate response of a cell to any change in water potential is a change in its water content or degree of hydration (Baver et al., 1972, p. 396).

As stated, the components of the water potential in a leaf cell are not completely independent. On a short-term basis, if the total water potential is specified, this determines both the osmotic and the turgor potential, as well as the degree of hydration. (We recognize that we can never *measure* a zero water potential in a leaf, even when fully hydrated. The water potential is always slightly negative.)

As the relative water content decreases, the solute concentration must increase proportionately, if the solute content remains constant (Gardner and Ehlig, 1965). This results in a decrease in the osmotic potential. If it is assumed that the relation between the turgor pressure and the cell volume is linear, then

$$\psi_p = e(\theta - \theta_o)/\theta_o, \tag{21.3}$$

where θ_o is the relative water content at which the turgor potential becomes zero and e is the modulus of elasticity. Substituting Equations 21.2 and 21.3 into Equation 21.1a, we get

$$\psi = \psi_s^{\circ}/\theta + e(\theta - \theta_o)/\theta_o + \psi_m(\theta), \tag{21.4}$$

in which ψ_m is now a function of θ. Equation 21.4 gives us a relation between the water potential and the relative water content of the cell. $\psi_m(\theta)$ represents the relation between the water content and the matric potential. Growth can be expected to cause some departure from the expression used in deriving Equation 21.4, but to the extent that the assumptions are valid, Equation 21.4 gives a unique relation between the total water potential and the relative water content of the leaf.

In practice, it is easier to make the measurements needed to test Equation 21.4 on tissue rather than on single cells. Therefore, Gardner and Ehlig (1965) used tissue [leaves of cotton (*Gossypium hirsutum* L.), bell pepper (*Capsicum frutescens* L.), sunflower (*Helianthus annuus* L.), and birdsfoot trefoil (*Lotus corniculatus* L.)]. The plants were grown in a greenhouse. To obtain different values of water potential, they withheld water from the plants until their leaves wilted to the desired extent. Water potential and osmotic potential were determined with thermocouple psychrometers. The relative water content was determined by using the method of Barrs and Weatherley (1962).

Figure 21.3 shows the relation between the relative water content and osmotic potential for the four plant species, as determined by Gardner and Ehlig (1965). The data are plotted on a logarithmic scale and the straight line has a slope of 45 degrees, as would be predicted if the solute content were to remain constant and the amount of bound water were negligible.

If a plant is growing in a saline soil solution, then over a period of time, the solute content of the cells tends to adjust accordingly. The rate of adjustment varies from species to species. Figure 21.4 shows the relation between the total water potential and the osmotic potential for bell pepper on both saline and nonsaline substrates (Ehlig et al., 1968).

If we neglect the matric and gravitational potentials, we can use Equation 21.1 to obtain the turgor potential by subtracting the osmotic potential from the total water potential. All three potentials are plotted as a

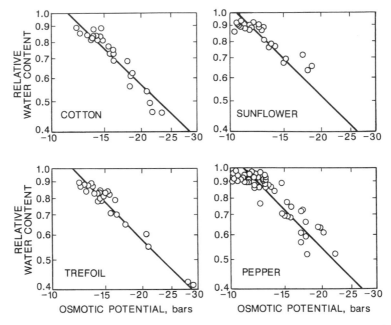

FIG. 21.3   Leaf relative water content as a function of the average osmotic potential in the plant leaf. The straight lines represent the relation expected if the solutes behave ideally and there is no bound water. (From Gardner, W.R., and Ehlig, C.F., Physical aspects of the internal water relations of plant leaves. Plant Physiology 40; 705–710, ©1965, American Society of Plant Physiologists. Reprinted by permission of the American Society of Plant Biologists, Rockville, Maryland.)

function of relative water content for nonsaline plants (Fig. 21.5). Of particular interest is the abrupt change in slope of the pressure potential (turgor potential) at a leaf relative water content of about 0.85. In pepper, for example, this corresponds to a total water potential of about −11 bars (−1.1 MPa) and coincides with the appearance of marked symptoms of visible wilting. The change in slope corresponds to a change in the elastic modulus of the leaf tissue and explains the wilting symptoms. This also corresponds roughly with the point at which the stomata are almost completely closed (Baver et al., 1972, p. 398).

III. ELASTICITY

Before we look at the data that Gardner and Ehlig (1965) calculated for the moduli of elasticity of the plants they studied, let us define *modulus of elasticity*. To do this, we refer to a college physics book (Schaum, 1961, pp. 90–91). *Elasticity* is defined as that property by virtue of which a body

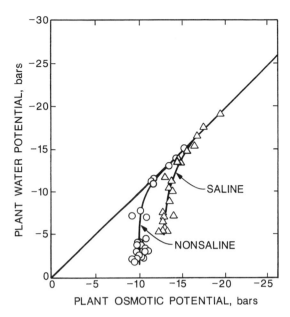

FIG. 21.4 Plant leaf water potential (total) as a function of the osmotic potential component. (From Ehlig, C.F., et al., Effect of salinity on water potential and transpiration in pepper (*Capsicum frutescens*). *Agronomy Journal* 60, 249–253, ©1968, American Society of Agronomy: Madison, Wisconsin. Reprinted by permission of the American Society of Agronomy.)

tends to return to its original size or shape after a deformation and when the deforming forces have been removed. *Stress* is measured by the force applied per unit area that produces or tends to produce deformation in a body. It is expressed in such units as lb/ft^2, $newton/m^2$, and $dynes/cm^2$.

Stress = force/(area of surface on which force acts) = F/A.   (21.5)

Strain is the fractional deformation resulting from a stress. It is measured by the ratio of the change in some dimension of the body to the total value of the dimension in which the change occurred. (A strain is a pure number and has no dimensions.) Thus, if a wire of initial length l experiences an elongation Δl when a force is applied to the wire, the *longitudinal strain* is

longitudinal strain = (change in length)/(initial length) = $\Delta l/l$.   (21.6)

The *elastic limit* is the smallest value of the stress required to produce permanent strain in the body. Within the elastic limit of any body, the ratio of the stress to the strain produced is a constant. This constant is called the *modulus of elasticity* of the material of the body.

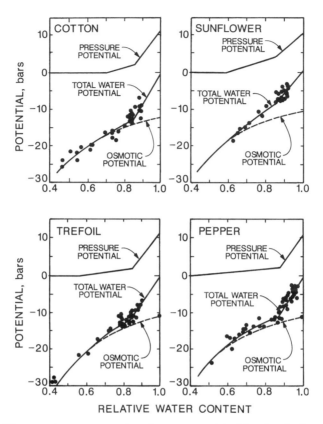

FIG. 21.5 The osmotic, pressure, and total water potential of the plant leaf as a function of the relative water content. The circles represent the experimentally determined values for the total water potential. The dashed line is the theoretically predicted osmotic potential. The osmotic and pressure potential components are added to give the calculated relation between total water potential and relative water content indicated by the smooth curve. (From Gardner, W.R., and Ehlig, C.F., Physical aspects of the internal water relations of plant leaves. Plant Physiology 40; 705–710, ©1965, American Society of Plant Physiologists. Reprinted by permission of the American Society of Plant Biologists, Rockville, Maryland.)

$$\text{Modulus of elasticity} = \text{stress required to produce unit strain}$$
$$= \text{stress/strain} \qquad (21.7)$$

Equation 21.7 is called *Hooke's law*. (For a biography of Hooke, see the Appendix, Section V.)

There are two types of elasticity: length elasticity and volume elasticity. We now define *length elasticity* or *Young's modulus*, Y. (For a biography of Young, see the Appendix, Section VI.) Consider that a wire or rod of length l and cross-sectional area A experiences an elongation Δl when a stretching force f is applied to it. Then

$$Y = \text{(longitudinal stress)/(longitudinal strain)} = (F/A)/(\Delta l/l) = (Fl)/(A\Delta l)$$
$$(21.8)$$

Y may be expressed in lb/in^2, newton/m^2, or dynes/cm^2. Y depends only on the material of the wire or rod and not on its dimensions.

We now define *volume elasticity* or *bulk modulus*, B. Consider that a body is subjected to a hydrostatic pressure, the same amount of force acting perpendicularly on each unit of surface area. The shape of the body remains the same but its volume decreases.

$$\text{Volume stress} = F/A = \text{normal force per unit area} = \text{pressure increase } \Delta p$$
$$(21.9)$$

$$\text{Volume strain} = \text{(volume decrease } \Delta V)/\text{(initial volume } V) = \Delta V/V$$
$$(21.10)$$

$$B = \text{volume stress/volume strain} = \Delta p/(\Delta V/V) = (V\Delta p)/\Delta V \quad (21.11)$$

The reciprocal of the bulk modulus of a substance is called the *compressibility* of the substance.

IV. ELASTICITY APPLIED TO PLANT LEAVES

Now let us return to the analysis of Gardner and Ehlig (1965). They wanted to determine the elasticity of plant cells. They first plotted turgor potential (pressure potential) versus relative water content (Fig. 21.6). If the relative water content is taken as a measure of average cell size (volume) (i.e., they are calculating bulk modulus), it is obvious that cell size is not a simple linear function of turgor pressure. However, the data can be represented reasonably well by two straight-line segments. One of the line segments is drawn so as to pass through the point of maximum turgor pressure corresponding to ψ_s°, as determined from Fig. 21.3, when $\theta = 1$. It appears that Hooke's law is obeyed reasonably well, if a distinction is made between a condition of high turgor pressure and one of low turgor pressure and with a different (bulk) modulus of elasticity for each range. The change in the elasticity occurs at about 2 bars for cotton, trefoil, and pepper, and at about 3.5 bars for sunflower.

Gardner and Ehlig (1965) then looked at the elastic properties of a leaf along different axes (in plane of leaf and perpendicular to plane of leaf) (length elasticity or Young's modulus). To investigate this, they determined the areas of the individual leaf disks as a function of relative water content

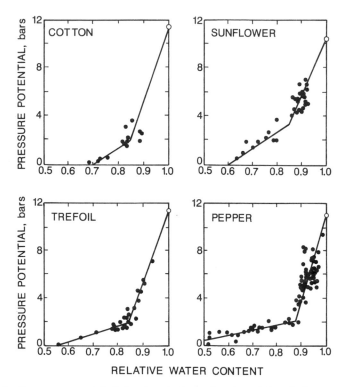

FIG. 21.6   Pressure potential of the plant leaf as a function of the relative water content. The pressure potential at a relative water content of unity was taken numerically equal to the osmotic potential at this water content in Fig. 21.3. (From Gardner, W.R., and Ehlig, C.F., Physical aspects of the internal water relations of plant leaves. Plant Physiology 40; 705–710, ©1965, American Society of Plant Physiologists. Reprinted by permission of the American Society of Plant Biologists, Rockville, Maryland.)

(Fig. 21.7). They divided the relative water content (as noted, an indication of volume) by the relative area to obtain the thickness of the leaf discs (Fig. 21.8). Note that the data in Fig. 21.8 fall on a straight line above a water content of about 0.4, but tend to curve toward the origin at lower water contents. The curvilinear part of the curve is explained by assuming that the water bound in the cell walls does not contribute to the expansion of the leaf. The straight-line portion of the curve is displaced upward because of this water. On extrapolating the curves in Fig. 21.8 back to zero relative water content, the quantity of water involved can be estimated. This turns out to be, for example, approximately 10% for sunflower, trefoil, and pepper, relative to the fully turgid condition. (In Fig. 21.8, read from the dashed line on the ordinate horizontally over to the solid line, and then read the corresponding relative water content on the abscissa.)

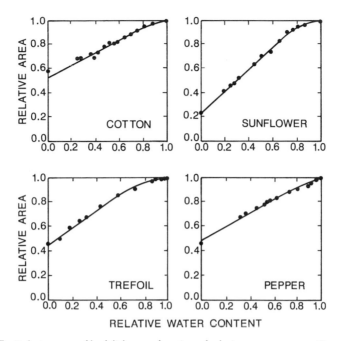

FIG. 21.7   Relative area of leaf disks as a function of relative water content. (From Gardner, W.R., and Ehlig, C.F., Physical aspects of the internal water relations of plant leaves. Plant Physiology 40; 705–710, ©1965, American Society of Plant Physiologists. Reprinted by permission of the American Society of Plant Biologists, Rockville, Maryland.)

The relative diameter and the relative thickness are plotted in Fig. 21.9 as a function of pressure potential (turgor pressure or turgor potential). Most of the increase in volume with increasing turgor pressure occurs in the leaf thickness with only a small increase occurring in the lateral dimensions of the leaf. All four species studied exhibited nearly the same moduli of elasticity in the high turgor pressure range, with more variation between species in the low pressure range. Values for the elastic moduli taken from the slopes of the lines in Figs. 21.6 and 21.9 are given in Table 21.1. Ordinarily, the elastic modulus is defined in terms of the increase in a dimension relative to that dimension when there is zero stress. However, it is much more difficult to fix precisely the point of zero turgor than the point of maximum turgor. For this reason, the moduli in Table 21.1 were calculated with respect to a relative water content of 1.0.

We can compare the values for moduli of elasticity in Table 21.1 to those of nonliving materials (Table 21.2). (To compare units in Tables 21.1 and 21.2, remember that 1 newton/m^2 = 10 dynes/cm^2 because 1 newton = 10^5 dynes and 1 m^2 = 10^4 cm^2. For example, brass has a bulk modulus of

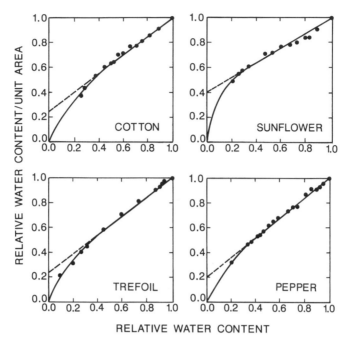

FIG. 21.8 Relative water content per unit area as a function of the relative water content. This ratio gives a measure of leaf thickness. (From Gardner, W.R., and Ehlig, C.F., Physical aspects of the internal water relations of plant leaves. Plant Physiology 40; 705–710, ©1965, American Society of Plant Physiologists. Reprinted by permission of the American Society of Plant Biologists, Rockville, Maryland.)

TABLE 21.1   Moduli of elasticity

Species	Turgor pressure range (bars)	Bulk modulus (dynes/cm²)	In plane of leaf (dynes/cm²)	Perpendicular to plane of leaf (dynes/cm²)
Cotton	>2	6.0×10^7	42.0×10^7	8.1×10^7
	<2	1.5×10^7	5.0×10^7	2.0×10^7
Sunflower	>3.4	4.7×10^7	46.5×10^7	7.9×10^7
	<3.4	1.4×10^7	3.3×10^7	2.3×10^7
Trefoil	>2	6.0×10^7	48.0×10^7	7.7×10^7
	<2	0.63×10^7	2.6×10^7	0.85×10^7
Pepper	>2	7.1×10^7	35.5×10^7	9.9×10^7
	<2	0.44×10^7	1.6×10^7	0.59×10^7

From Gardner, W.R. and Ehlig, C.F., Physical aspects of the internal water relations of plant leaves. *Plant Physiology* 40; 705–710, ©1965, American Society of Plant Physiologists. Reprinted by permission of the American Society of Plant Biologists, Rockville, Maryland.

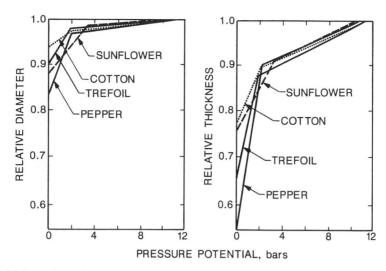

FIG. 21.9 Relative diameter and relative thickness of leaf disks as a function of the pressure potential. The slopes of these lines are proportional to the moduli of elasticity. (From Gardner, W.R., and Ehlig, C.F. Physical aspects of the internal water relations of plant leaves. Plant Physiology 40; 705–710, ©1965, American Society of Plant Physiologists. Reprinted by permission of the American Society of Plant Biologists, Rockville, Maryland.)

10×10^{10} N/m². This equals 10×10^{11} dynes/cm².) Comparing Tables 21.1 and 21.2, we see that the bulk modulus of turgid plants is about 10^4 times less than that of nonliving materials. The bulk modulus of wilted plants is about 10^5 times less than that of nonliving materials. The modulus in the plane of a leaf of turgid plants is about 10^3 times less than Young's modulus for nonliving materials. The modulus in the plane of a leaf of a wilted plant is about 10^4 times less than Young's modulus for nonliving materials. A dry cotton fiber has a Young's modulus of 1×10^{11} dynes/cm² (Nobel, 1974, p. 38). (A dry cotton fiber is almost entirely cellulose.) Young's modulus for cotton fibers is about 5% of that for steel. One can see that the moduli of elasticity for plants can be fairly large.

Some interesting conclusions concerning the phenomenon of wilting can be drawn from the data on elasticity (Table 21.1). It has generally been assumed that the permanent wilting point corresponds to zero turgor pressure in the plant leaf. The data (Fig. 21.6 and Table 21.1) indicate that visible wilting symptoms occur at a turgor pressure of 2 or 3 bars. Therefore, the visible wilting associated with the permanent wilting point is due to a marked change in the elastic properties of the cell when the turgor pressure drops below a critical value, rather than the complete absence of turgor. This is logical from a physical standpoint. Disregarding the support given

TABLE 21.2   Typical elastic constants

Material	Young's modulus N/m^2	Bulk modulus N/m^2
Aluminum	6.9×10^{10}	...[a]
Brass	9.0×10^{10}	10×10^{10}
Copper	11×10^{10}	14×10^{10}
Nickel	21×10^{10}	...
Steel	20×10^{10}	17×10^{10}
Tungsten	35×10^{10}	...
Glass	5.4×10^{10}	3.6×10^{10}
Ethyl ether	...	0.6×10^9
Ethyl alcohol	...	1.1×10^9
Water	...	2.1×10^9
Mercury	...	28×10^9

[a]Not given.
From Shortley, G., and Williams, D., *Elements of Physics*, 5th ed. p. 225, ©1971. Reprinted by permission of Pearson Education, Inc: Upper Saddle River, New Jersey.

to the leaf blade by the veins, the bending of a leaf is similar to the bending of a beam. The extent to which the leaf will flex under its own weight should be inversely proportional to the appropriate modulus of elasticity and to the cube of the blade thickness (one cubes a leaf dimension to get a volume). When the turgor pressure is above 2 bars, the thickness is relatively constant and little variation in flexure with varying turgor pressure is to be expected. When the turgor pressure is reduced below the critical pressure of about 2 bars, the elastic modulus decreases markedly, allowing the leaf to sag. As the turgor pressure is further reduced the reduction in leaf thickness tends to permit futher bending. The cotton leaf, on one hand, is relatively rigid and is well supported by the veins, so that it exhibits only modest wilting. The pepper, which, on the other hand, is quite elastic and undergoes a considerable change in thickness, shows extreme wilting as the turgor pressure approaches zero (Fig. 21.9). The critical turgor pressure at which this change in elasticity is observed corresponds to a water potential of about −11 to −13 bars. This is in good agreement with the traditionally accepted permanent wilting point, which is reasonably well correlated with a soil water potential of −15 bars.

V. APPENDIX: BIOGRAPHY OF ROBERT HOOKE

Robert Hooke (1635–1703) was an English experimental physicist, who discovered the first law of elasticity for solid bodies, known as Hooke's law. He was born July 18, 1635, at Freshwater, Isle of Wight (Preece, 1971a).

In 1654, Robert Boyle (1627–1691; English physicist and chemist) settled at Oxford, where he erected a laboratory, kept several operators at work, and engaged, in 1655, Robert Hooke as his chemical assistant. After reading of the air-pump of Otto von Guericke (1602–1686; German physicist), Boyle used Hooke's skill to make a less clumsy pump, which was completed in 1659 (Cajori, 1929, p. 78).

On November 12, 1662, Hooke was appointed curator of experiments to the Royal Society, of which he was elected a fellow in 1663, and filled the office during the remainder of his life. In 1665 he was appointed professor of geometry in Gresham college. He was secretary to the Royal Society between 1677 and 1683, publishing in 1681–1682 the papers read before that body under the title of *Philosophical Collections*.

Hooke's optical investigations led him to adopt in 1665 in an imperfect form the undulatory theory of light, which preceded the paper on the wave theory of light presented by Christian Huygens (1629–1695; Dutch physicist) at the meeting of the French Academy of Sciences in 1678 (Preece, 1971a). (Huygens was induced by Louis XIV to settle in Paris, where he remained from 1666 to 1681 and, like his great contemporaries Newton and Leibniz, Huygens never married.) Hooke was the first to state clearly that the motions of the heavenly bodies must be regarded as a mechanical problem, and he approached in a remarkable manner the discovery of universal gravitation (Preece, 1971a).

Hooke invented the wheel barometer, discussed the application of barometric indications to meteorologic forecasting, and originated the idea of using the pendulum as a measure of gravity. He is credited with the invention of the anchor escapement for clocks and of the application of spiral springs to the balances of watches (1676) (Preece, 1971a). Hooke died on March 3, 1703, in London. His principle writings are *Micrographia* (1665), *Lectiones Cutlerianae* (1674–1679), and *Posthumous Works*.

VI. APPENDIX: BIOGRAPHY OF THOMAS YOUNG

Thomas Young (1773–1829), an English physicist and physician, who gave his name to Young's modulus, was born at Milverton, Somersetshire, England, June 13, 1773 (Preece, 1971b). This great scientist had an extraordinary childhood (Cajori, 1929, p. 148). He could read with fluency at the age of two. When four years old he had read the Bible twice through; at the age of six he could repeat the whole of Goldsmith's *Deserted Village*. He devoured books, whether classical, literary, or scientific, in rapid succession. At about 16 he abstained from using sugar on account of his opposition to the slave trade. At 19 he entered upon a medical education, which was pursued first in London, then in Edinburgh (Scotland), Göttingen (Germany), and finally at Cambridge (England). He began medical practice in London in 1799 (Preece, 1971b). In 1801, he accepted the office of professor of natural philosophy in the Royal Institution, the metropolitan school of science established in the preceding year. He held this position for two years. In 1802, he was appointed foreign secretary of the Royal Society, and held this office for the remainder of his life. He was elected fellow of the Society in 1794.

Young's earliest studies were on the anatomic and optical properties of the eye. Then followed his first epoch of optical discovery, 1801–1804. In 1801, the paper that Young read before the Royal Society dealt with the color of thin plates, in which he supported the undulatory theory of light (Cajori, 1929, p. 149). He made crucial early researches that effectively established the wave theory and was the first to make a thorough application of it to sound and light. He gave the word *energy* its scientific significance (Preece, 1971b).

Young's observations were made with great exactness, but his mode of explaining them was condensed and somewhat obscure (Cajori, 1929, p. 149). His papers, containing the great principle of interference, constituted by far the most important publication on physical optics issued since the time of Newton, yet they made no impression on the scientific public. They were attacked by Lord Brougham in the *Edinburgh Review*. Young's articles were declared to contain "nothing which deserves the name either of experiment or discovery," to be "destitute of every species of merit." "We wish to raise our feeble voice," says Brougham, "against innovations that can have no other effect than to check the progress of science." After stating that the law of interference was "absurd" and "illogical," Brougham said, "We now dismiss, for the present, the feeble lucubrations of this author, in which we have searched without success for some traces of learning, acuteness, and ingenuity, that might compensate his evident deficiency in the powers of solid thinking, calm and patient investigation, and successful development of the laws of nature, by steady and modest observation of her operations." Young issued an able reply, published in the form of a pamphlet, which failed to turn public opinion in favor of this theory (Cajori, 1929, p. 150).

Because his wave theory was laughed at, Young proceeded to other studies. The 12 succeeding years after 1801 were given to medical practice and to the study of philology, especially the decipherment of Egyptian hieroglyphic writing. The Rosetta stone (black basalt, 114 cm long and

71 cm wide) is an ancient Egyptian stone bearing inscriptions in two languages and three scripts: hieroglyphics, demotic (another ancient Egyptian writing), and Greek. It was found in August, 1799, by a French man, whose name is given variously as Bouchard or Boussard, during the execution of repairs to the fort of St. Julien near the town of Rosetta, or Rashid, on the left bank of a branch of the Nile in the western delta, about 48 km from Alexandria. It passed into British hands with the French surrender of Egypt (1801) and is now in the British Museum, London. The inscription records the commemoration of the accession of Ptolemy V Ephiphanes to the throne of Egypt in the year 197–196 B.C. in the ninth year of his reign. The stone gave the key to the translation of Egyptian hieroglyphics hitherto undeciphered (Seton-Williams, 1971).

The decipherment of the hieroglyphic inscription was largely the work of Young and Jean François Champollion (1790–1832; French Egyptologist). Young discovered that the royal names were written within ovals known as cartouches, and he worked out the names of Ptolemy and Cleopatra. He also discovered in 1814 the way in which the hieroglyphic signs were to be read, by examining the direction in which the birds and animals in this pictorial script faced. The work of these two men established the basis for the translation of all hieroglyphic texts (Seton-Williams, 1971). One hieroglyph that we recognize today is the ankh, a cross with a loop at the top. It is the symbol for life.

When Augustin Fresnel (1788–1827; French physicist) began to experiment on light and to bring into prominence Young's theory, Young then resumed his early studies, and entered into his second great epoch of optical investigation (Cajori, 1929, p. 149). Young died in London on May 10, 1829.

REFERENCES

Barrs, H.D., and Weatherley, P.E. (1962). A re-examination of the relative turgidity technique for estimating water deficits in leaves. *Australian J Biol Sci* 15; 413–428.

Baver, L.D., Gardner, W.H., and Gardner, W.R. (1972). *Soil Physics*. 4th ed. Wiley: New York.

Bowes, B.G. (2000). *A Color Atlas of Plant Structure*. Iowa State University Press: Ames.

Cajori, F. (1929). *A History of Physics*. Macmillan: New York.

Ehlig, C.F., Gardner, W.R., and Clark, M. (1968). Effect of soil salinity on water potentials and transpiration in pepper (*Capsicum frutescens*). *Agronomy J* 60; 249–253.

Esau, K. (1965). *Plant Anatomy*. 2nd ed. Wiley: New York.

Esau, K. (1977). *Anatomy of Seed Plants*. 2nd ed. Wiley: New York.

Fahn, A., and Cutler, D.F. (1992). *Xerophytes: Encyclopedia of Plant Anatomy*. Vol. 13, Part 3. Gebrüder Borntraeger: Berlin.

Gardner, W.R., and Ehlig, C.F. (1965). Physical aspects of the internal water relations of plant leaves. *Plant Physiol* 40; 705–710.

Hatch, M.D., and Slack, C.R. (1966). Photosynthesis by sugar-cane leaves. A new carboxylation reaction and the pathway of sugar formation. *Biochem J* 101; 103–111.

Laetsch, W.M. (1974). The C_4 syndrome: A structural analysis. *Annu Rev Plant Physiol* 25; 27–52.

Mellor, R.S., and Jensen, R.G. (1986). Photosynthesis: Nature's big green machine. *Sci Food Agr* 4(1); 14–19.

Nobel, P.S. (1974). *Introduction to Biophysical Plant Physiology*. W.H. Freeman and Company: San Francisco.

Preece, W.E. (General Ed.) (1971a). Hooke, Robert. *Encyclopaedia Britannica* 11; 669.

Preece, W.E. (General Ed.) (1971b). Young, Thomas. *Encyclopaedia Britannica* 23; 909.

Rachidi, F., Kirkham, M.B., Stone, L.R., and Kanemasu, E.T. (1993). Soil water depletion by sunflower and sorghum under rainfed conditions. *Agr Water Manage* 24; 49–62.

Schaum, D. (1961). *Theory and Problems of College Physics*. 6th ed. Schaum: New York.

Seton-Williams, M.V. (1971). Rosetta stone. *Encyclopaedia Britannica* 19; 629.

Shortley, G., and Williams, D. (1971). *Elements of Physics*. 5th ed. Prentice-Hall: Englewood Cliffs, New Jersey.

Slavík, B. (1963). Relationship between the osmotic potential of cell sap and the water saturation deficit during the wilting of leaf tissue. *Biol Plant* 5; 258–264.

Torres, A.M., and Costello, W.L. (1963). *A Laboratory Manual for General Botany*. Wm. C. Brown: Dubuque, Iowa.

Woolhouse, H.W. (1978). Light-gathering and carbon assimilation processes in photosynthesis: Their adaptive modifications and significance for agriculture. *Endeavour* (new series) 2(1); 35–46.

22

Stomata and Measurement of Stomatal Resistance

The two main parts of a plant that control its water status are the roots, where water enters, and the stomata on the leaves, where water exits. We considered roots in Chapters 14 and 15. Here we consider stomata.

I. DEFINITION OF STOMATA AND THEIR DISTRIBUTION

The stomata are apertures in the epidermis, each bounded by two guard cells. In Greek, *stoma* means "mouth," and the term is often used with reference to the stomatal pore only. Esau (1965, p. 158) uses the term stoma to include the guard cells and the pore between them, and we will use her definition. The plural of stoma is *stomata*. There is no such word as "stomates."

Stomata occur in vascular plants. Vascular plants include the lower vascular plants such as horsetails (*Equisetum*), ferns (Class Filicinae), gymnosperms, and angiosperms. As noted before, the angiosperms are the flowering plants, and the group consists of the two large classes of Monocotyledoneae (monocotyledons) and Dicotyledoneae (dicotyledons) (Fernald, 1950).

By changes in their shape, the guard cells control the size of the stomatal aperture. The aperture leads into a substomatal intercellular space, the substomatal chamber, which is continuous with the intercellular spaces in the mesophyll. In many plants, two or more of the cells adjacent to the guard cells appear to be associated functionally with them and are morphologically distinct from the other epidermal cells. Such cells are called *subsidiary*, or *accessory*, cells (Esau, 1965, p. 158).

The stomata are most common on green aerial parts of plants, particularly the leaves. They also can occur on stems, but less commonly than on leaves. The aerial parts of some chlorophyll-free land plants (*Monotropa, Neottia*) and roots have no stomata as a rule, but rhizomes have such structures (Esau, 1965, p. 158). Stomata occur on some submerged aquatic plants and not on others. The variously colored petals of flowers often have stomata, sometimes nonfunctional. Fruits also can have stomata. Stomata are found on stamens and gynoecia.

Stomata can be distributed in the following ways on the two sides of a leaf:

- An *amphistomatous* leaf has stomata on both surfaces. Most plants have such a distribution.
- A *hypostomatous* leaf has stomata only on the upper surface. Many tree species are characterized by having hypostomatous leaves, such as horse chestnut (*Aesculus hippocastanum*) and basswood (*Tilia europea*) (Meidner and Mansfield, 1968; see their Table 1.1). The leaf of poplar (*Populus* sp.) is an exception. It has stomata on both surfaces and a petiole that allows the leaf to turn readily in the wind. These adaptations may allow its fast growth rate. The fast growth rate of poplar is one reason it is widely used in phytoremediation (use of plants to remove pollutants from soil).
- An *epistomatous* leaf has stomata only on the upper surface of the leaf. Some floating plants are epistomatous.
- A *heterostomatous* leaf has stomata that occur with more than twice the frequency on the abaxial surface than on the adaxial surface. An *isostomatous* leaf has stomata that occur with approximately equal frequencies on both surfaces.

The *stomatal ratio* is the ratio of stomatal frequency on the adaxial surface to that on the abaxial surface.

II. STOMATAL ANATOMY OF DICOTS AND MONOCOTS

Figure 22.1 shows how the stomata develop differently in broad-leaved plants (mainly dicotyledons), which have elliptical shapes, compared to grass species (monocotyledons), which have dumb-bell shapes. The most commonly occurring stomata are elliptical in shape and differentiate from a protodermal cell by division into two guard cells, which soon assume their typical shape—like a bean in surface view (Fig. 22.1, left). By separating slightly in the center, the guard cells form the stomatal pore between them. There is no radical change in shape of the guard cells as they grow in size except that the early rounded shape changes into a more elongated, elliptical one. Adjacent epidermal cells may or may not be distinctive in appearance, but they usually function as subsidiary cells (Meinder and Mansfield, 1968, p. 6).

In most members of the Poaceae (formerly Gramineae) (grass family) and Cyperaceae (sedge family), differentiation of a stoma begins with the division of two protoderm cells on either side of a stoma mother cell. The two daughter cells resulting from these divisions, which lie adjacent to the stoma mother cell, are the two future subsidiary cells. They are clearly distinguishable in shape from the other epidermal cells. The stoma mother cell divides next to form the guard cells, between which the stomatal pore appears. At this stage, the graminaceous stoma resembles the elliptical one in shape, but a further stage in its development results in an elongation of the guard cells which finally assume the characteristic dumb-bell shape (Fig. 22.1, right) (Meidner and Mansfield, 1968, pp. 6–8).

In leaves with parallel veins, such as those of monocotyledons and some dicotyledons, and in the needles of conifers, the stomata are arranged in parallel rows. In netted-veined leaves, which include most dicotyledons and a few monocotyledons, the stomata are scattered (Esau, 1965, p. 158). In leaves with parallel veins, which have the stomata in longitudinal rows, the developmental stages of the stomata are observable in sequence in the successively more differentiated portions of the leaf. This sequence is basipetal, that is, from the tip of the leaf downward. In the netted-veined leaves, the different developmental stages are mixed in mosaic fashion so that mature stomata occur side by side with immature ones (Esau, 1965, p. 166).

III. STOMATAL DENSITY

Esau (1965, p. 158) gives the density of stomata as between 100 and 300 per square millimeter for leaves of many species. The number of stomata is dependent on the species. Meidner and Mansfield (1968; see their Table 1.1) give the frequency of stomata on leaves of different species, including the lower vascular plants (ferns), gymnosperms, and angiosperms. Most plants have more stomata on the lower (abaxial) surface than on the upper (adaxial) surface, but wheat (*Triticum* sp.) is an exception. It has more stomata on

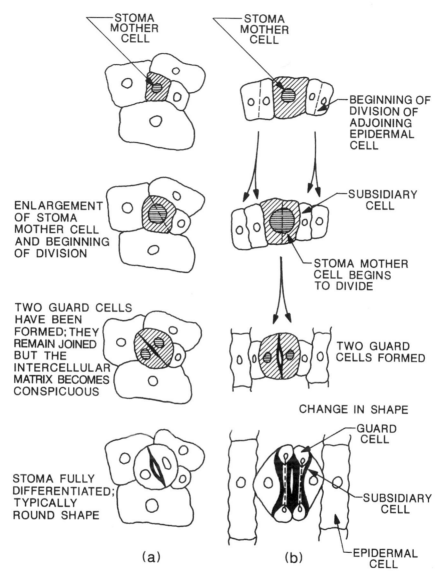

FIG. 22.1 Four stages in the differentiation of (a) elliptically shaped and (b) graminaceous stomata. (From Meidner, H., and Mansfield, T.A., *Physiology of Stomata*, p. 7, ©1968, McGraw-Hill Book Co: New York. This material is reproduced with permission of The McGraw-Hill Companies.)

the upper surface than on the lower surface. The number of stomata per unit area changes as a leaf grows. It tends to be higher in earlier stages of development than in later stages (Meidner and Mansfield, 1968, p. 6). Stomata may grow in size and change shape as the leaf blade expands.

At maturity of a leaf, the number of stomata per unit leaf area may not be constant. It can be affected by environmental factors. More stomata per unit area occur in sun leaves than in shade leaves. More stomata per unit area occur in leaves of plants growing in moist soil and high humidity than in dry conditions. Stomatal density can be affected by leaf position. Liang et al. (1975) measured stomatal density on the 15 uppermost leaves of six varieties of grain sorghum [*Sorghum bicolor* (L.) Moench] and their 15 F_1 hybrids. The second leaf from the top had the highest density, and leaf no. 15 had the lowest. Distribution also can vary with distance from the leaf base. Liang et al. (1975) found in their study with sorghum that stomatal density on the abaxial surface (which had more stomata than the adaxial surface) was highest at the basal portion of the leaves.

IV. DIFFUSION OF GASES THROUGH STOMATAL PORES

The distribution of stomata affects the diffusion of gases through them. Early important investigations on the diffusion of gases and liquids through small openings were carried out by Brown and Escombe (1900). These investigations proved that the rates of diffusion through small single apertures are proportional to the diameters and not to the areas of the openings. This agreed with results previously established by Stefan for the converse case of evaporation from circular surfaces of water (Maximov, 1929, p. 172). He compared evaporation from large surfaces (e.g., lakes) and small surfaces. For surfaces of small dimension, diffusion is more rapid at the edges than at the center, because at the margins the molecules of water vapor can diffuse fan-wise in all directions instead of only perpendicularly to the surface at the center. It follows that, in still air, the smaller the area of the evaporating surface, the more rapid the rate of evaporation. But for areas of such small dimensions as leaves or small bowls of water, it appears, as has been mathematically calculated by Stefan (1881), that evaporation is proportional not to the area of these objects but to their periphery or radius (Maximov, 1929, p. 136).

Brown and Escombe (1900) found that their "diameter law" holds good also for the case of diffusion through a number of small openings (i.e., through a "multiperforate septum"). From this it follows that more water vapor will diffuse in unit time through several small apertures than through a single larger opening with an area equal to the combined areas of the smaller ones. If, however, the perforations in a septum separating two mixtures of gases of different composition (e.g., dry and moist air) are very close together, the rate of diffusion is modified. The "lines of flow" of the diffusing molecules, which normally tend to diverge fan-wise as they issue from the apertures (Fig. 22.2), now interfere with one another and mutually hinder the spread of the diffusing particles, thus slowing down the rate of diffusion. Brown and Escombe (1900) showed experimentally that such interference begins when the distance between the apertures is somewhat less than ten times the diameter of the holes. The fact that the rate of diffusion through small openings is proportional not to the area, but to the diameter of the opening, greatly increases the possible amount of diffusion that can take place through a multiperforate septum (Maximov, 1929, p. 172–173).

V. GUARD CELLS

Guard cells may occur at the same level as the adjacent epidermal cells, or they may protrude above or be sunken below the surface of the epidermis (Fig. 22.3). In some plants, stomata are restricted to the epidermis that lines depressions in the leaf, the stomatal crypts. Epidermal hairs may also be prominently developed in such crypts. Stomata are level with the epidermal cells in most mesophytic plants and plants that grow in moist habitats. Plants that grow in dry habitats often have stomata that are situated below the level of the epidermal cells.

The guard cells are generally crescent-shaped with blunt ends (kidney-shaped) in surface view (Fig. 22.3D) and often have ledges of wall material

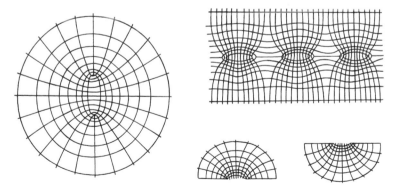

FIG. 22.2 Diagrammatic representation of the diffusion of water vapor through small openings. Left: Diffusion through a single opening in a vertical septum; the fanlike, diverging lines show the courses of the diffusing particles; perpendicular to them are concentric lines of equal vapor density. Right, above: Diffusion through a horizontal multiperforate septum (three openings are represented); Right, below: Diffusion through single openings of the same size. (From Maximov, N.A., *The Plant in Relation to Water*, p.173, ©1929, George Allen & Unwin, Ltd: London.)

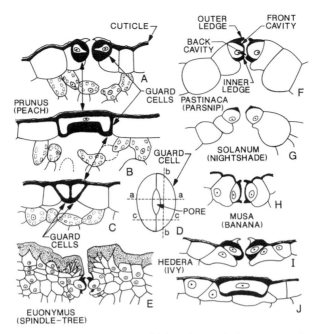

FIG. 22.3 Stomata in abaxial epidermis of foliage leaves. A-C, stomata and some associated cells from each leaf sectioned along planes indicated in **D** by the broken lines *aa, bb*, and *cc*. **E-I**, stomata from various leaves cut along the plane *aa*. **J**, one guard cell of ivy cut along the plane *bb*. The stomata are raised in **A, F, G**. They are slightly raised in **I**, slightly sunken in **H**, and deeply sunken in **E**. The hornlike protrusions in the various guard cells are sectional views of ledges. Some stomata have two ledges (**E, F, H**); others only one (**A, G, I**). Ledges are cuticular in **A, E, I**. The *Euonymus* leaf has a thick cuticle; epidermal cells are partly occluded with cutin. (From Esau, K., *Plant Anatomy*, 2nd ed., p. 159, ©1965, John Wiley & Sons, Inc: New York. This material is used by permission of John Wiley & Sons, Inc.)

on the upper and lower sides. In sectional views such ledges appear like horns (Fig. 22.3E, F, H). Sometimes a ledge occurs only on the upper side (Fig. 22.3A, G, I), or none is present. If two ledges are present, the upper delimits the front cavity above the stomatal pore, and the lower encloses the back cavity between the pore and the substomatal chamber (Fig. 22.3F). The ledges are more or less heavily cutinized (Esau, 1965, p. 159).

The walls of the guard cells can be differentially thickened. The change in shape of the guard cells occurs because the wall that is turned away from the stomatal aperture, the so-called back wall, is thin and apparently elastic (Fig. 22.3A, E-I). When the turgor increases, the thin wall bulges away from the aperture, while the front wall (facing the pore) becomes straight or concave. The whole cell appears to bend away from the aperture, and

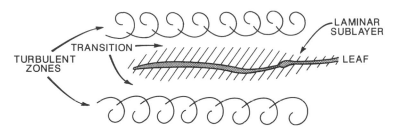

FIG. 22.4   Air structure near a small object (like a leaf) in an air stream. (From Rosenberg, N.J., *Microclimate: The Biological Environment*, p. 79, ©1974, John Wiley & Sons, Inc: New York. This material is used by permission of John Wiley & Sons, Inc.)

layer will be thicker the rougher the surface is. Roughness is nearly zero over very smooth surfaces, like open water on a calm day. Roughness increases with increasing height of objects sticking above the surface (Rosenberg, 1974, p. 104). The boundary layer is thinner the more windy the conditions. In growth chamber experiments, it is important to have fans circulating air in the closed chambers, so that the boundary layer (or more specifically, boundary-layer resistance) is reduced and gas exchange (in particular, carbon dioxide uptake) is similar to natural conditions in the open environment. Both boundary-layer resistance and stomatal resistance are important in controlling gas transport through leaves (see next section).

VIII. LEAF RESISTANCES

The resistances to water-vapor transport in a leaf are the epidermal resistance, made up of the stomatal resistance and the cuticular resistance, and the boundary-layer resistance. The resistances to carbon dioxide transport in a leaf are the same as for water vapor (stomatal, cuticular, and boundary-layer resistances), plus a fourth resistance called the mesophyll resistance, discussed later in this section.

Water vapor diffuses through two of the resistances in a leaf acting in *series*: the stomatal-aperture resistance (stomatal resistance) (r_s) and the boundary-layer (air) resistance (r_a), which results from the lengthening of the diffusion path outside of the stomata and which is an inverse function of wind and turbulence (Gale and Hagan, 1966). The resistance to cuticular water loss (r_c) is very large and is in *parallel* to r_s.

Let us now review our physics concerning resistors in series and parallel (Schaum, 1961, p. 156). In a series circuit (Fig. 22.5, left), resistance is as follows:

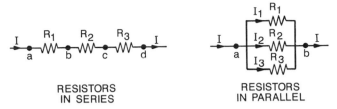

FIG. 22.5 Left: Resistors in series. Right: Resistors in parallel. (From Schaum, D., *Theory and Problems of College Physics*, p. 158, ©1961, Schaum Publishing Co: New York. This material is reproduced with permission of The McGraw-Hill Companies.)

$$R = R_1 + R_2 + R_3 + ..., \qquad (22.1)$$

where R = equivalent resistance of a series combination of conductors having resistances $R_1, R_2, R_3, ...$ The total *potential difference* across several resistors connected in series is equal to the sum of the potential differences across the separate resistors. Current in every part of the series circuit is the same.

In a parallel circuit (Fig. 22.5, right), resistance is as follows:

$$1/R = (1/R_1) + (1/R_2) + (1/R_3) + ..., \qquad (22.2)$$

where R = equivalent resistance of a parallel combination of conductors having resistances $R_1, R_2, R_3, ...$ R is always less than the smallest of the individual resistances. Connecting additional resistors in parallel decreases the joint resistance of the combination. The *potential difference* across several resistors in parallel is the same as that across each of the resistors. The potential difference is the same across all branches. The sum of the currents in the branches is equal to the value of the line current. Current values in the different branches vary inversely as the resistances of the different branches (Fig. 22.5, right) (Schaum, 1961).

The conductance via the cuticle r_c^{-1} is very small and may be neglected, unless r_s is large, as when the stomata closes. As noted, the epidermal resistance (r_e) is made up of the two resistances r_c and r_s in parallel (Waggoner, 1966):

$$r_e = 1/[(1/r_c) + (1/r_s)] = (r_s r_c)/(r_s + r_c). \qquad (22.3)$$

The stream of water, T (in units of g cm^{-2} s^{-1}, for example), transpired from a leaf is assumed in accordance with diffusion theory to be proportional to the difference ΔX in water concentration (g/cm^3) between the surfaces of the mesophyll cells and the free air outside (Waggoner, 1966):

$$T = \Delta X/[r_a + (r_s\, r_c)/(r_s + r_c)] = \Delta X/(r_a + r_e). \qquad (22.4)$$

Or Equation (22.4) may be shown as follows (Gale and Hagan, 1966):

$$T = \{[H_2O]_{int} - [H_2O]_{ext}\}/(r_s + r_a), \qquad (22.5)$$

where T is transpiration, defined above, $[H_2O]_{int}$ is the water vapor concentration at the mesophyll surface and $[H_2O]_{ext}$ is the vapor concentration of the air (cm^3 vapor/cm^3 air), and r_s and r_a are the resistances as defined above (s/cm). We are neglecting r_c.

The diffusion theory, upon which Equations 22.4 and 22.5 are based, is Fick's law. (For a biography of Fick, see the Appendix, Section XI.) In 1855, Adolf Fick discovered the linear flow law of diffusion, which is called Fick's law, to describe the diffusion of solutes in solution, and it is as follows (Kirkham and Powers, 1972, p. 75):

$$Q = DA(C_1 - C_2)/L, \qquad (22.6)$$

where Q is the quantity of solute per unit time, D is the diffusion coefficient, L is the length of the element through which the diffusion is occurring, A is the cross-sectional area of the element, and $(C_1 - C_2)/L$ *is the concentration gradient.*

Photosynthesis may be described similarly as a diffusion process of CO_2 from the outside air to the chloroplasts, but here a fourth resistance (in addition to r_s, r_c, r_a) to diffusion of CO_2 is present in the liquid phase from the mesophyll wall to the chloroplast ($r_m{}'$). In addition to liquid phase CO_2 diffusion resistance, $r_m{}'$ also includes all the metabolic factors that affect the photosynthetic rate. Thus, photosynthesis may be expressed as follows (Gale and Hagan, 1966):

$$P = \{[CO_2]_{ext} - [CO_2]_{int}\}/(r_s{}' + r_a{}' + r_m{}'), \qquad (22.7)$$

where P is the photosynthetic rate (cm^3 CO_2 cm^{-2} s^{-1}); $[CO_2]_{ext}$ is the concentration of the carbon dioxide in the outside air and $[CO_2]_{int}$ is the CO_2 concentration at the site of the CO_2 sink, that is, the chloroplast (cm^3 CO_2/cm^3 air); and $r_s{}'$, $r_a{}'$, and $r_m{}'$ are the resistances to CO_2 diffusion as defined above (s/cm). The primes denote resistance to flow of carbon dioxide, and no primes are used to denote resistance to flow of water vapor. Using Waggoner's (1966) analysis, we get:

$$P = \Delta X'/[r_a{}' + (r_s{}'r_c{}')/(r_s{}' + r_c{}') + r_m{}'] = \Delta X'/(r_a{}' + r_e{}' + r_m{}'), \qquad (22.8)$$

where P is the photosynthetic rate, as defined above, and $\Delta X'$ is the decrease in carbon dioxide concentration between the air and the site of chemical combination of carbon dioxide with a receptor.

The fact that there is a fourth resistance (mesophyll resistance) for carbon dioxide transport, which is not present for water-vapor transport, has been the theoretical basis for the use of antitranspirants. When an antitranspirant is applied to a leaf, transpiration (T; Equation 22.5) should be reduced more than photosynthesis (P; Equation (22.7) is reduced. However, in practice, an antitranspirant reduces both T and P tremendously, so that photosynthesis is essentially stopped until the antitranspirant is removed. Figures 22.6 and 22.7 show circuits that illustrate

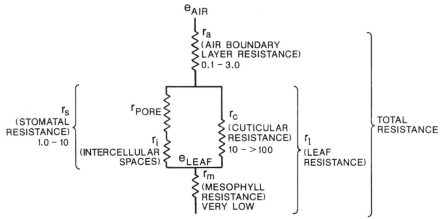

FIG. 22.6 Diagram showing resistances in seconds per centimeter to diffusion of water vapor from a leaf. Stomata and cuticular resistances vary widely among species and with leaf hydration and atmospheric humidity. The rate of transpiration is proportional to Δe, the water vapor pressure gradient, e_{leaf} to e_{air}, and inversely proportional to the resistances in the pathway. (From Kramer, P.J., *Water Relations of Plants*, p. 296, ©1983, Academic Press: New York. Reprinted by permission of Academic Press.)

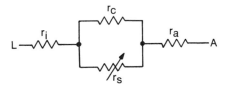

FIG. 22.7 Resistances encountered by a water molecule diffusing from a leaf cell (L) into the surrounding air (A). r_i is the resistance of the intercellular spaces, r_c the resistance of the cuticle, r_s the variable resistance of the stomata, and r_a the resistance of the boundary layer of unstirred air at the leaf surface through which water molecules must diffuse. (From Baker, D.A., Water relations. In *Advanced Plant Physiology* (M.B. Wilkins, Ed.), pp. 297–318, ©1984, Pitman Publishing Limited: London. Reprinted by permission of Pearson Education Limited, Essex, United Kingdom and by permission of Dennis A. Baker.)

resistances encountered in a leaf, as conceived by plant physiologists (Kramer, 1983; Baker, 1984).

The upper surface of a leaf (usually the adaxial surface) and the lower surface of a leaf (usually the abaxial surface) each have a resistance associated with them. If a leaf has no stomata on a surface, then there will be no stomatal resistance for that surface. Resistances of adaxial and abaxial stomata are assumed to act in parallel (Kramer, 1983, p. 302), or

$$1/R_{total} = (1/R_{abaxial}) + (1/R_{adaxial}). \qquad (22.9)$$

However, this assumption means that the potential on the abaxial surface of the leaf is the same as the potential on the adaxial surface of the leaf (see the preceding paragraphs concerning resistance in a parallel circuit). But this is not the case for plants (Kirkham, 1986). The adaxial surface of a leaf has a different water potential than the abaxial surface. So the limitation of Equation 22.9, as applied to the total resistance of a plant leaf, should be recognized. However, it is the only equation we have to get the total resistance of a leaf, when the resistances on each surface are known. So we use it.

IX. MEASUREMENT OF STOMATAL APERTURE AND STOMATAL RESISTANCE

Because water is lost mainly through the stomata on the surfaces of leaves, it is critical to know the extent of stomatal opening, to evaluate how much water a plant is losing. Slavík (1971), Kanemasu (1975a), Willmer (1983), and Weyers and Meidner (1990) enumerate different methods used to assess stomatal aperture. These methods include the following:

1. Observation under a microscope (Hsiao and Fischer, 1975a; Schoch and Silvy, 1978; Willmer and Beattie, 1978; Omasa et al., 1983; Martin et al., 1983; Weyers and Meidner, 1990, p. 106–116).

2. Use of cobalt-chloride paper (Teare et al., 1973; Kanemasu and Wiebe, 1975). Cobalt-chloride paper is prepared by dipping filter paper in a solution of $CoCl_2.6\,H_2O$ and then drying it. The paper is blue when dry, but pink when moist. The dry, blue paper, when placed on a leaf, covered with plexiglass, and held firmly by a small spring clamp, will turn pink from water vapor escaping from the leaf surface. This method can be used to compare the rates of transpiration from upper and lower leaf surfaces and from leaves of different plants under different environmental conditions.

3. Determination of resistance from leaf-chamber (cuvette) transpiration, which also can incorporate the capability to monitor CO_2 assimilation

for photosynthesis. For early work, see the following: Davenport (1975); Syber and Moldau (1977); Blacklow and Maybury (1980); Bloom et al. (1980); Bell and Incoll (1981a, 1981b); Griffiths and Jarvis (1981); Kohsiek (1981); Rawson and Love (1982); Schulze et al. (1982); Daley et al. (1984). All work with the two models of the portable photosynthetic systems of Li-Cor, Inc. (Lincoln, Nebraska) (Model LI-6200; Model LI-6400) report data from cuvette measurements. In 1987, Model LI-6200 was put on the market and is a closed system; subsequently, Model LI-6400 was developed and it is an open system.

4. Mass-flow porometry. When stomata close, the permeability of the leaf to various gases (porosity) is greatly reduced. In mass-flow porometry, air is forced under pressure through the leaf, and the rate of flow or leaf resistance to flow is indicative of porosity (Fig. 22.8) (Hsiao and Fischer, 1975b). The mass-flow porometer developed by Gregory and Pearse (1934) is the basis for most mass-flow porometers. Amphistomatous leaves are needed to use mass-flow porometers, because air must enter one side of the leaf and exit from the other side. If the resistance of one epidermis is high, the reading obtained with the porometer reflects mainly the opening of that epidermis. When using mass-flow porometers, it is assumed that the mesophyll resistance is constant and small compared to the resistance offered by the epidermis of a leaf with closed stomata. Mass-flow porometers are not available commercially.

5. Diffusion porometry. Diffusion porometers measure diffusion of water vapor from the substomatal cavities through the stomata. Diffusion porometry includes both transient (dynamic)-state and steady-state methods (Kanemasu, 1975a; Knof, 1980). In the steady-state porometer, dry gas is passed over an enclosed leaf at a known flow rate and the humidity of the exhaust gas is measured (Fig. 22.9) (Campbell, 1975). In the transient-state porometer, a sensor responsive to a change in humidity is clamped to a leaf (van Bavel et al., 1965; Kanemasu et al., 1969; Ehrler, 1975; Kanemasu, 1975b). Tan and Black (1978) describe a diffusion porometer for use on conifer needles.

Day (1977) and Parkinson and Day (1980) give theory associated with the steady-state porometer, and Chapman and Parker (1981) supply theory for the transient-state porometer. Of all the methods used to measure stomatal resistance only (i.e., photosynthetic rate is not measured), diffusion porometers (transient and steady-state types) are most widely used for quantitative measurements (Livingston et al., 1984). They are commercially available (Figs. 22.9 and 22.10). In the United States, only the steady-state porometer

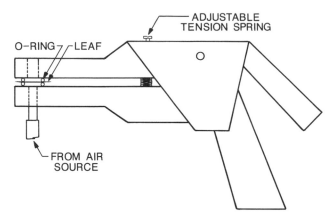

FIG. 22.8 A mass flow porometer. The basic structural material consists of plexiglass cemented together. The critical aspect in construction is alignment of the two O-rings, both horizontally and vertically. Alignment of the two arms of the cup (cut from 1.3-cm-thick plexiglass sheets) is ensured by fixing, with a close-fitting metal pin, the upper arm snugly between the two large parallel trapezoidal plates glued to the lower arm. One O-ring is glued to an arm first. The other arm is then sanded to ensure good horizontal alignment of the O-rings. Vertical alignment is effected when the second O-ring is glued onto the arm. (From Hsiao, T.C., and Fischer, R.A., Mass flow porometers. In *Measurement of Stomatal Aperture and Diffusive Resistance* (Kanemasu, E.T., Ed.) ©1975 Bulletin 809, College of Agriculture Research Center, Washington State University: Pullman, Washington. Reprinted by permission of the Director of the Washington State University Agricultural Research Center, Pullman, Washington.)

is made (by Li-Cor, Inc., Lincoln, Nebraska) (Fig. 22.9). However, the transient-state one is made in Cambridge, England, by Delta-T Devices, Ltd. (Squire et al., 1981), and imported for sale by Decagon Devices (Pullman, Washington) and Dynamax, Inc. (Houston, Texas) (Fig. 22.10).

Early diffusion porometers monitored the diffusion of various gases (hydrogen, nitrous oxide, radioactive argon) through the leaf (Kanemasu and Wiebe, 1975). Moreshet and Falkenflug (1978) described a stomatal diffusion porometer that measures the diffusion of radioactive krypton through leaves. Most porometers in use today, however, measure the diffusion of water vapor. The water-vapor sensors in the porometers usually contain lithium chloride. A humidity sensor supplied commercially by Vaisala (Helsinki, Finland) has a faster response time than that of the lithium-chloride sensor (Visscher et al., 1978), and it is used in porometers (Squire et al., 1981), including the one shown in Fig. 22.10. Accuracy of sensors can be increased by preventing sorption of water vapor on the walls surrounding the sensor (Gandar and Tanner, 1976) and by taking into

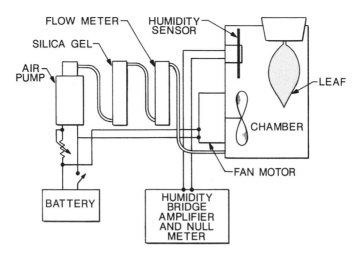

FIG. 22.9 Block diagram of steady state porometer showing components and interconnections. (From Campbell, G.S., Steady-state diffusion porometers. In *Measurement of Stomatal Aperture and Diffusive Resistance* (Kanemasu, E.T., Ed.), pp. 20–23, ©1975 Bulletin 809, College of Agriculture Research Center, Washington State University: Pullman, Washington. Reprinted by permission of the Director of the Washington State University Agricultural Research Center, Pullman, Washington, and Gaylon S. Campbell.)

account changes in the resistance of the sensor with changes in temperature (Berkowitz and Hopper, 1980). Commercially available diffusion porometers (e.g., Fig. 22.10) measure temperature along with stomatal resistance.

The transient and steady-state diffusion porometers have been compared. Bell and Squire (1981) found a systematic difference of 20 to 30% between the measurements made with the two instruments. They felt that the difference was due either to a systematic error in one or both of the instruments or to the different principles of operation. Gay (1983) reported that transient diffusion porometers had a greater accuracy than steady-state porometers, but that some overestimated low resistances. In contrast, Johnson (1981) found that at low resistance, the two types of porometers produced a linear and nearly equal response.

Kanemasu (1975a) gives a summary of the five methods used to measure stomatal aperture and diffusive resistance and points out the strengths and weaknesses of each method.

X. THEORY OF MASS-FLOW AND DIFFUSION POROMETERS

As noted in Section IV, if stomata are spaced so that diffusion from one does not interfere with another, stomata can be considered to conduct

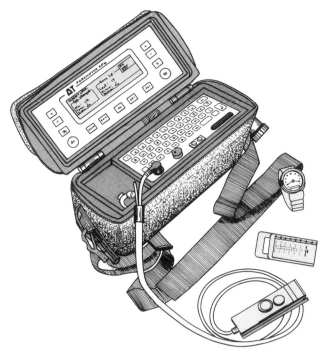

FIG. 22.10   A commercially available transient porometer. Automatic cycling ensures consistent results by repeating the measurement cycle (in typically 3 to 10 seconds) so that as soon as a repeatable value has been reached—usually after about 4 or 5 cycles—the next leaf can be sampled. The relative humidity level at which the instrument cycles can be set between 20% and 70% to match the ambient relative humidity as closely as possible, to avoid upsetting the stomata. The sensor head, shown in the lower right of the figure, weighs 80 grams and incorporates a window for checking the alignment of the leaf with the sampling area—a slot 2.5 × 22.5 mm. The calibration plate, shown above the sensor head in the figure, has six values of diffusion resistance in the range 0 to 30 s/cm. The porometer comes with a rechargeable battery, padded carrying case, and a strap so one can put it around the neck while taking measurements in the field. The size of the porometer is 350 × 200 × 100 mm and it weighs 3.2 kg. (From a Dynamax, Inc, Houston, Texas, brochure. Reprinted by permission of Dynamax, Inc., Houston, Texas.)

water vapor more or less independently of each other. Hence, stomatal mass-flow and diffusive resistances per unit leaf area are inversely proportional to the number of stomata in that area (i.e., to stomatal frequency) (Hsiao, 1975).

The simplest physical model of a stomatal pore is that of a cylinder (Hsiao, 1975). Therefore, the relation between mass-flow resistance (in the mass-flow porometer) and stomatal opening tends to take on a form similar to that of the Poiseuille equation: resistance to flow is inversely proportional to the fourth power of the radius of the opening. This approximation becomes invalid, however, in the case of nearly closed stomata, because the term for interaction with the path wall becomes large and must be considered. In most cases, the stomatal pore is not circular and the length of the pore (normal to the conducting path) does not necessarily vary with the width of the pore. The mass-flow resistance then becomes inversely proportional to the third or even lower power of the width (Hsiao, 1975).

For diffusive resistance, the simplest approach is to apply Fick's law of diffusion to an assumed simple pore geometry. The result is that, to the first approximation, stomatal diffusive resistance is inversely proportional to the total pore area. For circular stomatal pores (Hsiao, 1975),

$$r_s = (A/nD)(4L_s/\pi d^2), \tag{22.10}$$

where A is the leaf area being studied, n is the number of stomata, D is the diffusivity of water vapor in air, L_s is the depth of the stomatal tube (i.e., of the stomatal pore), and d is the stomatal pore diameter. If the so-called "end correction" (Monteith, 1973, p. 145) is applied to one end (outer end) of the stomatal tube, a factor, $1/2\ d$, is added, and the resulting equation is (Kanemasu, 1975b):

$$r_p = (A/nD)[(4t\ /\ (\pi d^2 + 1/2d)], \tag{22.11}$$

where r_p is the resistance of a calibration plate, t is the thickness of the plate, A is the aperture area, n is the number of holes, D is the diffusivity of water vapor in air, and d is the diameter of the holes. Equation 22.11 is identical to the equation used for calculating the resistance of the calibration plate for the diffusion porometer (Kanemasu, 1975b).

XI. APPENDIX: BIOGRAPHY OF ADOLF FICK

Adolf Eugen Fick (1829–1901), a physiologist, was born in Kassel, Hesse, Germany, on September 3, 1829, the son of Friedrich and Marianne (Spousel) Fick. He got his M.D. at the University of Marburg in 1851 and married Emile von Cölln in 1862. He was an assistant to Carl Ludwig (1816–1895; German physiologist) in Zurich in 1852. Fick was a professor of physiology in Zurich beginning in 1862 and was a professor at the University of Würzburg from 1868. He was the author of *Die medizinische Physik* (1856) and *Untersuchungen über elektrischen nervenreizung* (1864).

He made important discoveries in every branch of physiology. He proved that carbohydrates rather than albumin are the source of muscle energy. He constructed the first pletysmography, which measured the pulse rate. In about 1864, he invented the myotonograph for measuring and recording muscle tension. In 1870, he developed a method to determine cardiac output by gasometry (Marquis Who's Who, 1968). He discovered the linear flow law of diffusion, named after him, to describe diffusion of solutes in solution, as in animal tissue (Fick, 1855). (He was a prosector; Kirkham and Powers, 1972, p. 429.) He died in Blankenberghe, Belgium, on August 21, 1901.

REFERENCES

Assmann, S.M. (2001). From proton pump to proteome. Twenty-five years of research on ion transport in higher plants. *Plant Physiol* 125; 139–141.

Assmann, S.M., and Wang, X.-Q. (2001). From milliseconds to millions of years: Guard cells and environmental responses. *Current Opinion Plant Biol* 4; 421–428.

Baker, D.A. (1984). Water relations. In *Advanced Plant Physiology* (Wilkins, M.B., Ed.), pp. 297–318. Pitman: London.

Bell, C.J., and Incoll, L.D. (1981a). A handpiece for the simultaneous measurement of photosynthetic rate and leaf diffusive conductance. I. Design. *J Exp Bot* 32; 1125–1134.

Bell, C.J., and Incoll, L.D. (1981b). A handpiece for the simultaneous measurement of photosynthetic rate and leaf diffusive conductance. II. Calibration. *J Exp Bot* 32; 1135–1142.

Bell, C.J., and Squire, G.R. (1981). Comparative measurements with two water vapour diffusion porometers (dynamic and steady-state). *J Exp Bot* 32; 1143–1156.

Berkowitz, G.A., and Hopper, N.W. (1980). A method of increasing the accuracy of diffusive resistance porometer calibrations. *Ann Bot* 45; 723–727.

Blacklow, W.M., and Maybury, K.G. (1980). A battery-operated instrument for non-destructive measurements of photosynthesis and transpiration of ears and leaves of cereals using $^{14}CO_2$ and a lithium chloride hygrometer. *J Exp Bot* 31; 1119–1129.

Bloom, A.J., Mooney, H.A., Björkman, O., and Berry, J. (1980). Materials and methods for carbon dioxide and water exchange analysis. *Plant Cell Environ* 3; 371–376.

Brown, H.T., and Escombe, F. (1900). Static diffusion of gases and liquids in relation to the assimilation of carbon and translocation in plants. *Phil Trans Roy Soc London* 193B; 223–291.

Campbell, G.S., Davenport, D.C., Ehrler, W.L. (1975). Steady-state diffusion porometers. In *Measurement of Stomatal Aperture and Diffusive Resistance* (Kanemasu, E.T., Ed.), pp. 20–23. Bulletin 809, College Agriculture Research Center, Washington State Univ: Pullman, Washington.

Chapman, D.C., and Parker, R.L. (1981). A theoretical analysis of the diffusion porometer: Steady diffusion through two finite cylinders of different radii. *Agr Meteorol* 23; 9–20.

Daley, P.F., Cloutier, C.F., and McNeil, J.N. (1984). A canopy porometer for photosynthesis studies in field crops. *Can J Bot* 62; 290–295.

Davenport, D.C. (1975). Stomatal resistance from cuvette transpiration measurements. In *Measurement of Stomatal Aperture and Diffusive Resistance* (Kanemasu, E.T., Ed.), pp. 12–15. Bulletin 809, College Agriculture Research Center, Washington State Univ: Pullman, Washington.

Day, W. (1977). Stomatal resistance in different gases. *J Appl Ecol* 14; 643–647.

Ehrler, W.L. (1975). The porometer of van Bavel, Nakayama, and Ehrler. In *Measurement of Stomatal Aperture and Diffusive Resistance* (Kanemasu, E.T., Ed.), pp. 15–17. Bulletin 809, College Agriculture Research Center, Washington State Univ: Pullman, Washington.

Esau, K. (1965). *Plant Anatomy*. 2nd ed. Wiley: New York.

Fernald, M.L. (1950). *Gray's Manual of Botany*. 8th ed. American Book: New York.

Fick, A. (1855). Über Diffusion. *Annalen der Physik* (Leipzig) 170; 59–86. (Cited by Kirkham and Powers, 1972, p. 458).

Fischer, R.A. (1968). Stomatal opening: Role of potassium uptake by guard cells. *Science* 160; 784–785.

Fujino, M. (1967). Adenosinetriphosphate and adenosinetriphosphatase in stomatal movement. *Sci Bull Fac Educ Nagasaki Univ* 18; 1–47.

Gale, J., and Hagan, R.M. (1966). Plant antitranspirants. *Annu Rev Plant Physiol* 17; 269–282.

Gandar, P.W., and Tanner, C.B. (1976). Water vapor sorption by the walls and sensors of stomatal diffusion porometers. *Agronomy J* 68; 245–249.

Gay, A.P. (1983). Transit time diffusion porometer calibration: An analysis taking into account temperature differences and calibration non-linearity. *J Exp Bot* 34; 461–469.

Gregory, F.G., and Pearse, H.L. (1934). The resistance porometer and its application to the study of stomatal movement. *Proc Roy Soc* B144; 477–493.

Griffiths, J.H., and Jarvis, P.G. (1981). A null balance carbon dioxide and water vapour porometer. *J Exp Bot* 32; 1157–1168.

Heath, O.V.S., and Mansfield, T.A. (1969). The movements of stomata. In *The Physiology of Plant Growth and Development* (Wilkins, M.B., Ed.), pp. 301–332. McGraw-Hill: New York.

Hsiao, T.C. (1975). Relationships among aperture, mass-flow resistance and diffusive resistance. In *Measurement of Stomatal Aperture and Diffusive Resistance* (Kanemasu, E.T., Ed.), pp. 24–25. Bulletin 809, College Agriculture Research Center, Washington State Univ: Pullman, Washington.

Hsiao, T.C., and Fischer, R.A. (1975a). Microscopic measurements. In *Measurement of Stomatal Aperture and Diffusive Resistance* (Kanemasu, E.T., Ed.), pp. 2–5. Bulletin 809, College Agriculture Research Center, Washington State Univ: Pullman, Washington.

Hsiao, T.C., and Fischer, R.A. (1975b). Mass flow porometers. In *Measurement of Stomatal Aperture and Diffusive Resistance* (Kanemasu, E.T., Ed.), pp. 5–11. Bulletin 809, College Agriculture Research Center, Washington State Univ: Pullman, Washington.

Imamura, S. (1943). Untersuchungen über den Mechanismus der Turgorschwankung der Spaltöffnungsschliesszellen. *Japanese J Bot* 12; 251–346.

Johnson, J.D. (1981). Two types of ventilated porometers compared on broadleaf and coniferous species. *Plant Physiol* 68; 506–508.

Kanemasu, E.T. (Ed.) (1975a). *Measurement of Stomatal Aperture and Diffusive Resistance*. Bulletin 809, College Agriculture Research Center, Washington State Univ: Pullman, Washington.

Kanemasu, E.T. (1975b). The porometer of Kanemasu, Thurtell, and Tanner. In *Measurement of Stomatal Aperture and Diffusive Resistance* (Kanemasu, E.T., Ed.), pp. 17–20. Bulletin 809, College Agriculture Research Center, Washington State Univ: Pullman, Washington.

Kanemasu, E.T., and Wiebe, H.H. (1975). Other methods. In *Measurement of Stomatal Aperture and Diffusive Resistance* (Kanemasu, E.T., Ed.), pp. 23–24. Bulletin 809, College Agriculture Research Center, Washington State Univ: Pullman, Washington.

Kanemasu, E.T., Thurtell, G.W., and Tanner, C.B. (1969). Design, calibration and field use of a stomatal diffusion porometer. *Plant Physiol* 44; 881–885.

Kirkham, D., and Powers, W.L. (1972). *Advanced Soil Physics*. Wiley: New York.

Kirkham, M.B. (1986). Water relations of the upper and lower surfaces of maize leaves. *Biol Plant* 38; 249–257.

Knof, G. (1980). Ein registrierendes Transpirationsporometer zur Messung des Diffusions-widerstandes an Pflanzen in Feldversuchen. *Arch. Acker-u. Pflanzenbau u Bodenkd* 24; 647–654. (In German, no English summary.)

Kohsiek, W. (1981). A rapid-circulation evaporation chamber for measuring bulk stomatal resistance. *J Appl Meteorol* 20; 42–52.

Kramer, P.J. (1983). *Water Relations of Plants*. Academic Press: New York.

Li, J., Wang, X.-Q., Watson, M.B., and Assmann, S.M. (2000). Regulation of abscisic acid-induced stomatal closure and anion channels by guard cell AAPK kinase. *Science* 287; 300–303.

Liang, G.H., Dayton, A.D., Chu, C.C., and Casady, A.J. (1975). Heritability of stomatal density and distribution on leaves of grain sorghum. *Crop Sci* 15; 567–570.

Livingston, N.J., Black, T.A., Beames, D., and Dunsworth, B.G. (1984). An instrument for measuring the average stomatal conductance of conifer seedlings. *Can J For Res* 14; 512–517.

Macallum, A.G. (1905). On the distribution of potassium in animal and vegetable cells. *J Physiol* 52; 95–128.

Marquis Who's Who. (1968). *Who's Who in Science from Antiquity to the Present*. 1st ed. Marquis-Who's Who: Chicago.

Martin, E.S., Donkin, M.E., and Stevens, R.A. (1983). *Stomata*. Edward Arnold: London.

Maximov, N.A. (1929). *The Plant in Relation to Water*. George Allen & Unwin: London.

Meidner, H., and Mansfield, T.A. (1968). *Physiology of Stomata*. McGraw-Hill: New York.

Monteith, J.L. (1973). *Principles of Environmental Physics*. American Elsevier: New York.

Moreshet, S., and Falkenflug, V. (1978). A krypton diffusion porometer for the direct field measurement of stomatal resistance. *J Exp Bot* 29; 267–275.

Omasa, K., Hashimoto, Y., and Aiga, I. (1983). Observation of stomatal movements of intact plants using an image instrumentation system with a light microscope. *Plant Cell Physiol* 24; 281–288.

Parkinson, K.J., and Day, W. (1980). Temperature corrections to measurements made with continuous flow porometers. *J Appl Ecol* 17; 457–460.

Peaslee, D.E., and Moss, D.N. (1968). Stomatal conductivities in K-deficient leaves of maize (*Zea mays* L.). *Crop Sci* 8; 427–430.

Rawson, H.M., and Love, D.C. (1982). A chamber for rapid measurement of cereal leaf gas exchange. *Photosynthetica* 16; 67–70.

Romano, L.A., Jacob, T., Gilroy, S., and Assmann, S.M. (2000). Increases in cytosolic Ca^{2+} are not required for abscisic acid-inhibition of inward K^+ currents in guard cells of *Vicia faba* L. *Planta* 211; 209–217.

Rosenberg, N.J. (1974). *Microclimate: The Biological Environment*. Wiley: New York.

Schaum, D. (1961). *Theory and Problems of College Physics*. 6th edition. Schaum: New York.

Schoch, P.G., and Silvy, A. (1978). Méthode simple de numération des stomates et des cellules de l'épiderme des végétaux. *Ann. Amelior Plantes* 28; 455–461. (In French, English sum.)

Schulze, E.-D., Hall, A.E., Lange, O.L., and Walz, H. (1982). A portable steady-state porometer for measuring the carbon dioxide and water vapour exchanges of leaves under natural conditions. *Oecologia* 53; 141–145.

Slavík, B. (1971). Determination of stomatal aperture. In *Plant Photosynthetic Production. Manual of Methods* (Šesták, Z., Čatský, J., and Jarvis, P.G., Eds.), pp. 556–565. Dr W. Junk N.V. Pub: The Hague, The Netherlands.

Squire, G.R., Black, C.R., and Gregory, P.J. (1981). Physical measurements in crop physiology. II. Water relations. *Exp Agr* 17; 225–242.

Stefan, J. (1881). Versuche über die Verdampfung. *Sitzungsber dK Akad d Wiss Wien*, Abt. II, 85; 943 (cited by Maximov, 1929, p. 429).

Syber, A. Yu., and Moldau, Kh. A. (1977). An apparatus with separate conditioning of the plant and an individual leaf for determination of stomate resistance and leaf water content. *Soviet Plant Physiol* 24 (Part 2, No. 6); 1049–1054.

Tan, C.S., and Black, T.A. (1978). Evaluation of a ventilated diffusion porometer for the measurement of stomatal diffusion resistance of Douglas-fir needles. *Arch Met Geoph Biokl* Ser. B, 26; 257–273.

Taylor, A.R., and Assmann, S.M. (2001). Apparent absence of a redox requirement for blue light activation of pump current in broad bean guard cells. *Plant Physiol* 125; 329–338.

Teare, I.D., Mohan Rao, M.R., and Kanemasu, E.T. (1973). Correlation of transpiration rates by cobalt chloride method and stomatal-diffusion porometer. *Indian J Agr Sci* 43; 639–642.

van Bavel, C.H.M., Nakayama, F.S., and Ehrler, W.L. (1965). Measuring transpiration resistance of leaves. *Plant Physiol* 40; 535–540.

Visscher, G.J.W., Griffioen, H., and van Leeuwen, C.H. (1978). Investigations on a diffusion porometer with a fast humidity sensor. *Neth J Agr Sci* 26; 366–372.

Waggoner, P.E. 1966. Decreasing transpiration and the effect upon growth. In *Plant Environment and Efficient Water Use* (Pierre, W.H., Kirkham, D., Pesek, J., and Shaw, R., Eds.), pp. 49–72. American Society of Agronomy and Soil Science Society of America: Madison, Wisconsin.

Weyers, J., and Meidner, H. (1990). *Methods in Stomatal Research*. Longman Scientific and Technical: Harlow, Essex, England.

Willmer, C.M. (1983). *Stomata*. Longman: London.

Willmer, C.M., and Beattie, L.N. (1978). Cellular osmotic phenomena during stomatal movements of *Commelina communis*. I. Limitations of the incipient plasmolysis technique for determining osmotic pressures. *Protoplasma* 95; 321–332.

Yamashita, T. (1952). Influence of potassium supply upon various properties and movement of the guard cell. *Sieboldia Acta Biol* 1; 51–70.

23

Solar Radiation, Black Bodies, Heat Budget, and Radiation Balance

The sun is the battery that drives processes on earth, including evaporation, transpiration, and the ascent of water in plants. Therefore, it is essential to understand the sun's energy. In this chapter we study solar radiation and the laws associated with a black body, because the sun can be considered to be a black body. We calculate the heat budget at the surface of the earth and define the radiation balance.

I. SOLAR RADIATION

All bodies emit radiant energy in the form of electromagnetic waves when they are at a temperature above absolute zero ($-273.16°C$ or $-459.69°F$ = hypothetical point at which a substance would have no molecular motion and no heat). The source of this *thermal radiation* or *temperature radiation* is the incessant molecular motion. During collisons, or more generally as a result of interactions between molecules, part of their energy is transformed into radiation. Conversely, radiation can be absorbed by the molecules and converted into kinetic and potential energy, thereby raising the temperature of the body (van Wijk and Scholte Ubing, 1966, p. 62). (We ignore radiation from radioactive materials. This is another type of radiation.)

Solar radiation reaches the outer surface of the earth's atmosphere with an almost constant intensity of about 1400 W m^{-2} or 2.0 cal cm^{-2} min^{-1} measured perpendicularly to the solar beam (Johnson, 1954). About 98% is contained in the wavelength interval 0.2 to 4.5 μ, including about 40 to 45% in the 0.4 to 0.7 μ range (visible) (Fig. 23.1), and about 2% at wavelengths shorter and longer than these limits. The distribution of the incident flux with wavelength can be regarded as comparatively smooth, with few major gaps and a peak at the wavelength of green light (about 0.5 μ). Throughout the main region, the distribution of incident flux with wavelength corresponds roughly with that expected from radiation theory for a perfect absorber and emitter at a temperature of 6,000°K (Fig. 23.2) (Slatyer, 1967, p. 28). (We soon will define a perfect absorber and emitter.)

II. TERRESTRIAL RADIATION

In contrast to solar radiation, the earth's temperature is roughly 300°K. The black body radiation (which we will define soon), corresponding to this temperature, has its maximum spectral intensity at approximately 10 μ and 98% of its energy is contained in the wavelength interval 0.5 to 80 μ (van Wijk and Scholte Ubing, 1966, p. 74, 92; Slatyer, 1967, p. 29). In consequence, the spectral range of solar radiation and terrestrial thermal radiation, although overlapping slightly, can be considered as completely separate (Fig. 23.3). The former (solar radiation), although containing some infrared radiation, is commonly called short-wave radiation, and the latter (terrestrial thermal radiation) is called long-wave radiation (Slatyer, 1967, p. 29).

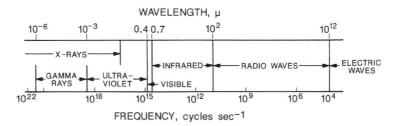

FIG. 23.1 Electromagnetic spectrum on logarithmic wavelength and frequency scales. (From Rosenberg, N.J., *Microclimate: The Biological Environment*, p. 6, ©1974, John Wiley & Sons, Inc. New York. This material is used by permission of John Wiley & Sons, Inc.)

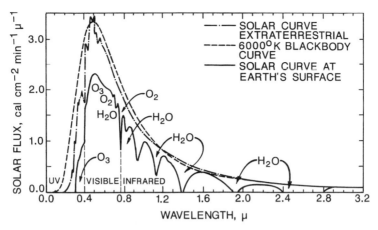

FIG. 23.2 Theoretical and actual spectra of solar radiation at the top of the atmosphere and the actual spectrum at the earth's surface. (From Rosenberg, N.J., *Microclimate: The Biological Environment*, p. ©1974, John Wiley & Sons, Inc: New York. This material is used by permission of John Wiley & Sons, Inc.)

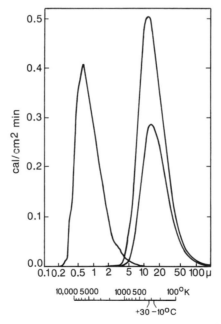

FIG. 23.3 Distribution of intensity of two bands of atmospheric radiation, according to wavelength. The curve on the left is for shortwave (solar) radiation and the two curves on the right are for longwave (terrestrial) radiation. The taller curve on the right corresponds to earth temperature of 30°C and the curve nested inside it is for earth temperature of −10°C. (From Geiger, R., *The Climate Near the Ground*. Rev. ed. Translated by Scripta Technica, Inc. p. 8, ©1965, Harvard University Press: Cambridge, Massachusetts. This material is used by the permission of the legal successor to Rudolph Geiger, Prof. Dr. Walter Geiger, Perlschneider Str. 18, 81241 Munich, Germany.)

III. DEFINITION OF A BLACK BODY

Before we continue further, let us define black-body radiation, using the description of Shortley and Williams (1971, pp. 323–326). All materials at temperatures above absolute zero are continually emitting radiation. As the temperature of a solid is increased, the energy radiated from the solid increases rapidly. The amount of radiant power emitted by a solid depends significantly on the character of the surface of the solid. In beginning the discussion of radiation, it is helpful to define a "perfect radiator," whose rate of radiation is the maximum possible for its temperature. That such a maximum exists can be shown by considering the inverse process, *absorption*.

Figure 23.4 shows in cross section a solid object maintained at a uniform temperature T throughout. Within this solid there are two identical evacuated spherical cavities containing opaque bodies of the same size but of different materials; for example, body 1 may be made of wood and body 2 may be made of polished metal. It is found by experiment that as a result of radiative interchanges of heat, the temperatures of bodies 1 and 2 eventually become equal to the temperature T of the enclosing walls and remain at that temperature, in accordance with the general principle of thermal equilibrium.

We assume that the inner walls of the cavities are perfectly absorbing. Then the only radiation reaching bodies 1 and 2 is that *radiated* by the

FIG. 23.4 Two spherical bodies of the same size but of different materials suspended within evacuated spherical cavities. (From Shortley, G. and Williams, D., *Elements of Physics*, 5th ed., p. 323, ©1971. Reprinted by permission of Pearson Education, Inc: Upper Saddle River, New Jersey.)

inner walls of the cavities; these cavity walls do not reflect any radiation and return it to the bodies. Under these circumstances, the radiant energy incident per second on unit area of each body will be the same; call it E, in W/m^2. Of the incident radiation E, a certain fraction will be reflected and the remainder will be absorbed. As indicated in Fig. 23.5, let ρ denote the fraction of the incident radiation that is reflected and α denote the fraction that is absorbed; ρ is called the *reflectance* and α the *absorptance*. These quantities are dimensionless and their sum is unity for the surface of any opaque body; $\alpha + \rho = 1$. The product $\rho_1 E$ gives the radiant power reflected from unit area of body 1, while $\rho_2 E$ gives the radiant power reflected from unit area of body 2, in W/m^2. Similarly, $\alpha_1 E$ and $\alpha_2 E$ give the radiant power absorbed per unit area of bodies 1 and 2.

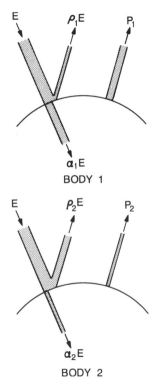

FIG. 23.5 Radiant energy incident per second on unit area, E (W/m^2), of the two bodies illustrated in Fig. 23.4. Of the incident radiation E, a certain fraction will be reflected, ρ, and the remainder will be absorbed, α. The radiated power per unit area (W/m^2) is P_1 for body 1 and P_2 for body 2. The figure shows that the rate of absorption equals the rate of emission. (From Shortley, G., and Williams, D., *Elements of Physics*, 5th ed., p. 324, ©1971. Reprinted by permission of Pearson Education, Inc: Upper Saddle River, New Jersey.)

Now, let the radiated power per unit area, in W/m^2, be P_1 for body 1 and P_2 for body 2. If the temperatures of the bodies in Figs. 23.4 and 23.5 are to remain constant, as much energy must be lost per second by radiant emission as is gained by absorption and we may write

rate of absorption = rate of emission,

or

$$\alpha_1 E = P_1 \qquad (23.1)$$

and

$$\alpha_2 E = P_2. \qquad (23.2)$$

Dividing the first equation by the second, we find that

$$\alpha_1/\alpha_2 = P_1/P_2, \text{ or } P_1/\alpha_1 = P_2/\alpha_2. \qquad (23.3)$$

Equation 23.3, and the observed temperature equality, give *Kirchhoff's principle of radiation*, which states: *The ratio of the rates of radiation of any two surfaces at the same temperature is equal to the ratio of the absorptances of the two surfaces.* Qualitatively, we can say that good radiators are good absorbers. (For a biography of Kirchhoff, see the Appendix, Section XIII.)

Now we return to the problem of defining a perfect radiator. There is a maximum value of the absorptance α. Because no surface can absorb *more* than *all* of the incident radiation, the maximum value α can have is unity. In view of Equation 23.3, we may say that a surface having the maximum rate of radiation is one that has the maximum absorptance and is, therefore, one that absorbs all radiation incident upon it. Such a surface is *black* to all types of radiation. Therefore, we may define a perfect radiator as follows: A *perfect radiator* is a body that absorbs all incident radiation and is, therefore, called a *black body*. A perfect radiator is a perfect absorber.

IV. EXAMPLE OF A BLACK BODY

No material surface absorbs all of the radiation incident upon it. Even lampblack reflects about 1% of the incident radiation. In practice, a perfectly black surface can be most closely approximated by a very small opening in the wall of a large cavity such as the one shown schematically in Fig. 23.6. Radiation may enter or leave the cavity through the opening. Of the radiation entering through the opening, a part is absorbed by the interior walls of the cavity and a part is reflected. Of the part reflected, only a small fraction escapes through the opening and the remainder is again partially absorbed and partially reflected by the walls. After repeated reflections, all of the entering radiation is absorbed except for the small portion that escapes through the opening. The opening, therefore, approximates a *black surface* or *perfect absorber*.

The inside walls of the cavity are radiating as well as absorbing, and a part of this radiation escapes through the opening. It can be shown that if the walls are at a uniform temperature T, the radiation that escapes is almost identical with the radiation that would be emitted by a perfect radiator at temperature T. The hole closely approximates the surface of a black body emitting so-called *black-body radiation*. To indicate the accuracy of this approximation, we note that computation shows that even if the interior surface of a sphere has an absorptance of only 1/2, a 25-cm sphere with a 5-cm hole will absorb 99 percent of diffuse radiation (coming equally from all directions) incident on the hole, and hence will radiate through the hole 99 percent of the radiation of a perfect radiator. A smaller hole will do correspondingly better (Shortley and Williams, 1971, p. 325).

V. TEMPERATURE OF A BLACK BODY

The total radiation emitted from the surface of a body increases rapidly as the temperature of the surface is increased. The quantitative relation

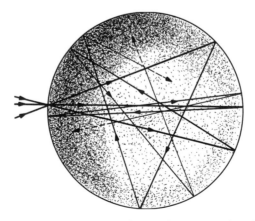

FIG. 23.6 A small hole in the wall of an enclosure, showing complete absorption of several representative rays. (From Shortley, G., and Williams, D., *Elements of Physics*, 5th ed., p. 325, ©1971. Reprinted by permission of Pearson Education, Inc: Upper Saddle River, New Jersey.)

between rate of radiation and surface temperature of an ideal radiator or black body is given by the *Stefan-Boltzmann law* and has the form

$$P_{Black} = \sigma T^4 \text{ (black body)}, \tag{23.4}$$

where P is the radiated power per unit area. The rate of radiation increases as the *fourth power* of the absolute temperature T. The proportionality constant σ is called the Stefan-Boltzmann constant and has the value 5.670 $\times$ 10^{-8} W m^{-2} K^{-4}. (For a biography of Stefan, see the Appendix, Section XIV, and for that of Boltzmann, see Section XV.)

VI. GRAY BODY

The total radiation from many surfaces that are definitely not black also is very nearly proportional to the fourth power of the absolute tempera-ture. This is true of surfaces composed of platinum, iron, tungsten, carbon, and many other materials. In every case, however, the propor-tionality constant is less than that for an ideal-radiator surface. Such a radiator is called a *gray body*. Because the absorptance α of a gray body is independent of its temperature, we see, by comparision with a black body in Equations 23.3 and 23.4, that its rate of radiation is (Shortley and Williams, 1971, p. 326)

$$P = \alpha P_{Black} = \alpha \sigma T^4. \tag{23.5}$$

Because of this relation, α also is called the *emissivity* of the surface.

VII. SPECTRUM OF A BLACK BODY

The Stefan-Boltzmann law gives the total rate of radiation of a perfect radiator (black body) at absolute temperature T, but gives no information concerning the *spectrum* of a perfect radiator (Shortley and Williams, 1971, p. 843). The spectrum of a perfect radiator is continuous. To discuss the relative amounts of energy in radiation of different wavelengths in the spectrum, we introduce the quantity P_λ, which gives the radiated power per unit area in a unit wavelength range at wavelength λ. This quantity, called the spectral radiance, can be determined by means of a spectrome-ter. What is observed in actuality is the amount of radiant power con-tained in a short wavelength interval between λ and $\lambda + \Delta\lambda$. The radiant power per unit area of source, emitted in this wavelength range, is given by $P_\lambda\Delta\lambda$, represented by the shaded area in Fig. 23.7. The unit in which P_λ is measured is W/m^2 per unit wavelength interval; for example, W/m^2 per nanometer.

Plots of the distribution of power in the spectrum of a black body at different temperatures are shown by the solid lines in Fig. 23.8. These curves all have two basic similarities in form:

1. They do not cross; the curves for higher temperatures are above the curves for lower temperatures at all wavelengths.
2. The maxima of the curves are displaced toward shorter wavelengths as the temperature of the black body is increased.

The progressive shift of maximum toward the violet end of the spectrum accounts for the observed change in color of a radiating metal body from red through white to blue as its temperature is increased. Sunlight has the characteristics of black-body radiation corresponding to a temperature of about 6000K and serves to define "white." Incandescent-lamp filaments are much cooler (about 3000°K) and give light that is more orange than daylight. Certain stars, such as Vega (12,000°K) are much hotter and appear blue. These points regarding color are illustrated by the broken curves in Fig. 23.8, in which the ordinates of the 3000- and 12,000-degree curves have been scaled so that all three curves are plotted with the same maximum.

The wavelength λ_M of the maximum of the curve (see Fig. 23.7) is found experimentally to vary inversely as the absolute temperature, according to the law

$$\lambda_M = A/T \qquad (23.6)$$

where A is a constant whose value is $A = 2.8978 \times 10^6$ nm·°K. This relation is called *Wien's displacement law*. (For a biography of Wien, see the

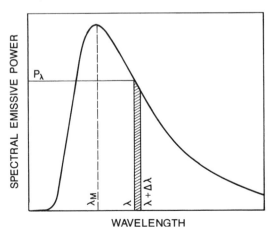

FIG. 23.7 Power radiated per unit area as a function of wavelength; definition of P_λ. (From Shortley, G., and Williams, D., *Elements of Physics*, 5th ed., p. 844, ©1971. Reprinted by permission of Pearson Education, Inc: Upper Saddle River, New Jersey.)

Appendix, Section XVI.) Thus with each doubling of temperature (see Fig. 23.8), the value of λ_M is divided by two (Shortley and Williams, p. 843–845).

VIII. SUN'S TEMPERATURE

With our knowledge about black bodies, the Stefan-Boltzmann law, and Wien's displacement law, we now return to solar radiation. The sun's surface temperature T_s can be calculated with the Stefan-Boltzmann law. The Stefan-Boltzmann constant can be expressed as 8.26×10^{-11} cal cm^{-2} min^{-1} K^{-4} (Geiger, 1965, p. 6). Geiger (1965, p. 7) says, Since radiation decreases with the square of the distance, and the sun's radius is $s = 695,560$ km, the earth's radius R is negligible in comparison with the distance M of the sun from the earth,

$$(\sigma T_s^{4})/k = M^2/s^2, \qquad\qquad (23.7)$$

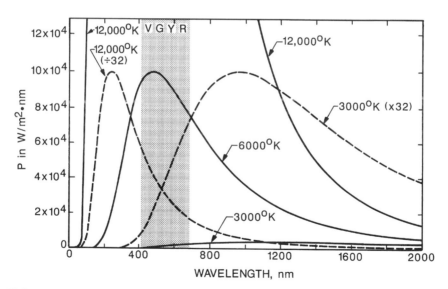

FIG. 23.8 The solid lines show plots of black-body radiation curves for temperatures of 3000, 6000, and 12,000°K. Broken lines show the 3000°K curve with ordinates multiplied by 32 and the 12,000°K curve with ordinates divided by 32; this adjustment brings the maxima of these curves to the same value as the maximum of the 6000°K curve. (From Shortley, G., and Williams, D., *Elements of Physics*, 5th ed., p. 844, ©1971. Reprinted by permission of Pearson Education, Inc: Upper Saddle River, New Jersey.)

from which the value of T_s is found to be 5793°K" (Geiger, 1965, p. 7) ($M = 150 \times 10^6$ km; $R = 6370$ km).

IX. EARTH'S TEMPERATURE

The Stefan-Boltzmann law also provides a conclusion about the earth's mean temperature, based on the assumption that the earth radiates like a black body. Because the earth's temperature is subject to variations in time, but remains unchanged on the whole over thousands of years, the amount radiated by the surface of the sphere of area $4\pi R^2$ must be equal to the quantity received by the cross-sectional area πR^2 multiplied by the solar constant k. The mean surface temperature of the earth T_E, calculated from the equation

$$(\sigma T_E) \, (4\pi R^2) = k\pi R^2 \qquad (23.8)$$

is found to be 278K = 5°C. The surface temperature of the earth observed near the ground is higher (14°C), because of the protective effect of the atmosphere, which is correspondingly cooler at higher levels (−50 to −80°C) (Geiger, 1965, p. 7).

X. COMPARISON OF SOLAR AND TERRESTRIAL RADIATION

Even if the quantity of radiation received from the sun is equal to that radiated by the earth, the two types of radiation are fundamentally different in quality. The total intensity of solar radiation is spread over a wide range of wavelengths. According to Wien's displacement law (Equation 23.6), the product of the temperature T of a radiating body and the wavelength corresponding to maximum intensity of radiation, λ_M, is constant. With T in degrees Kelvin and λ_M in microns (Geiger, 1965, p. 7)

$$T\lambda_M = 2880 \, (°K, \mu). \qquad (23.9)$$

As noted above, the higher the temperature of a body, the farther the radiation maximum is displaced toward shorter wavelengths. For the surface temperature of 5793°K (calculated temperature of sun; see above), λ_M is 0.50 μ; the observed maximum is 0.47 μ, which means a higher temperature of the sun. The difference shows that the sun radiates only approximately as a black body. In either case, the most intense solar radiation occurs in the blue-green range of visible light. The wavelength of maximum intensity of radiation for the earth's actual surface temperature of 14°C or 287°K is about 10.0 μ, which is well into the invisible infrared (Geiger, 1965, p. 7) (Fig. 23.1).

Distribution of intensities over the spectrum is so asymmetric that 25% of the total radiation lies below λ_M, in the short-wavelength range, and 75% is above λ_M. It is therefore appropriate to introduce a wavelength λ_s as a center of balance, such that 50% of the total intensity lies on either side of it; then $T\lambda_s$ = 4100 (°K, μ). In Fig. 23.3, below the abscissa, which has a logarithmic wavelength scale, is shown a scale of temperature determined by this equation. For solar radiation, λ_s is 0.7 μ, in the visible red. Forty percent of solar radiation lies within the infrared part of the spectrum (Geiger, 1965, p. 8). On the left in Fig. 23.3 is the curve of solar radiation from observations. The area enclosed by the curve represents the total intensity, hence the solar constant, reduced to one quarter for comparison with the earth's radiation. The two distribution curves on the right correspond to earth temperatures of +30°C and −10°C. The figure makes it clear that in meteorology it is correct to distinguish between two fundamentally different streams of radiation. Solar radiation and diffuse sky radiation are in the range from 0.3 to 2.2 μ. Radiation emitted by the earth and its atmosphere lies between 6.8 and about 100 μ. The intervening range from 2.2 to 6.8 μ is used by both types of radiation to the extent of less than 5%. Hence there is a marked division between the two kinds of radiation, which we shall refer to as short-wavelength and long-wavelength radiations (Geiger, 1965, pp. 8–9).

XI. HEAT BUDGET

Now let us turn to the heat budget at the surface of the earth. We shall use Geiger's description (1965, pp. 9–10). We assume an ideal case in which the earth's surface is entirely horizontal and extensive. In this case, the boundary between ground and atmosphere is a plane. The plane contains no heat, but under normal circumstances a considerable exchange of heat occurs across it. The quantities that determine this heat exchange will now be discussed.

Radiation S is the first (major) factor of heat exchange. Heat arrives at the earth's surface from the sun, the sky, and the atmosphere (insolation). Heat is sent back into space (outgoing or terrestrial radiation). Factors that add heat to the surface of the ground are considered positive; those that subtract heat from it are negative. The sum of insolation and outgoing radiation, that is, the balance, decides in individual cases whether S is positive or negative. In Geiger's (1965) book, the unit for S, as for all factors in the heat budget, is cal cm^{-2} min^{-1}, also called in English the langley per minute, abbreviated ly min^{-1}. These are not SI units. The SI unit is W/m^2 (one watt = 1 joule/s = 0.239 cal/s).

The second factor B is determined by the flow of heat from the ground to the surface or in the reverse direction. During a cold winter night heat flows upward through the ground and B is, therefore, positive; on a summer afternoon, B is negative because heat is transported downward from the surface.

Third, the air above the ground plays a part in the exchange of heat L. This factor also may be positive or negative. Transport of heat to or from the ground depends not only on physical heat conduction, as within the ground, but also on mass exchange (eddy diffusion) because of the great mobility of the air.

Fourth, there is the effect of evaporation V. This is measured, like all the other heat-economy factors, in calories per square centimeter per minute (Geiger, 1965) or W/m^2. The quantity of heat in calories required to evaporate 1 g of water is called the latent heat of vaporization and varies with temperature. At 25°C, it is 583 cal gram^{-1}. If a round figure of 600 cal gram^{-1} is used for temperatures above 0°C, then V in cal cm^{-2} min^{-1} corresponds to the evaporation of a certain depth of water in millimeters per hour. Normally V is negative, but positive values are possible, as when dew or hoarfrost form on the surface and heat of condensation or sublimation is released.

From the surroundings of the area under consideration, there can flow warmer or colder, moister or drier, air, a process that is called *advection*. This advection process has an effect on the heat economy of the area and upsets the assumption on which the previous discussion was based, namely that horizontal counter-influences are absent. We introduce the additional advection process by defining the factor Q.

Precipitation may entail a gain or a loss of heat for the ground, depending on its temperature, and this is given by the symbol N. Over oceans, lakes, and rivers, the factor W, for the exchange of heat between water and its surface, is used instead of B.

The complete equation for the heat exchange at a flat vegetation-free ground surface is:

$$S + B \ (or \ W) + L + V + Q + N = 0. \tag{23.10}$$

Note: Geiger was a German, so the letters stand for German words, as follows: S = die Sonne (the sun); B = der Boden (the soil); W = das Wasser (the water); L = die Luft (the air); V = die Verdunstung or die Verdampfung (the evaporation); Q = die Quer (*in die Quer* in German means "crosswise"); N = der Niederschlag (the rain).

XII. RADIATION BALANCE

Now let us consider specifically the factor S in Equation 23.10, because it is the most important factor taking part in the heat exchange at the surface of the earth. We continue with Geiger's (1965, p. 12–13) analysis.

The symbol S means the radiation balance or net radiation. If insolation is greater than outgoing (terrestrial) radiation, the balance is positive; if it is less, the balance is negative. A negative balance is described as a net loss of radiation. Sometimes the term "outgoing radiation" is used to designate the Stefan-Boltzmann radiation loss and sometimes for the negative radiation balance. Geiger (1965) uses the term "effective outgoing radiation" and avoids the ambiguous term "outgoing radiation."

The radiation balance consists of two radiation streams of different spectral ranges (Fig. 23.3). There is a short-wavelength part only as long as the sun shines, that is, during the daytime. Radiation reaching the surface of the earth consists of that part of direct solar radiation I that is not reflected by clouds, absorbed by the atmosphere, or scattered diffusely, and also that part of the nondirectional sky radiation H that represents diffusely scattered radiation that has reached the ground and provides "daylight" within the visible spectrum. The value of $I + H$ reaching a horizontal surface is called *global radiation*. Part of this radiation is reflected by the earth's surface. This short-wavelength reflected radiation R depends on the nature of the ground, in contrast to $I + H$. The reflection factor or *albedo* is the ratio of the reflected to the incident radiation, usually expressed as a percentage (Table 23.1).

Incoming long-wavelength radiation is of no significance in the radiation balance of the earth as a planet. It is, however, of great importance for the radiation balance of the earth's surface. The atmosphere of the earth contains water vapor, ozone, and other gases (Fig. 23.2), all of which absorb radiation and emit it according to Kirchhoff's law (Equation 23.3). This long-wavelength atmospheric radiation G is termed counterradiation, because it counteracts the terrestrial radiation loss. It occurs both by day and by night, and, in fact, is somewhat greater during the day, because it is dependent on temperature.

It might be expected that part of this long-wavelength radiation would also be lost through reflection by the ground. However, the earth's natural surface cover can be considered to resemble a black body. In general, the albedo of natural surfaces is less than 5%. Snow cover, which reflects so strongly within the visible spectrum that newly fallen snow produces a

TABLE 23.1 Albedo of various surface for total solar radiation with diffuse reflection (both short wavelength)

Surface	Percent reflected
Fresh snow cover	75–95
Compressed snow	70
Melting snow	30–65
Dense cloud cover	60–90
Old snow cover	40–70
Clean firm snow	50–65
Dry salt cover	50
Light sand dunes, surf	30–60
Clean glacier ice	30–46
Dirty firm snow	20–50
Lime	45
Granite	15
Quartz sand	35
Sandy soil	15–40
Meadows and fields	12–30
Prairie, wet	22
Prairie, dry	32
Stubble fields	15–17
Grain crops	10–25
Pine, spruce wood	10–14
Deciduous wood	16–37
Yellow leaves (autumn)	33–36
Desert, midday	15
Desert, low solar altitude	35
Bare fields	12–25
Wet plowed fields	5–14
Densely built-up areas	15–25
Woods	5–20
Grass, green	16–27
Grass, dried	16–19
Dark clay, wet	2–8
Dary clay, dry	16
Sand, wet	9
Sand, dry	18
Dark cultivated soil	7–10
Water surfaces, sea	3–10
Water, 0 to 30°C	2
Water, 60°C	6
Water, 85°C	58

From Geiger, 1965, p. 15, and van Wijk and Scholte Ubing, 1966, p. 87. This material is used by the permission of Dr. Walter Geiger, legal successor to Prof. Dr. Rudolf Geiger, Perlschneider Str. 18, 81241 Munich, Germany.

striking improvement in light conditions, is practically an ideal black body for long waves, reflecting at the most 0.5% of the incident radiation.

According to the Stefan-Boltzmann law, the radiation emitted by the soil surface by day and by night would be exactly σT^4 (T is the surface temperature), if the ground were a black body. As just stated, this condition is largely fulfilled by natural surfaces. To the extent that it is not fulfilled, the outgoing radiation will be reduced according to Kirchhoff's law. But at the same time, the amount of outgoing long-wavelength reflected radiation would be increased, and it is not possible to distinguish it instrumentally from the terrestrial radiation.

The radiation balance S is, therefore, given by the equation

$$S = I + H + G - \sigma T^4 - R \ (\text{cal cm}^{-2} \text{ min}^{-1} \text{ or W/m}^2) \qquad (23.11)$$

The last two factors in Equation 23.11 depend on the nature of the ground surface, while the first three on the right-hand side of the equation are independent of it. Figure 23.9 shows the magnitude of these factors for a summer day and a summer night.

XIII. APPENDIX: BIOGRAPHY OF GUSTAV KIRCHHOFF

Gustav Robert Kirchhoff (1824–1887), the German physicist who established spectroscopy on a sound theoretical basis and studied complex electrical circuits as well as radiation, was born at Königsberg (Kaliningrad) on March 12, 1824. He was educated at the university of his native town. After acting as *Privatdozent* in Berlin (1847–1850), he became extraordinary professor of physics in Breslau in 1850. Four years later he was appointed professor of physics in Heidelberg, and in 1875 he was transferred to Berlin, where he remained for the rest of his life.

Kirchhoff's contributions to experimental and mathematical physics were numerous and important. In his work in electricity, he modified the resistance bridge, brought to public attention by Wheatstone (see Chapter 20), and developed a theorem that gives the distribution of currents in a network. Kirchhoff extended Ohm's theory for a linear conductor (see Chapter 20) to the case of conductors in three dimensions, and so generalized the equations dealing with the flow of electricity in conductors. Another important piece of work was the demonstration that an electric disturbance is propagated along wire with the same velocity as light is propagated in free space (Preece, 1971a).

His name is best known for the researches, in conjunction with the great German chemist Robert Wilhelm Bunsen (1811–1899), on the devel-

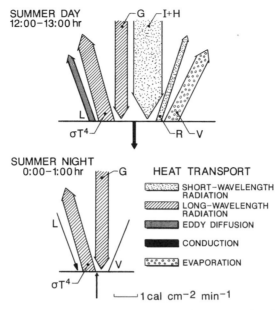

FIG. 23.9 The importance of radiation as compared with the other factors in the heat budget. *I*, direct solar radiation; *H*, diffusely scattered radiation; *G*, long-wave atmospheric radiation; σT^4, radiation emitted by soil surface; *R*, short-wave reflected radiation; *L*, heat; *V*, evaporation; *I + H*, global radiation. (From Geiger, R., *The Climate Near the Ground*. Rev ed. Translated by Scripta Technica, Inc., p. 14, ©1965. Harvard University Press: Cambridge, Massachusetts. This material is used by the permission of the legal successor to Rudolf Geiger Prof. Dr. Walter Geiger, Perlschneider Str. 18, 81241 Munich, Germany.)

opment of spectrum analysis. The rich period of Kirchhoff's life was the twenty years he taught in Heidelberg and worked with Bunsen. It was during the years 1859–1862 that these great investigators together made the outstanding discoveries of spectrum analysis. At the time the physical laboratory in Heidelberg was unpretentious and was located in a house, the "Riesengebäude," then 150 years old. The memorable researches were carried on in a small room. In 1857 Bunsen and Henry E. Roscoe first described the Bunsen burner. This new burner furnished Bunsen and Kirchhoff with a nonluminous gas flame of fairly high temperature, in which chemical substances could be vaporized and a spectrum could be obtained, due purely to the luminous vapor (Cajori, 1929, p. 168).

To Kirchhoff belongs the merit of having enunciated a complete account of the theory of spectrum analysis. He established the method on a solid basis. He gave the explanation of the Fraunhofer lines and thus opened up to investigation a new field in spectrum analysis applied to the composition of celestial bodies (Preece, 1971a). Although spectrum analysis, as a terrestrial science, was due equally to Kirchhoff and Bunsen, its celestial applications belong to Kirchhoff alone. Kirchhoff's explanation of the Fraunhofer lines was epoch-making. Said Helmholtz (1821–1894; German physicist), "It has in fact most extraordinary consequences of the most palpable kind, and has become of the highest importance for all branches of natural science. It has excited the admiration and stimulated the fancy of men as hardly any other discovery has done, because it has permitted an insight into worlds that seemed forever veiled for us." In this connection, Kirchhoff frequently related the following story. The question of whether or not Fraunhofer's lines reveal the presence of gold in the sun was being investigated at the time. Kirchhoff's banker remarked on this occasion: "What do I care for gold in the sun if I cannot fetch it down here?" Shortly afterwards Kirchhoff received from England a medal for his discovery, and its value in gold. While handing it over to his banker, he observed, "Look here, I have succeeded at last in fetching some gold from the sun." (Cajori, 1929, p. 169).

Kirchhoff's researches concerning radiation played a leading role. He defined a perfect black body, and, as to the experimental realization of it, he suggested a closed box with black walls inside, kept at a constant temperature and having a very small opening through which radiation may pass from the inside to the outside. He died in Berlin on October 17, 1887 (Preece, 1971a).

XIV. APPENDIX: BIOGRAPHY OF JOSEF STEFAN

Josef Stefan (1835–1893), the Austrian physicist who made original contributions to the kinetic theory of gases, hydrodynamics, and radiation, was born on March 24, 1835, at St. Peter near Klagenfurt. He was educated at the University of Vienna, where he became doctor of philosophy in 1858; then *Privatdozent* in mathematical physics; in 1863 professor ordinarius of physics; and in 1866 director of the Physical Institute. He was a distinguished member of the Vienna Academy of Sciences, of which he was appointed secretary in 1875. Before Stefan's work, Kirchhoff had already described the perfect radiator as the perfectly black body, namely, one that absorbed all the radiation that fell on it and reflected none, but emitted radiation of all wavelengths. Stefan showed empirically in 1879 that the radiation of such a body was proportional to the fourth power of its absolute temperature, a relationship since known as Stefan's law or as the Stefan-Boltzmann law after it had been deduced by Ludwig Boltzmann in

1884 from thermodynamic considerations. Stefan's law was one of the first important steps leading to the understanding of black-body radiation from which the quantum idea of radiation sprang. Stefan died on January 7, 1893, in Vienna (McKie, 1971).

XV. APPENDIX: BIOGRAPHY OF LUDWIG BOLTZMANN

Ludwig Boltzmann (1844-1906), Austrian physicist, made important contributions to many branches of physics. His greatest achievements were the development of statistical mechanics and the statistical explanation of the second law of thermodynamics. He was born in Vienna on February 20, 1844, and studied at the university there, receiving his doctorate in 1866. He held professorships in mathematics (Vienna, 1873–1876), experimental physics (Graz, 1876–1889), and theoretical physics (Graz, 1869–1873; Munich, 1889–1893; Vienna, 1894–1900; Leipzig, 1900–1902; Vienna, 1902–1906). Despite his several professorships, theoretical physics was his real vocation (Klein, 1971).

In 1905, when he was professor of theoretical physics at the University of Vienna, Boltzmann was invited to give a course of lectures in the summer session at the University of California in Berkeley. His recollections of that summer survive in his popular essay, "Reise eines deutschen Professors ins Eldorado." An abridged translation in presented in *Physics Today* (Boltzmann, 1905). Boltzmann's great sense of humor is evident in this writing.

When Boltzmann began his scientific work, he attacked the problem, until then unconsidered, of explaining the second law of thermodynamics on the basis of the atomic theory of matter. In a series of papers published during the 1870s, Boltzmann showed that the second law could be understood by combining the laws of mechanics, applied to the motions of the atoms, with the theory of probability. In this way he made clear that the second law is an essentially statistical law and that a system will approach a state of thermodynamic equilibrium, because the equilibrium state is overwhelmingly the most probable state. The entropy function of thermodynamics, whose behavior shows the trend to equilibrium and whose maximum value characterizes the equilibrium state, is itself a measure of the probability of the macroscopic state. (The equation relating entropy and probability is engraved on the monument at Boltzmann's grave in Vienna.) He built much of the structure of statistical mechanics, a structure later elaborated by the U.S. mathematical physicist Josiah Willard Gibbs (1839–1903) (Klein, 1971).

Apart from Boltzmann's work on statistical mechanics, he made extensive calculations in the kinetic theory of gases. He was also one of the first Europeans to recognize and to expound on the importance of James Clerk Maxwell's (1831–1879; Scottish physicist) theory of electromagnetism, a subject on which he published a two-volume treatise. Boltzmann also derived, using thermodynamics, Stefan's law for black-body radiation, a derivation that Hendrik Antoon Lorentz (1853–1928, Dutch physicist who got the Nobel prize in physics in 1902) called "a true pearl of theoretical physics" (Klein, 1971).

Boltzmann's work in statistical mechanics was strongly attacked by Wilhelm Ostwald (1853–1932; German chemist who received the Nobel Prize in chemistry in 1909) and the energeticists who did not believe in atoms and wanted to base all of physical science on energy considerations only. Boltzmann also suffered from misunderstandings, on the part of others, about his ideas on the nature of irreversibility. They did not fully grasp the statistical nature of his reasoning. He was fully justified against both sets of opponents by the discoveries in atomic physics, which began shortly before 1900 and by the fluctuation phenomena, such as Brownian motion, which could be understood only by statistical mechanics (Klein, 1971). Cercignani (1998) gives an in-depth discussion of the scientific world in which Boltzmann lived.

Depressed by the criticism of his work, Boltzmann took his own life by hanging on September 5, 1906, at Duino, near Trieste, Italy (Klein, 1971).

XVI. APPENDIX: BIOGRAPHY OF WILHELM WIEN

Wilhelm Wien (1864–1928), German physicist and Nobel Prize winner, was born January 13, 1864, at Gaffken, East Prussia. He studied at the universities of Göttingen, Heidelberg, and Berlin, and in 1890 entered the Physicotechnical Institute near Berlin as assistant to Helmholtz. In 1896, he was appointed professor at the technical high school in Aachen. In 1899 he went to Giessen; in 1900 to Würzburg; and in 1920 to Munich. He wrote on optical problems; on radiation, especially black-body radiation, for which in 1911 he was awarded the Nobel Prize; on water and air currents, on discharge through rarefied gases, cathode rays, and X rays. Wien's most important contributions to black-body radiation are contained in three laws named after him, the most famous of these being known as Wien's displacement law. His autobiography was published posthumously under the title *Aus dem Leben und Wirken eines Physikers* (1930). Wien died on August 30, 1928, in Munich (Preece, 1971b).

REFERENCES

Boltzmann, L. (1905). A German professor's trip to El Dorado. *Physics Today* 45(1); 44–51, 1992. (Translated by Bertram Schwarzschild).

Cajori, F. (1929). *A History of Physics*. Macmillan: New York.

Cercignani, C. (1998). *Ludwig Boltzmann: The Man Who Trusted Atoms*. Oxford University Press: New York.

Geiger, R. (1965). *The Climate Near the Ground*. Rev ed. Translated by Scripta Technica, Inc. Harvard University Press: Cambridge, Massachusetts.

Johnson, F. S. (1954). The solar constant. J. *Meteorol* 11; 431–439.

Klein, M.J. (1971). Boltzmann, Ludwig. *Encyclopaedia Britannica* 3; 893.

McKie, D. (1971). Stefan, Josef. *Encyclopaedia Britannica* 21; 198.

Preece, W.E. (General Ed.) (1971a). Kirchhoff, Gustav. *Encyclopaedia Britannica* 13; 383.

Preece, W.E. (General Ed.) (1971b.) Wien, Wilhelm. *Encyclopaedia Britannica* 23; 499.

Rosenberg, N. J. (1974). *Microclimate: The Biological Environment*. Wiley: New York.

Shortley, G., and Williams, D. (1971). *Elements of Physics*. 5th ed. Prentice-Hall: Englewood Cliffs, New Jersey.

Slatyer, R. O. (1967). *Plant-Water Relationships*. Academic Press: London.

Van Wijk, W. R., and Scholte Ubing, D.W. (1966). Radiation. In *Physics of Plant Environment* (van Wijk, W.R., Ed.), pp. 62–101. Second ed. North-Holland: Amsterdam.

24

Measurement of Canopy Temperature with Infrared Thermometers

Plant temperature and water use are related because, if a plant is well watered, the stomata are open, transpirational cooling occurs, and canopy temperature is cool. Conversely, as a plant becomes water stressed, stomata close, transpiration is reduced, and canopy temperature increases. Consequently, one can use canopy temperature to characterize the water status of a crop (Kirkham et al., 1983; 1984; 1985). In the 1970s, portable, commercially available infrared thermometers that measure thermal radiation were developed and refined (Jackson et al., 1980). They provide a means to measure remotely plant canopy temperatures, and measurements with them are easy because the instruments are hand-held and lightweight (Jackson et al., 1977). (Jackson and colleagues at the U.S. Water Conservation Laboratory in Phoenix, Arizona, did pioneering experiments with portable infrared thermometers. For a biography of Jackson, see the Appendix, Section VII.) In this chapter we consider the theory and use of infrared thermometers.

I. INFRARED THERMOMETERS

Infrared thermometers have the advantage of measuring many leaves at one time. Before their development, it was difficult to determine the magnitude of the temperature difference either between plants or between plants and air, because there was no way of defining the temperature of a group of leaves. A leaf with the surface normal to incident solar radiation has a higher temperature than a leaf that has a surface parallel to the sun's rays or one that is shaded (Tanner, 1963). Severe sampling problems exist if one can make only a few measurements on individual leaves, such as one does when using thermocouples. And the temperature that is measured depends on the location of the thermocouple (for example, base of leaf versus tip of leaf). Tanner (1963) said, "There is no single temperature value that represents the plant and which has been demonstrated to be useful for any given research problem."

The developments in infrared thermometry have provided instruments that surmount the sampling problem. The thermal radiation from all plant surfaces in the field of view (F.O.V.) of the instrument is integrated into a single measurement. A temperature measurement with an infrared thermometer gives a temperature with a particular definition: the black-body temperature that would produce the radiation entering the instrument from plant parts in the field of view (Tanner, 1963). Because the thermal-radiation emissivity of green plants is high (0.95 to 0.97) (Tanner, 1963), the measured (apparent) radiation temperature can be converted to the plant temperature with little error. Most natural surfaces have high emissivities, ranging between 0.90 and 0.98 (Campbell, 1977, p. 49). Measurements made with infrared thermometers are particularly useful in studies of transpiration (water loss from plants), because the temperature measured with the instrument (radiated from the upper part of the plant) gives weight to the plant portions participating most actively in transpiration (Tanner, 1963).

As we noted in Chapter 23, a good approximation of a black body is a small hole in the wall of a hollow body (Fig. 23.6). A beam of radiation that enters the hole and hits the inside wall is partly reflected to another part of the wall, where again a fraction is absorbed and so on. After a number of reflections, little radiant energy is left and the chance that some of it is reflected outwards through the hole is exceedingly small. For similar reasons, a dense vegetative cover in which part of the leaves are seen on edge when viewed from above is much darker (i.e., has a lower reflection factor) than the surface of a single leaf (van Wijk and Scholte Ubing, 1966, p. 66).

II. DEFINITIONS

In Chapter 23, we defined black body and emissivity. Here we define other terms that are used in association with radiation and in the literature dealing with infrared thermometers. We shall use the definitions provided by van Wijk and Scholte Ubing (1966, pp. 62–63). *Radiant energy* is the energy traveling in the form of electromagnetic waves. It has the dimension of energy so that it is measured in joule, erg, calorie, or an equivalent quantity. The amount of radiant energy emitted, transferred, or received per unit time is called *radiant flux* Φ (Greek letter, capital phi). It has the dimension of energy per unit time. In physical literature, the watt (W) (1 watt = 1 joule s^{-1}) or the erg s^{-1} are commonly used as units; in meteorology, the unit cal min^{-1} is frequently employed. *Radiant flux density $H = d\Phi/dA$* is the flux per unit of surface; it is expressed as W m^{-2}, erg cm^{-2} s^{-1}, cal cm^{-2} min^{-1} (= langley min^{-1}) (one langley = 1 cal cm^{-2}), or equivalent units. The units of radiant flux density (van Wijk and Scholte, 1966, p. 63) are the same as those for radiated power per unit area (Shortley and Williams, 1971, p. 324). When it is desired to point out that the radiant flux is directed toward the surface of observation, the term *irradiancy* or *irradiance* is used. If one wants to stress that the radiation is emitted by a source, the radiant flux density is sometimes called *radiancy* or *emittancy*. Emittancy is also called *radiant emittance* (Campbell, 1977, p. 48).

As noted in Chapter 23, the amount of radiant energy contained in thermal radiation depends strongly on the wave length λ of the radiation that is emitted or received. It is often necessary to consider the energy, intensity, flux, etc. per unit of wave length interval. Such quantities are called spectral quantities. They will be indicated by the subscript lambda (λ).

III. PRINCIPLES OF INFRARED THERMOMETRY

Let us now turn to the basic principles of infrared thermometers. We follow the analysis of Perrier (1971, p. 654). The energy emitted by a body that is not perfectly black is given by Equation 23.5, $P = \alpha P_{Black} = \alpha\sigma T^4$. For a perfect black body, $\alpha = 1$ and Equation 23.5 reduces to Equation 23.4, the Stefan-Boltzmann law. We apply Equation 23.5 to the surface temperature, T_s, of leaves. If P is in units of W m^{-2}, the surface temperature will be in $^\circ K$, and the Stefan-Boltzmann constant σ will be 5.67×10^{-8} W m^{-2} K^{-4} (Campbell, 1977, p. 49). The term α (emissivity) is dimensionless. The emissivity, α, is sometimes abbreviated ε, the abbreviation used by

Perrier (1971). Surface temperature can be calculated from Equation 23.5 if surface emissivity is known and the flux of thermal radiation emitted is measured. [Perrier (1971, p. 654) uses the term "emittance," but most publications use the term "emissivity" for ε.]

A radiometer has a sensor that receives energy from the measured surface through the optics of the radiometer, which define the field of view by use of a diaphragm, lens, and sometimes a mirror, and bring localized surface areas into focus. It is necessary to select the waveband of thermal energy emitted by the surface from the total energy received by the sensor and originating at the surface. Therefore, a filter with a sharp bandpass in the infrared region (Table 24.1) is used generally to eliminate the short-wave radiation. The bandpass of 8 to 14 μm is particularly suitable (Fig. 24.1). This selected bandpass includes the peak of black body emission at normal temperature (9 to 10 μm) so that the maximum energy is measured. Moreover, water does not absorb radiation in this band; thus, the effect of water vapor on the measurement is minimized. But that part of the long-wave radiation emitted by the surroundings and reflected by the surface in this waveband cannot be eliminated directly (Fig. 24.2; ϕ_r).

Theoretically, a simple integration of the relation representing the response of infrared thermometers can be written as follows:

$$A = \varepsilon\sigma T_s^4 + (1 - \varepsilon)B \ [A \text{ in units of W m}^{-2}] \qquad (24.1)$$

where A is the flux of long-wave radiation from the surface; T_s is the real temperature of the surface (leaves); B is the total incident long-wave (or thermal) radiation in units of W m^{-2}; $(1 - \varepsilon)B$ is the reflected component influencing the thermometer output; ε is the surface emissivity; and σ is the Stefan-Boltzmann constant.

TABLE 24.1   The electromagnetic spectrum

Type of radiation	Frequency range (cycles/sec)	Wavelength range (cm)
Electric waves	0 to 10^4	Infinity to 3×10^6
Radio waves	10^4 to 10^{11}	3×10^6 to 0.3
Infrared	10^{11} to 4×10^{14}	0.3 to 7.6×10^{-5}
Visible	4×10^{14} to 7.5×10^{14}	7.6×10^{-5} to 4×10^{-5}
Ultraviolet	7.5×10^{14} to 3×10^{18}	4×10^{-5} to 10^{-8}
X rays	16×10^{16} to 3×10^{22}	10^{-6} to 10^{-12}
Gamma rays	3×10^{18} to 3×10^{21}	10^{-8} to 10^{-11}

From Rosenberg, N.J., *Microclimate: The Biological Environment*, p. 5, ©1974, John Wiley & Sons. New York. This material is used by permission of John Wiley & Sons, Inc.

WAVELENGTH, μm ⟶

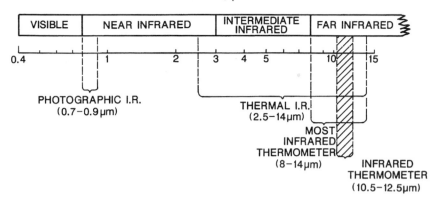

FIG. 24.1 A portion of the electromagnetic spectrum relating photographic infrared, thermal infrared, and infrared thermometer ranges to the visible and infrared regions. (From Jackson, R.D., Pinter, Jr., P.J., Reginato, R.J., and Idso, S.B., p. 5, 1980. Hand-Held Radiometry. *Agr Rev Manuals ARM-W-19,* United States Department of Agriculture, Science and Education Administration, Western Region: Oakland, California.)

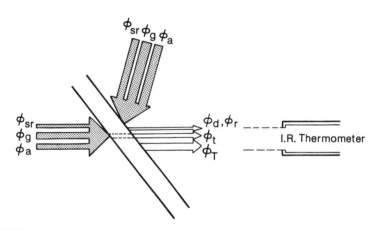

FIG. 24.2 Scheme of fluxes of energy on a surface like a leaf. ϕ_T: energy emitted by the surface; ϕ_g: part of global radiation received by the surface; ϕ_a: part of long-wave radiation emitted by sky and received by the surface; ϕ_{sr}: radiation from the surroundings received by the surface; ϕ_r: reflected part of all these radiations; ϕ_d: diffused part of all these radiations; ϕ_t: transmitted part. (From Perrier, A., Leaf temperature measurement. In *Plant Photosynthetic Production. Manual of Methods,* p. 632–671, ©1971, Z. Šesták, J. Čatský, and P.G. Jarvis, Eds. Fig. 17.7c, p. 656. Dr W. Junk N.V.: The Hague, The Netherlands. With kind permission of Kluwer Academic Publishers and Professor Alain Perrier.)

It is supposed that ε is independent of T_s over a narrow range (−15 to 60°C) and is independent of wavelength over a narrow waveband (8 to 14 μm). This condition is important only in the second term $(1 - \varepsilon)B$. It is supposed also that the filter function is practically independent of the temperature T_s (0 to 40°C) and the temperature of the filter is constant. As a first approximation, since ε and B are known, by assuming that ε is close to unity (generally, for leaves $0.94 < \varepsilon < 0.98$), the real surface temperature (T_s) can be estimated (T) from the relation

$$A = \sigma T^4. \tag{24.2}$$

In Equation 24.2 (compare with Equation 24.1), the calculated surface temperature (T) will be overestimated because the term containing B is neglected and also underestimated because ε is overestimated. This compensation between the two opposed deviations leads to a small overall error in the calculation of the surface temperature (T). Experience shows that B is most often less than A, and the maximum of B is reached in the evening (more scattering and reflection) or under a cloudy sky. The error generally varies between 0.5 and 1.5°C.

IV. USE OF A PORTABLE INFRARED THERMOMETER

Now let us turn to field use of infrared thermometers, following the description of Jackson et al. (1980, p. 52). To obtain representative canopy temperatures, it is desirable to point the infrared thermometer so that a maximum amount of vegetation is viewed by the sensor. This can be accomplished by viewing the target obliquely and at right angles to any structures which might be present in the field. It is best to take readings looking in several directions to minimize effects that the sun's angles (Kimes et al., 1980) (altitude angle; azimuth angle) may have on target temperature. The viewing angle used by Kirkham et al. (1984) was 30 degrees. Jackson et al. (1980) take measurements 1 to 2 hours following solar noon, a time when a maximum difference between canopy and air temperature usually occurs. Routine weather observations, such as cloud cover, windspeed, precipitation, target conditions, and wet- and dry-bulb air temperatures, are recorded whenever canopy temperatures are measured. Wet- and dry-bulb air temperatures are essential in determining the Crop Water Stress Index (see Chapter 25). It is best to take measurements on cloud-free days to minimize errors due to reflection and scattering from clouds.

V. CALIBRATION OF INFRARED THERMOMETERS

Jackson et al. (1980) found that the readout temperature on most factory calibrated instruments is not an accurate representation of apparent blackbody temperatures. They calibrated all instruments under standardized conditions. Jackson et al. (1980) and Perrier (1971, p. 655) describe the calibration of an infrared thermometer. Let us use Perrier's description.

The unique relationship between the data supplied by the infrared thermometer (A_o) and the flux of long-wave radiation A (Equation 24.1) reaching the apparatus from the surface is obtained in the laboratory by measuring A_o for many different surface temperatures (T) of a reference black body. The temperature (T) gives the flux A (Equation 24.2), so that it is possible to draw the curve relating A_o to A or directly to T. For these measurements, the infrared thermometer is put either close to the surface of a sphere immersed in a temperature-controlled bath (Fig. 24.3a) (thus obtaining a very good black body at known temperature T) or at the top of a perfectly reflecting cone placed on a reference surface (anodized aluminum) (Fig. 24.3b), the temperature of which is controlled and varied.

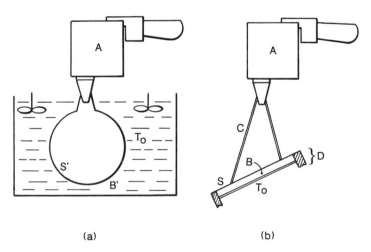

(a)   (b)

FIG. 24.3 Schematic diagram of infrared thermometer being calibrated either using (a) a controlled temperature bath, B', or (b) an aluminum block, B. Other abbrevations: A, radiation thermometer; T_o, controlled temperature; S', spherical surface (black surface); S, anodized aluminum surface; C, cone (reflecting surface); D, reference surface system. (From Perrier, A., Leaf temperature measurement. In *Plant Photosynthetic Production. Manual of Methods*, p. 632–671, ©1971, Z. Šesták, J. Čatský, and P.G. Jarvis, Eds. Fig. 17.7c, p. 656. Dr W. Junk N.V.: The Hague, The Netherlands. With kind permission of Kluwer Academic Publishers and Professor Alain Perrier.)

Such calibration curves relating A_o to T are reproducible to within a range of 0.3°C. Some manufacturers provide a black-body plate with a thermometer imbedded in it to perform checks of the calibration. Stigter et al. (1982) also describe calibration of infrared thermometers.

VI. ADVANTAGES OF INFRARED THERMOMETERS

Infrared thermometers have three advantages. First, they are easy to use. The infrared thermometer is pointed at the canopy and a readout on the back of the instrument, facing the viewer, immediately displays the temperature. The instruments can give either the temperature of the canopy or the difference in temperature between the air and the canopy. The latter temperature usually is preferred, because it indicates how stressed a crop is. Canopies with temperatures below ambient temperature are less water stressed than those with temperatures above ambient temperature. (The air temperature can be measured separately with a thermometer.) Infrared thermometers have been used to schedule irrigations of crops such as corn (Clawson and Blad, 1982). In such work it is important to measure the canopy temperature of a well-watered control for a standard, local reference.

A second advantage of infrared thermometers is that they can rapidly measure temperatures remotely and nondestructively. A third advantage is that they can integrate temperatures over an area (the field of view) and thus avoid the sampling problem of single-point measurements made, for example, with thermocouples. Note that thermocouples used to measure leaf temperature touch the leaf directly and measure a different temperature than that determined with an infrared thermometer. The infrared thermometer measures a black-body temperature. Measurements made with thermocouples and infrared thermometers cannot be compared directly.

Canopies must be well developed and covering the soil before data can be collected with commercially available infrared thermometers (Fig. 24.4). Measurements cannot be made on individual plants, such as those in pots in controlled environments. Amiro et al. (1983) describe a small infrared thermometer that can be used on broad or narrow leaves grown under controlled-environment or field conditions. A focusing system must be implemented for narrow leaves.

Of all the instruments available to measure water in plants (thermocouple psychrometers, pressure chambers, diffusion porometers, infrared thermometers), the infrared thermometer might have the most immediate, practical value. It provides an easy way to measure canopy temperature

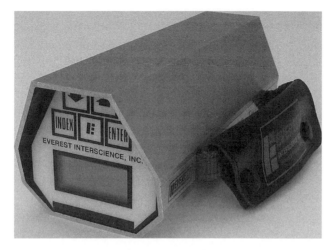

FIG. 24.4 A commercially available, hand-held infrared thermometer. (From a brochure of Everest Interscience, Tucson, Arizona. Picture courtesy of Everest Interscience.)

and to schedule irrigations. An elevated canopy temperature indicates water stress and a need for irrigation. Producers might use the infrared thermometer on crops to detect water stress before damage occurs. This would be particularly important on high-value crops, such as those grown by horticulturists. Canopy-temperature measurements can be made at different locations in a field to identify stressed areas. Consequently, infrared thermometers are valuable as a means to determine remotely spatial variability due to drought or any other stress that reduces the transpiration rate.

VII. APPENDIX: BIOGRAPHY OF RAY JACKSON

Ray Dean Jackson, a soil physicist at the U.S. Water Conservation Laboratory (retired), was born in Shoshone, Idaho, on September 28, 1929. He married in 1952 and 1968 and has seven children (*American Men and Women of Science*, 1994). He served in the U.S. Marine Corps before receiving his B.S. degree at Utah State University in 1956. He earned an M.S. in soil physics from Iowa State University in 1957, and a Ph.D. from Colorado State University in 1960. From 1957 to 1960 he was a soil scientist with the Soil and Water Conservation Research Division, Agriculture Research Service (ARS) of the United States Department of Agriculture (USDA), in Colorado. In 1960, he joined the U.S. Water Conservation Laboratory of the USDA in Phoenix, Arizona, as a research physicist, where he worked until retirement in 1992. He was research leader and the technical advisor for soil-plant-atmosphere relations for the Western Region of the ARS. He was an adjunct professor of soil and water science at the University of Arizona, Tucson. In the summer of 1964 he was an OECD (Organization for Economic Cooperation and Development) fellow at Rothamsted Experimental Station, England.

Jackson published on subjects relating to diffusion in porous media, soil-water evaporation, soil-water movement, heat transfer, simultaneous heat and water transfer, atmospheric radiation, and infrared thermometry. He was perhaps the first researcher to publish measured soil-water diffusivity data for the relatively dry water contents of the western United States, a region where water-vapor diffusion predominates, and he showed that diffusion theory held at these low water contents. This work formed the basis for the development of the "desert survival skill" developed by Jackson and van Bavel (1965). They demonstrated that a transparent plastic sheet covering a hole in the soil could be used to collect potable water from desert soils and plants. This technique is taught in survival courses worldwide (American Society of Agronomy, 1975).

Jackson received the Superior Service Unit Award from the USDA in 1963. He is a fellow of the American Society of Agronomy, Soil Science Society of America, and the American Association for the Advancement of Science (*American Men and Women of Science*, 1994). In 1992, he won the Outstanding Scientist of the Year Award from the USDA-ARS. The award recognized his leadership skills and his research, which resulted in the commercialization of hand-held infrared thermometers to measure remotely plant leaf temperatures for determination of a crop's water needs (American Society of Agronomy, 1992).

REFERENCES

American Men and Women of Science. (1994). 19th ed. Vol. 4, J–L. R.R. Bowker: New Providence, New Jersey.

American Society of Agronomy. 1975. Ray D. Jackson. *Agronomy J* 67; 103.

American Society of Agronomy. 1992. People. USDA-ARS. *Agronomy News*, December issue, p. 10.

Amiro, B.D., Thurtell, G.W., and Gillespie, T.J. (1983). A small infrared thermometer for measuring leaf temperature in leaf chambers. *J Exp Bot* 34; 1569–1576.

Campbell, G.S. (1977). *An Introduction to Environmental Biophysics*. Springer-Verlag: New York.

Clawson, K.L., and Blad, B.L. (1982). Infrared thermometry for scheduling irrigation of corn. *Agronomy J* 74; 311–316.

Jackson, R.D., and van Bavel, C.H.M. (1965). Solar distillation of water from soil and plant materials: Simple desert survival technique. *Science* 149; 1377–1379.

Jackson, R.D., Reginato, R.J., and Idso, S.B. (1977). Wheat canopy temperature: A practical tool for evaluating water requirements. *Water Resources Res* 13; 651–656.

Jackson, R.D., Pinter, Jr., P.J., Reginato, R.J., and Idso, S.B. (1980). Hand-Held Radiometry. *Agr Rev Manuals ARM-W-19.* United States Department of Agriculture, Science and Education Administration, Western Region: Oakland, California.

Kimes, D.S., Idso, S.B., Pinter, Jr., P.J., Reginato, R.J., and Jackson, R.D. (1980). View angle effects in the radiometric measurement of plant canopy temperatures. *Remote Sensing Environ* 10; 273–284.

Kirkham, M.B., Johnson, D.E., Kanemasu, E.T., and Stone, L.R. (1983). Canopy temperature and growth of differentially irrigated alfalfa. *Agr Meteorol* 29; 235–246.

Kirkham, M.B., Suksayretrup, K., Wassom, C.E., and Kanemasu, E.T. (1984). Canopy temperature of drought-resistant and drought-sensitive genotypes of maize. *Maydica* 29, 287–303.

Kirkham, M.B., Redelfs, M.S., Stone, L.R., and Kanemasu, E.T. (1985). Comparison of water status and evapotranspiration of six row crops. *Field Crops Res* 10; 257–268.

Perrier, A. (1971). Leaf temperature measurement. In *Plant Photosynthetic Production. Manual of Methods* (Šesták, Z., Čatský, J., and Jarvis, P.G., Eds.), pp. 632–671. Dr W. Junk N.V.: The Hague, The Netherlands.

Rosenberg, N.J. (1974). *Microclimate: The Biological Environment.* John Wiley and Sons: New York.

Shortley, G., and Williams, D. (1971). *Elements of Physics.* 5th ed. Prentice-Hall: Englewood Cliffs, New Jersey.

Stigter, C.J., Jiwaji, N.T., and Makonda, M.M. (1982). A calibration plate to determine performance of infrared thermometers in field use. *Agr Meteorol* 26; 279–283.

Tanner, C.B. 1963. Plant temperatures. *Agronomy J* 55; 210–211.

van Wijk, W.R., and D.W. Scholte Ubing. 1966. Radiation. In *Physics of Plant Environment* (van Wijk, W.R., Ed.), pp. 62–101. 2nd ed. North-Holland: Amsterdam, The Netherlands.

25

Stress-Degree-Day Concept and Crop-Water-Stress Index

The stress-degree-day (SDD) procedure and crop-water-stress index (CWSI) are popular methods to evaluate water stress in plants. They were developed by scientists at the United States Department of Agriculture (USDA) Water Conservation Laboratory in Phoenix, Arizona. Their work has resulted in many papers. See, for example, Jackson et al. (1977); Idso et al. (1977); Ehrler et al. (1978a, 1978b); Idso et al. (1978, 1979, 1980, 1981a, 1981b, 1981c, 1981d); Jackson et al. (1981); Idso (1982a, 1982b); Jackson (1982); Idso et al. (1982); Idso (1983); Sharratt et al. (1983); Idso et al. (1984). We now define stress-degree-day and crop-water-stress index and show their application.

I. STRESS-DEGREE-DAY PROCEDURE

The work of the Phoenix scientists began in 1976 (Dean, 1976). The stress-degree-day concept was developed before the crop-water-stress index. Let us follow the description of its development by Jackson et al. (1977). The water status of a plant is a primary determinant of grain yield. A means for evaluating water status by remote measurement could open the way to improved yield predictions and, in irrigated areas, to improved scheduling times. The temperature of a plant canopy can be measured remotely with lightweight, hand-held infrared thermometers (see Chapter 24). The difference between the temperature of a plant canopy and the temperature of the surrounding air $(T_c - T_a)$ may be an indicator of the water status of a crop because water stress causes partial stomatal closure, thus reducing transpiration and allowing sunlit leaves to warm above ambient air temperature. The Phoenix scientists introduced the concept of a *stress-degree-day (SDD)*. SDD is a daily value of $T_c - T_a$ measured at the time of maximum surface temperature (generally 1 to 1.5 hours after solar noon). SDD is defined as follows:

$$SDD = {}_{n=i}\Sigma^N (T_c - T_a)_n, \qquad (25.1)$$

which is the plant canopy temperature T_c minus the air temperature T_a 150 cm above the soil, summed over N days beginning at day i.

[The *SDD* concept is similar to the growing-degree-day concept: $GDD = \Sigma\{[(T_M + T_m)/2] - T_t\}$, where GDD is growing degree day; T_t is the threshold temperature for growth; T_M and T_m are the daily maximum and minimum air temperatures, respectively; and the GDD values are summed over N days, the number of days under consideration (Lowry, 1969, p. 194). The threshold temperature varies with different crops.]

Jackson et al. (1977) evaluated water stress in plants (durum wheat, *Triticum durum* Desf. var. Produra) by using the stress-degree-day concept. They differentially irrigated the wheat. Plot 1 was the dry treatment (only enough irrigation-water added to permit survival, a stressful condition in arid Phoenix, Arizona, where crops are usually amply watered, so they will grow). Plot 6 was the wet plot and was overwatered. Plots 2 through 5 received amounts of irrigation water that varied between the amounts added to Plots 1 and 6. Figure 25.1 shows their results. The greater the stress (lack of water), the greater was the value of the stress-degree-day. They began the summation of the SDD on day 83 (February 24, 1976), the day on which differential irrigation treatments were started. They ended the SDD summation on the day of harvest. Plot 6, which received an excessive amount of irrigation water, had canopy temperatures that were consistently less than air temperatures, and the SDD became less than −100 during the latter part of the season.

In general, if a plant has adequate water, $T_c - T_a$ will be near zero or negative; if it is water stressed, $T_c - T_a$ will be greater than zero. Thus, the sum of the positive values of $T_c - T_a$ may serve as an index of when to irrigate. Jackson et al. (1977) defined a positive SDD as follows:

$$SDD_{pos} = {}_{n=i}\Sigma^N (T_c - T_a)_n \qquad (25.2)$$

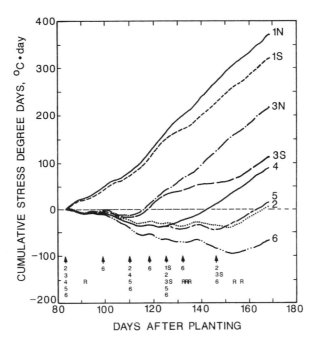

FIG. 25.1 Stress-degree-days versus days after planting. Plot numbers are shown at the right. Arrows indicate irrigations, the numbers of the plots receiving the irrigation being shown below the arrows. R indicates rain. (From Jackson, R.D., Reginato, R.J., and Idso, S.B., Wheat canopy temperature: A practical tool for evaluating water requirements. *Water Resources Research* 13(3); 651–656, ©1977, American Geophysical Union. Reproduced by permission of American Geophysical Union.)

in which values of $T_c - T_a$ less than zero are set equal to zero. The index i is the first day after irrigation, and N is the number of days required for SDD_{pos} to reach a prescribed value.

Figure 25.2 shows SDD_{pos} and soil water depletion (measured using a neutron probe) for two of the plots in the experiment of Jackson et al. (1977). Cloudiness and other climatic conditions can cause abrupt changes in the slope of the SDD_{pos} versus time graph (Fig. 25.2) during the first few days after irrigation. As water depletion increases, $T_c - T_a$ is always positive, the slope rapidly increases, and the effect of climatic factors diminishes.

Jackson et al. (1977) proposed an SDD_{pos} 10 as an index for the time to irrigate wheat in Arizona. They recognized that this value is somewhat dependent on the means used to measure T_c and T_a (for example, the height at which T_a is measured), that it may be soil and crop specific (they used a loam soil), and that it may be different under other climatic conditions.

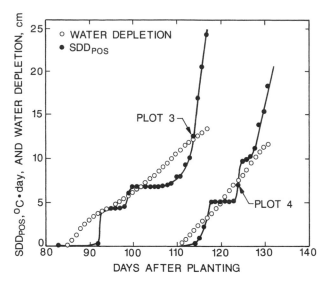

FIG. 25.2 Positive stress-degree-days and water depletion for two plots, beginning after the last irrigation. Numerical values on the ordinate are the same for both factors. (From Jackson, R.D., Reginato, R.J. and Idso, S.B., Wheat canopy temperature: A practical tool for evaluating water requirements. *Water Resources Research* 13(3); 651–656, ©1977, American Geophysical Union. Reproduced by permission of American Geophysical Union.)

Nevertheless, the SDD_{pos} appears to provide a possible means to develop irrigation scheduling based on remotely sensed plant-canopy temperatures.

II. CANOPY-MINUS-AIR TEMPERATURE AND EVAPOTRANSPIRATION

Let us now turn to the relation between $T_c - T_a$ and evapotranspiration. One approach to estimating the amount of water depleted from the root zone is to use an evapotranspiration equation based on the temperature difference $T_c - T_a$, such as the following equation (Jackson et al., 1977):

$$ET = R_n - G - f(u)C(T_c - T_a), \tag{25.3}$$

in which ET is evapotranspiration, R_n is net radiation, G is soil heat flux, $f(u)$ is a function of wind speed, and C is the volumetric heat capacity of air. This equation is a reliable predictor of crop evapotranspiration (Stone and Horton, 1974). Stone and Horton (1974) used Equation 25.3 for the same purpose that Jackson et al. (1977) were concerned with: to develop a method of predicting water use over large areas by using remotely sensed parameters.

To use Equation 25.3, Jackson et al. (1977) made some simplifying assumptions. They found that for their experimental conditions, wind was not of major importance in the calculation of ET using Equation 25.3. (This may not be true for locations with persistent winds and higher wind speeds than those recorded in Phoenix, Arizona.) They were not concerned with hourly values of ET, but wanted to calculate daily values of actual ET, using a minimum of input data and a one-time-of-day measurement of $T_c - T_a$. For 24-hour periods, it is safe to assume that the soil heat flux G is negligible. With Jackson et al.'s (1977) simplifying assumptions, Equation 25.3 becomes

$$ET = R_n - B(T_c - T_a), \qquad (25.4a)$$

in which B is a composite constant that must be determined.

The parameter B in Equation 25.4a was evaluated by using daily values for ET from a lysimeter, daily values of R_n over the lysimeter, and one-time-of-day (taken between 13:30 and 14:00 hours) measurements of $T_c - T_a$ for every day for which ET, R_n, and $T_c - T_a$ data were available, from day 60 until harvest of the wheat. These data are shown in Fig. 25.3. Figure 25.3A

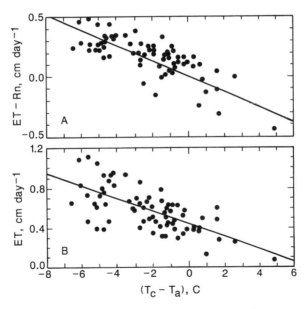

FIG. 25.3 Evapotranspiration and net radiation as a function of canopy-air temperature difference. (From Jackson, R.D., Reginato, R.J. and Idso, S.B., Wheat canopy temperature: A practical tool for evaluating water requirements. *Water Resources Research* 13(3); 651–656, ©1977, American Geophysical Union. Reproduced by permission of American Geophysical Union.) See text for explanation of parts A and B.

shows the relation for $ET - R_n$ versus $T_c - T_a$. A statistical value for B was obtained by forcing Equation 25.4a through the origin, since Equation 25.4a indicates that for $T_c - T_a = 0$, $ET - R_n = 0$. This yielded

$$ET = R_n - 0.064 \, (T_c - T_a). \tag{25.4b}$$

In Fig. 25.3B, the dependence of ET on $T_c - T_a$ alone was determined. The relation is

$$ET = 0.438 - 0.064 \, (T_c - T_a). \tag{25.5}$$

The constants in Equation 25.4b and Equation 25.5 were evaluated by using ET data from lysimeters. To test their applicability, ET was calculated by using R_n and $T_c - T_a$ data from the wheat plots. Water depletion was also calculated. The measured and calculated data are compared in Fig. 25.4. In Fig. 25.4A, ET was calculated from Equation 25.4b by using net radiation measured over the north sides of each plot. In Fig. 25.4B, the net radiation was averaged over the six plots for each day, and the average was used in Equation 25.4b. In Fig. 25.4C, the seasonal average

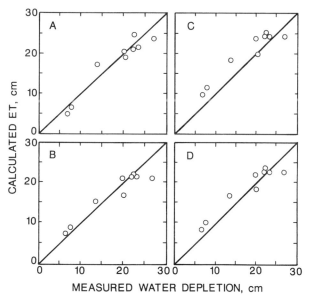

FIG. 25.4 Calculated evapotranspiration and measured water depletion. The lines indicate a 1:1 relation. See text for explanation of different parts of figure. (From Jackson, R.D., Reginato, R.J., and Idso, S.B., Wheat canopy temperature: A practical tool for evaluating water requirements. *Water Resources Research* 13(3); 651–656, ©1977, American Geophysical Union. Reproduced by permission of American Geophysical Union.)

of R_n was used in Equation 25.4b, whereas in Fig. 25.4D, R_n was taken as the statistically derived constant from Fig. 25.3B (i.e., from Equation 25.5). The data in Figs. 25.4A and 4B indicate that if daily estimates of R_n are available, water use can be estimated reasonably well by using Equation (25.4a).

The data of Jackson et al. (1977) indicate that air temperature could be determined on the ground and airborne scanners could measure T_c, enabling water use by crops to be evaluated over large areas. In sum, the work by Jackson et al. (1977) showed that: 1) the *SDD* concept can be used as an indicator for determining the times and amounts of irrigation; and 2) because predicted *ET*, from an expression relating *ET* to net radiation and $T_c - T_a$, and measured water used agreed reasonably well, the expression may be useful in determining amounts of irrigation water to apply.

III. CROP-WATER-STRESS INDEX

Now let us consider the crop-water-stress index, which was developed by the Phoenix scientists four years after the stress-degree-day concept was developed (Idso et al., 1982). The crop-water-stress index is also called the plant-water-stress index. Only the difference between canopy temperature and air temperature is considered in the stress-degree-day concept. However, stress-degree-day may be influenced by factors such as air vapor pressure, net radiation, and wind speed (Idso et al., 1981c). It is important to determine the significance of these other factors and to devise a means for adjusting for them. Consequently, the Phoenix workers developed a plant-(crop-)water-stress index that essentially normalizes the stress-degree-day value.

The basis for the plant-water-stress index was established by the work of Ehrler (1973). He used thermocouples to measure the leaf temperature of four varieties of cotton (*Gossypium hirsutum* L. "Deltapine SL," "Deltapine-16," and "Hopicala" and *Gossypium barbadense* L. "Pima-S4"). He found that for clear, sunny days, the difference between leaf and air temperature from 08:00 to 18:00 hours was a linear function of air vapor pressure deficit (*VPD*), as long as the plants were well supplied with water (Fig. 25.5).

Working with infrared thermometers, Idso et al. (1981c) extended Ehrler's (1973) data to include alfalfa (*Medicago sativa* L.), soybeans (*Glycine max* L. Merr.), and squash (*Curcurbita pepo* L.). They plotted values of $T_c - T_a$ versus *VPD* (Figs. 25.6, 25.7, 25.8) and found that

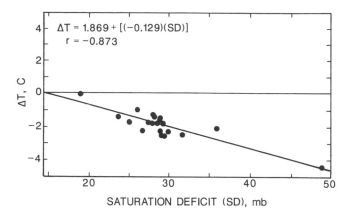

FIG. 25.5   The regression of cotton leaf-air temperature difference (ΔT, °C) on the saturation deficit of the air (mb). Air temperature and vapor pressure were measured 1 m above the cotton crop. The data are restricted to the period 08:00–18:00 hours (Mountain Standard Time) on predominantly sunny days when the crop was fully hydrated, i.e., from 2 to 6 days after a heavy irrigation. (From Ehrler, W.L., Cotton leaf temperatures as related to soil water depletion and meteorological factors. *Agronomy Journal* 65; 404–409, ©1973, American Society of Agronomy: Madison, Wisconsin. Reprinted by permission of the American Society of Agronomy.)

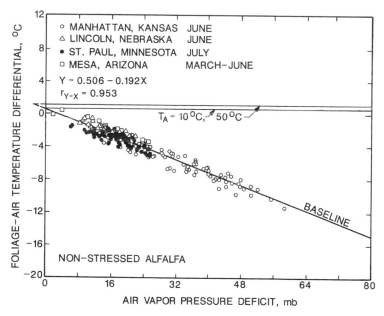

FIG. 25.6   Foliage-air temperature differential vs. air vapor pressure deficit for well-watered alfalfa grown at the specified sites and dates during 1980. (From Idso, S.B., Jackson, R.D., Pinter, Jr., P.J., Reginato, R.J., and Hatfield, J.L., Normalizing the stress-degree-day parameter for environmental variability. *Agricultural Meteorology* 24; 45–55, ©1981, Elsevier Scientific Publishing Company: Amsterdam. Reprinted by permission of Elsevier, Amsterdam.)

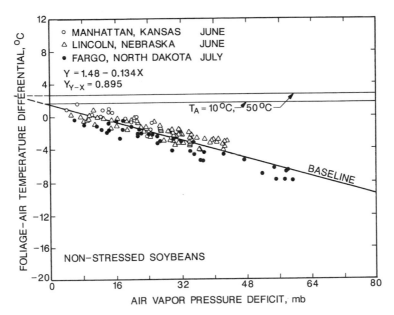

FIG. 25.7 Foliage-air temperature differential vs. air vapor pressure deficit for well-watered soybeans grown at the specified sites and dates. (From Idso, S.B., Jackson, R.D., Pinter, Jr., P.J., Reginato, R.J., and Hatfield, J.L., Normalizing the stress-degree-day parameter for environmental variability. *Agricultural Meteorology* 24; 45–55, ©1981, Elsevier Scientific Publishing Company: Amsterdam. Reprinted by permission of Elsevier, Amsterdam.)

crop-specific linear relationships prevailed throughout the greater portion of the daylight period (i.e., from about two to three hours after sunrise to about two to three hours before sunset). They also found these relationships to be essentially undisturbed by variations in other environmental parameters, such as wind speed or the normal course of insolation through the day. Only shading by clouds seemed to have a significant influence, reducing foliage (canopy) temperature relative to that of the air by several degrees (Idso, 1982b).

Figure 25.9 provides a generalized representation of these results and a framework for describing the development of the plant-water-stress index (Idso, 1982b). The lower limit of this graph, which represents a state of potential evaporation, is referred to as the *non-water-stressed baseline*. It is crop specific and must be obtained by experimentation as described in the preceding paragraphs. Once established, it is used to define the other limiting condition that prevails when water stress is a maximum and transpiration completely suppressed, which is accomplished as follows (Idso, 1982b).

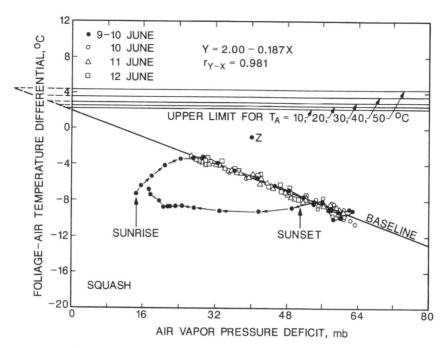

FIG. 25.8 Foliage-age temperature differential vs. air vapor pressure deficit for well-watered squash grown at Tempe, Arizona, in June, 1980. (From Idso, S.B., Jackson, R.D., Pinter, Jr., P.J., Reginato, R.J., and Hatfield, J.L., Normalizing the stress-degree-day parameter for environmental variability. *Agricultural Meteorology* 24; 45–55, ©1981, Elsevier Scientific Publishing Company: Amsterdam. Reprinted by permission of Elsevier, Amsterdam.)

Consider a well-watered plant transpiring at the potential rate. A plot of $T_c - T_a$ versus VPD (T_c, canopy temperature, is also called T_f, foliage temperature) for this plant will fall somewhere on the non-water-stressed baseline; and as the air VPD decreases to zero, it will move along this baseline to achieve the $T_f - T_a$ value representative of the linear relationship's intercept. If this term is positive, as it has proven to be (Idso, 1982b) (value a in Fig. 25.9), there will still be a small evaporative flux from the plant to the air, even though the air at that point is saturated, due to the positive vapor-pressure gradient (VPG) that exists between the plant and the air as a result of the plant's higher temperature.

This driving force for evaporation is easily evaluated as $VPG = \rho_s(T_f) - \rho_s(T_a)$, where $\rho_s(T_f)$ is the saturated vapor pressure at the temperature of the foliage and $\rho_s(T_a)$ is the saturated vapor pressure at the temperature of the air; for transpiration to be reduced to zero, it must be reduced to zero. One way by which this may be accomplished is to supersaturate the air, that is, to create a negative VPD equivalent in absolute magnitude

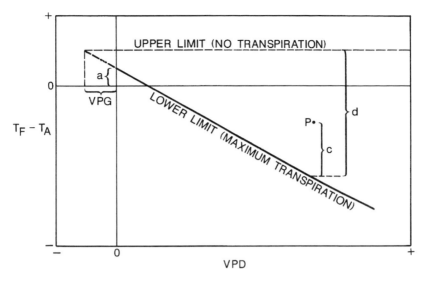

FIG. 25.9 The general form of the relationship between foliage-air temperature differential $(T_f - T_a)$ and air vapor pressure deficit (VPD) for a stand of vegetation sufficiently supplied with water to transpire at the potential rate, i.e., the lower limit (maximum transpiration line), plus an illustration of how the upper limit (no transpiration line) is derived using the vapor pressure gradient (VPG). The values c and d are used to define the plant water stress index. See text for explanation. (From Idso, S.B., Reginato, R.J., and Farah, S.M., Soil- and atmosphere-induced plant water stress in cotton as inferred from foliage temperatures. *Water Resources Research* 18(4); 1143–1148, ©1982, American Geophysical Union. Reproduced by permission of American Geophysical Union.)

to the VPG. Then, following the non-water-stressed baseline back into the negative VPD region by this amount will specify the upper limit to which $T_f - T_a$ may rise *at the particular air temperature in question*. This latter point of emphasis is made to underscore the fact that there is not a unique upper limit for a given species, as is the case with the non-water-stressed baseline, but rather a variety of limits corresponding to the variety of air temperatures that may prevail. For plants with a small baseline intercept (i.e., less than 0.5°C), this upper limit dependency on air temperature is weak and can sometimes be ignored (Idso, 1982b).

Consider now a data point representative of a stressed plant that locates it at position P in Fig. 25.9. In this format, Idso et al. (1981c) defined the *plant-water-stress index* ($PWSI$) (or *crop-water-stress index*, $CWSI$) as the ratio of the vertical distance between the data point and the non-water-stressed baseline and the total vertical distance between the baseline and the upper limit (i.e., $PWSI = c/d$). Thus defined, it can be seen that as a plant goes from a condition of maximum transpiration to one of no transpiration, the index goes from a value of zero to unity; Jackson et al. (1981) have demonstrated that actual transpiration (E) at any point P in this range is specified as $E = E_p(1–PWSI)$, where E_p is the potential evaporation rate that could be sustained in the given circumstances, but with a nonlimiting supply of soil moisture (Idso, 1982b). (The PWSI or CWSI has sometimes been referred to as the IJ index after Idso and Jackson, the two scientists who developed the concept. For a biography of Idso, see the Appendix, Section VII. A biography of Jackson appears in Chapter 24, Section VII.)

IV. HOW TO CALCULATE THE CROP-WATER-STRESS INDEX

Let us now take a specific example, which shows how to obtain the PWSI (or the CWSI). Let us refer to Fig. 25.8 (Idso et al., 1981c). Suppose at a time when the air VPD is 40 mb, the value of $T_f - T_a$ is –1°C, so that the point Z on Fig. 25.8 represents the status of the crop, which in this case is squash. Now, if the crop had been sufficiently supplied with water to evaporate at the potential rate, $T_f - T_a$ would have been –5.5°C, as obtained from intersecting the non-water-stressed baseline at VPD = 40 mb. Conversely, if the crop had not been transpiring at all, and T_a was 30°C (for example), then $T_f - T_a$ would be expected to have been about 3°C. With this information, we can define the PWSI (or CWSI) to be the ratio of the vertical distance above the non-water-stressed baseline that the point Z has conceptually traveled in falling below the potential evaporation rate to the total possible distance that it could conceptually travel, which in this example is –1°C – (–5.5°C) divided by 3°C – (–5.5°C) or 4.5°C/8.5°C = 0.53. Thus, we see that as the ratio of actual to potential evaporation goes from 1 to 0, the crop-water-stress index goes from 0 to 1.

V. CROP-WATER-STRESS INDEX FOR ALFALFA, SOYBEANS, AND COTTON

Idso et al. (1981c) did not determine $T_c - T_a$ versus VPD for water-stressed squash, so no measured points in Fig. 25.8 lie around point Z in the figure. They did, however, determine $T_c - T_a$ versus VPD for water-stressed alfalfa and soybeans (Figs. 25.10 and 25.11). Note that for water-stressed alfalfa and soybeans points lie between the baseline or lower limit (maximum transpiration) and the upper limit (no transpiration). In Fig. 25.12,

Idso et al. (1981c) have converted the data from Figs. 25.10 and 25.11 into the format of the crop-water-stress index. The soybeans, in this instance, were still fairly young, and covered only about 10% of the ground. Thus, with their rather limited rooting volume, they experienced a dramatic rate of stress development as the hot and dry day, on which the data were obtained, progressed (maximum air temperature was 39°C and minimum relative humidity was 17%). But the alfalfa, with its well-developed root system, showed a much greater buffering capacity to stress development, although it too showed a significant increase in stress in the afternoon. Maximum stress for both crops occurred about one to two hours after solar noon, indicating that this was a good time for a once-a-day measurement, as was used by Jackson et al. (1981), to quantify the stress history of several differently irrigated wheat plots. Figure 25.13 shows the crop-(plant-) water-stress index plotted for mildly stressed cotton (lower line) and moderately stressed cotton (upper line) before and after irrigation (Idso et al., 1982).

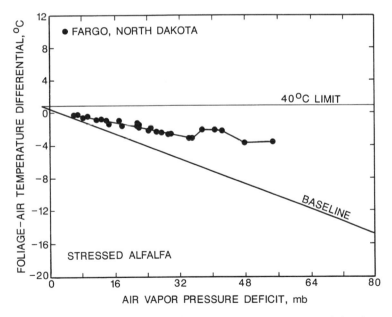

FIG. 25.10 Foliage-air temperature differential vs. air vapor pressure deficit for stressed alfalfa growing at Fargo, North Dakota. (From Idso, S.B., Jackson, R.D., Pinter, Jr., P.J., Reginato, R.J., and Hatfield, J.L., Normalizing the stress-degree-day parameter for environmental variability. *Agricultural Meteorology* 24; 45–55, ©1981, Elsevier Scientific Publishing Company: Amsterdam. Reprinted by permission of Elsevier, Amsterdam.)

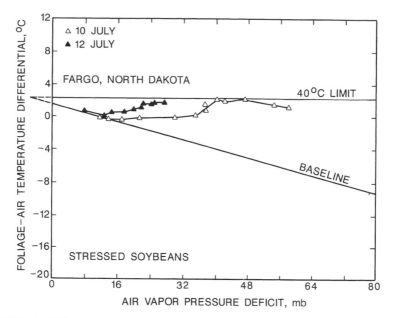

FIG. 25.11 Foliage-air temperature differential vs. air vapor pressure deficit for stressed soybeans growing at Fargo, North Dakota. (From Idso, S.B., Jackson, R.D., Pinter, Jr., P.J., Reginato, R.J., and Hatfield, J.L., Normalizing the stress-degree-day parameter for environmental variability. *Agricultural Meteorology* 24; 45–55, ©1981, Elsevier Scientific Publishing Company: Amsterdam. Reprinted by permission of Elsevier, Amsterdam.)

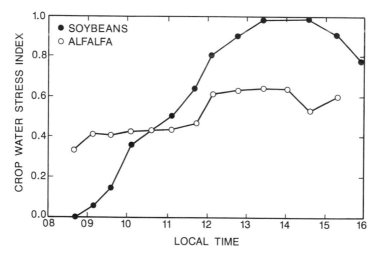

FIG. 25.12 The crop-water-stress index as a function of time for severely stressed soybeans and less severely stressed alfalfa at Fargo, North Dakota. (From Idso, S.B., Jackson, R.D., Pinter, Jr., P.J., Reginato, R.J., and Hatfield, J.L., Normalizing the stress-degree-day parameter for environmental variability. *Agricultural Meteorology* 24; 45–55, ©1981, Elsevier Scientific Publishing Company: Amsterdam. Reprinted by permission of Elsevier, Amsterdam.)

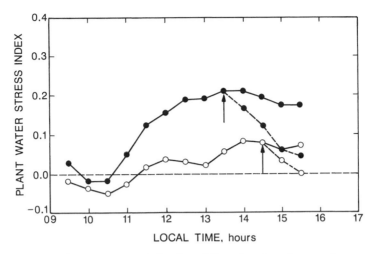

FIG. 25.13 The plant water stress index for mildly and moderately stressed cotton preceding and following irrigations shown by arrows. (From Idso, S.B., Reginato, R.J., and Farah, S.M., Soil- and atmosphere-induced plant water stress in cotton as inferred from foliage temperatures. *Water Resources Research* 18(4); 1143–1148, ©1982, American Geophysical Union. Reproduced by permission of American Geophysical Union.)

VI. IMPORTANCE OF A WIDE RANGE OF VAPOR-PRESSURE DEFICIT VALUES

Plots of $T_c - T_a$ versus vapor-pressure deficit for well-watered plants appear to yield a unique linear relationship under a specific climatic condition. The Phoenix scientists postulate that the existence of such linear relationships provides a simple criterion for identification of a potential evaporation (Idso et al., 1981c). Their findings also provide a means for normalizing the stress-degree-day value for environmental variability, by converting it into the crop-water-stress index. It is evident, however, that defining stress in this fashion limits the ability to quantify, with confidence, the crop-water-stress index under conditions of low vapor-pressure deficit, where the variability of $T_c - T_a$ approaches the degree of scatter inherent in the data (see Figs. 25.6 and 25.7). Therefore, it is important to have a wide range of vapor-pressure-deficit values to obtain meaningful crop-water-stress indexes.

VII. APPENDIX: BIOGRAPHY OF SHERWOOD IDSO

Sherwood B. Idso was born June 12, 1942, in Thief River Falls, Minnesota, where he lived until graduating from high school in 1960. He then enrolled in the Institute of Technology at the University of Minnesota, where he received a bachelor's degree in physics with distinction in 1964, an M.S. degree in 1966, and a Ph.D. in 1967. He moved to the U.S. Water Conservation Laboratory of the USDA in Phoenix, Arizona, in 1967, where he has since worked as a research physicist. He also is an adjunct professor in the Departments of Geology and Geography at Arizona State University in Tempe. Idso and his wife have seven children (Idso, 1982c).

Idso has published numerous scientific papers. He has studied heat and moisture transfer in the soil-plant-atmosphere continuum. He developed methods for evaluating evaporative water losses from soil, plants, and open water, along with a number of techniques for the remote sensing of soil- and plant-water status. He has an abiding interest in severe weather phenomena and is a dedicated investigator of dust storms and dust devils (Idso, 1982c). He is well known for his writings related to climate change.

In 1977, Idso received the Arthur S. Flemming Award "for his innovative research into fundamental aspects of agricultural-climatological interrelationships affecting food production and the identification of achievable research goals whose attainment could significantly aid in assessment and improvement of world food supplies." The Flemming Award is presented annually to people under the age of 40 who work in civilian or military capacities in the federal government. The Downtown Jaycees of Washington, D.C., sponsor the award (American Meteorological Society, 1978).

REFERENCES

American Meteorological Society. (1978). About our Members. *Bull Amer Meteorol Soc 59*; 747.

Dean, J.P. (1976). Sensing soil moisture—remotely. *Agr Res 25*(4); 7–9.

Ehrler, W.L. (1973). Cotton leaf temperatures as related to soil water depletion and meteorological factors. *Agronomy J 65*, 404–409.

Ehrler, W.L., Idso, S.B., Jackson, R.D., and Reginato, R.J. (1978a). Wheat canopy temperature: Relation to plant water potential. *Agronomy J 70*; 251–256.

Ehrler, W.L., Idso, S.B., Jackson, R.D., and Reginato, R.J. (1978b). Diurnal changes in plant water potential and canopy temperature of wheat as affected by drought. *Agronomy J 70*; 999–1004.

Idso, S.B. (1982a). Humidity measurement by infrared thermometry. *Remote Sensing Environ 12*; 87–91.

Idso, S.B. (1982b). Non-water-stressed baselines: A key to measuring and interpreting plant water stress. *Agr Meteorol 27*; 59–70.

Idso, S.B. (1982c). *Carbon Dioxide: Friend or Foe?* IBR Press: Tempe, Arizona. (Available from the Institute for Biospheric Research, Inc., 631 E. Laguna Drive, Tempe, Arizona 85282)

Idso, S.B. (1983). Stomatal regulation of evaporation from well-watered plant canopies: A new synthesis. *Agr Meteorol* 29; 217.

Idso, S.B., Jackson, R.D, and Reginato, R.J. (1977). Remote-sensing of crop yields. *Science* 196; 19–25.

Idso, S.B., Jackson, R.D., and Reginato, R.J. (1978). Extending the "degree day" concept of plant phenological development to include water stress effects. *Ecology* 59; 431–433.

Idso, S.B., Hatfield, J.L., Jackson, R.D., and Reginato, R.J. (1979). Grain yield prediction: Extending degree-day approach to accommodate climatic variability. *Remote Sensing Environ* 8; 267–272.

Idso, S.B., Reginato, R.J., Hatfield, J.L., Walker, G.K., Jackson, R.D., and Pinter, Jr., P.J. (1980). A generalization of the stress-degree-day concept of yield prediction to accommodate a diversity of crops. *Agr Meteorol* 21; 205–211.

Idso, S.B., Reginato, R.J., Jackson, R.D., and Pinter, Jr., P.J. (1981a). Measuring yield-reducing plant water potential depressions in wheat by infrared thermometry. *Irrigation Sci* 2; 205–212.

Idso, S.B., Reginato, R.J., Jackson, R.D., and Pinter, Jr., P.J. (1981b). Foliage and air temperatures: Evidence for a dynamic "equivalence point." *Agr Meteorol* 24; 223–226.

Idso, S.B., Jackson, R.D., Pinter, Jr., P.J., Reginato, R.J., and Hatfield, J.L. (1981c). Normalizing the stress-degree-day parameter for environmental variability. *Agr Meteorol* 24; 45–55.

Idso, S.B., Reginato, R.J., Reicosky, D.C., and Hatfield, J.L. (1981d). Determining soil-induced plant water potential depressions in alfalfa by means of infrared thermometry. *Agronomy J* 73; 826–830.

Idso, S.B., Reginato, R.J., and Farah, S.M. (1982). Soil- and atmosphere-induced plant water stress in cotton as inferred from foliage temperatures. *Water Resources Res* 18; 1143–1148.

Idso, S.B., Reginato, R.J., Jackson, R.D., and Pinter, Jr., P.J. (1984). Reply to Paw U's comments on "Foliage and Air Temperatures: Evidence for a Dynamic 'Equivalence Point.'" *Agr Meteorol* 31; 87–88.

Jackson, R.D. (1982). Canopy temperature and crop water stress. In *Advances in Irrigation* (Hillel, D., Ed.), Vol. 1, pp. 43–85. Academic Press: New York.

Jackson, R.D., Reginato, R.J., and Idso, S.B. (1977). Wheat canopy temperature: A practical tool for evaluating water requirements. *Water Resources Res* 13; 651–656.

Jackson, R.D., Idso, S.B., Reginato, R.J., and Pinter, Jr., P.J. (1981). Canopy temperature as a crop water stress indicator. *Water Resources Res* 17; 1133–1138.

Lowry, W.P. (1969). *Weather and Life: An Introduction to Biometeorology.* Academic Press: New York.

Sharratt, B.S., Reicosky, D.C., Idso, S.B., and Baker, D.G. (1983). Relationships between leaf water potential, canopy temperature, and evapotranspiration in irrigated and nonirrigated alfalfa. *Agronomy J* 75; 891–894.

Stone, L.R., and Horton, M.L. (1974). Estimating evapotranspiration using canopy temperatures: Field evaluation. *Agronomy J* 66; 450–454.

26

Potential Evapotranspiration

The term *potential evapotranspiration* must be defined when we talk of evapotranspiration. The concept of potential evapotranspiration was put forth by Thornthwaite (1948) and Penman (1948). We will use the definition of Rosenberg (1974, p. 172), noting that his definition is similar to that of Penman. (For a biography of Penman, see the Appendix, Section V.)

I. DEFINITION OF POTENTIAL EVAPOTRANSPIRATION

Rosenberg says that potential evapotranspiration (abbreviated as ETP by him, but as PET by most others) is "the evaporation from an extended surface of [a] short green crop which fully shades the ground, exerts little or negligible resistance to the flow of water, and is always well supplied with water. Potential evapotranspiration cannot exceed free water evaporation under the same weather conditions."

In fact, we know that real (actual) evapotranspiration differs from the potential under most circumstances. The reasons for these differences are best explained by reference to the conditions imposed by the definition of potential evapotranspiration and by an analysis of the reality of these conditions. We follow the discussion of Rosenberg (1974, pp. 172–178), even though we could use other references (e.g., Chang, 1968, pp. 129–144).

II. FACTORS THAT AFFECT POTENTIAL EVAPOTRANSPIRATION

First, let us look at the influence of *extended surfaces* (or what is called the influence of *fetch*) on potential evapotranspiration. An extended surface has great (if unspecified) fetch. Fields should be at least 20 m from their centers in any direction from which the wind blows. However, in some experiments, it has been found that temperature profiles to 5 m, even at 200 m from the edge of a field, are not fully adjusted to the new surface. Any visible difference in plant growth along the border of a field is evidence of inadequate field size for measuring PET within the meaning of the phrase "extended surface" (Rosenberg, 1974, p. 172).

Second, let us look at the influence of crop height. Many of the important crops grown worldwide are not short: corn, sorghum, winter and spring grains, cotton, and trees. The taller the crop, the more effective is its exchange of energy with the ambient air. Alfalfa (called lucerne in England) should fit the definition of a "short crop," but studies have shown that the quantities of water evaporated by this crop increase with increasing crop height (Fig. 26.1). We know also that the type of leaf influences the evapotranspiration rate and that, all things being equal, broad-leaved plants will transpire more than will the grasses (Rosenberg, 1974, p. 172).

Third, let us look at the influence of crop cover. Row crops do not normally shade the ground fully except in some cases at advanced stages in

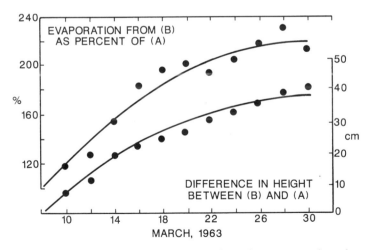

FIG. 26.1 Evapotranspiration from lysimeters, with similar exposures, but where crop of alfalfa (lucerne) was kept clipped short in one (**A**), but allowed to grow in height in other (**B**). Evaporation rates from (**B**) increased to more than double those from (**A**) as crop in (**B**) grew to 42 cm higher than in (**A**). Both lysimeters were surrounded by areas of alfalfa of similar heights to crops growing in them. (From Chang, J.-H., *Climate and Agriculture: An Ecological Survey,* Fig. 81, p. 144, ©1968. Aldine Publishing Co: Chicago. Reproduced by permission of Dr. Jen-Hu Chang.)

their development. Nor do the broadcasted crops such as alfalfa shade the ground for some time after periodic cuttings. We know that water use may continue to increase with increasing leaf area, even when leaf area great enough to shade the ground completely has been achieved. Brun et al. (1972), for example, showed that in soybean and sorghum fields the proportion of water lost, as transpiration increases, is closely correlated to leaf area index, with transpiration being approximately 50% of the total evapotranspiration at a leaf area index of 2. This proportion increases to 95% of the total evapotranspiration at a leaf area index of 4. [*Leaf area index* is defined as the area of leaves above a unit area of ground taking only one side of each leaf into account (Monteith, 1973, p. 52).] Figures 26.2 and 26.3 show relationships between evapotranspiration and leaf area index (Chang, 1968, p. 130; Ritchie, 1972).

Fourth, let us look at the influence of the internal plant resistance to water flow. The concept of potential evapotranspiration assumes that plants behave passively as wicks for the transport of water from the soil to the air. However, plants can close their stomata and increase the resistance to water flow. Under well-watered conditions, some plants appear to have very low resistances. Alfalfa is one of these plants. Cold weather, however, has an interesting effect on the resistance of the alfalfa crop, as shown in Figs. 26.4 and 26.5 for two days, April 21 and April 22, respectively. Evapotranspiraton (*LE*) was greatly reduced on April 22 (Fig. 26.5) after a cold night. Thus, even if alfalfa is well watered, its stomata appear to close when temperatures fall (Rosenberg, 1974).

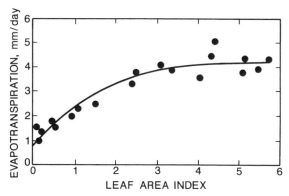

FIG. 26.2   Relationship between leaf area index and evapotranspiration. (From Chang, J.H., *Climate and Agriculture: An Ecological Survey*, Fig. 71, p. 130,©1968. Aldine Publishing Co: Chicago. Reproduced by permission of Dr. Jen-Hu Chang.)

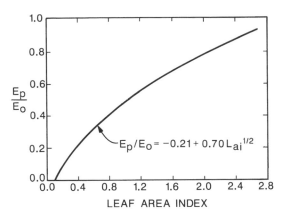

FIG. 26.3 The plant evaporation, E_p, relative to the potential evaporation, E_o, as influenced by leaf area index, when the soil water is not limited. (From Ritchie, J.T., Model for predicting evaporation from a row crop with incomplete cover. *Water Resources Research* 8(5); 1204–1213, ©1972, American Geophysical Union. Reproduced by permission of American Geophysical Union.)

Fifth, let us look at the influence of soil-water availability. Obviously, in the case of range-land and dry-land agriculture, plants are not always well supplied with water. This can be the case in irrigated agriculture also, inadvertently or by intention. When the water supply becomes limited, it is important to get the greatest yield per unit of water expended. For this reason, water is often added at critical stages of growth. We know that strategic irrigation at certain periods in the growth cycle of a crop may lead to great increases in yield.

The definition of PET states that the crop is "well supplied with water." Therefore, with decreasing soil moisture availability, evapotranspiration will be reduced below the potential (Figs. 26.6, 26.7, 26.8, 26.9, 26.10). Van Bavel (1967) suggested that the transpiration rate in alfalfa begins to diminish after a soil water potential of about −4 bars is reached and cites other works in which this breaking point has ranged from −0.2 to −10 bars for corn and cotton, respectively. Ritchie (1972) said that the evaporation from the soil surface is a two-stage process. The first stage is the constant-rate stage in which the evaporation is limited only by the supply of energy to the surface of the soil. The second stage is the falling-rate stage in which water movement to the evaporating sites near the surface is controlled by the hydraulic properties of the soil. These studies relating soil water to evapotranspiration show that the potential rate of evapotranspiration cannot prevail unless the soil is kept well supplied with water.

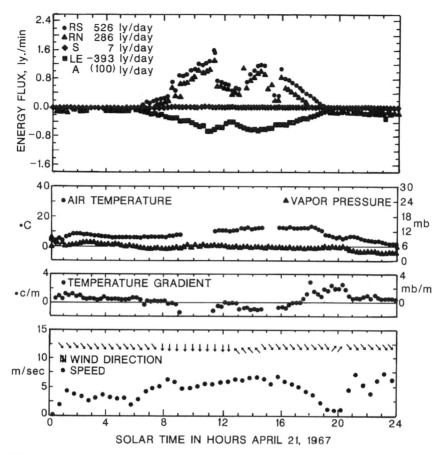

FIG. 26.4 Energy balance with lysimetrically measured evapotranspiration from alfalfa at Mead, Nebraska, April 21, 1967. Air temperature and vapor pressure measured at 100 cm; gradients between 45 and 100 cm; wind speed at 200 cm. *RS* = solar radiation; *RN* = net radiation; *S* = soil heat flux; *LE* = evaporation; *A* = sensible heat flux. (From Rosenberg, N.J., *Microclimate: The Biological Environment*, p. 174, ©1974, John Wiley & Sons, Inc: New York. This material is used by permission of John Wiley & Sons, Inc.)

Sixth, let us look at the relation of free water evaporation and plant water use (Rosenberg, 1974, p. 176). The condition that potential evapotranspiration cannot exceed free water evaporation under the same weather conditions probably applies well in humid regions. For example, the amount of water used by rye and fescue grass is about 80% of that evaporated from open water in evaporation pans, except when winds are strong and the air is hot and dry. Then the ratio drops, apparently because hot and dry conditions cause an increase in stomatal resistance.

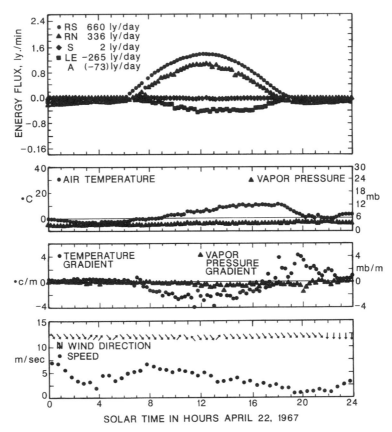

FIG. 26.5 Same as Fig. 26.4 except for April 22, 1967. (From Rosenberg, N.J., *Microclimate: The Biological Environment*, p. 175, ©1974, John Wiley & Sons, Inc: New York. This material is used by permission of John Wiley & Sons, Inc.)

However, in the Great Plains of the United States and in yet more arid regions, well-watered crops that exert little canopy resistance and that are tall or aerodynamically rough can consume more energy and transpire more water than is evaporated from free water surfaces. If the free water surfaces are extensive and the crop areas are not, the differences may be pronounced. A case in point occurred during the period of strong regional advection of sensible heat into eastern Nebraska during May, 1967 (Rosenberg, 1974, p. 177). Evaporation from evaporation pans with land exposures and with lake exposures showed lower daily evaporation than

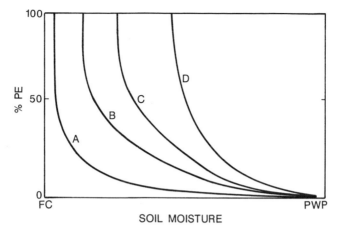

FIG. 26.6 Adjustment of potential evapotranspiration for soil dryness and rooting depth of crops. Curves A to D correspond to increases in rooting depth of crop. FC = field capacity; PWP = permanent wilting point. (From Chang, J.-H., *Climate and Agriculture: An Ecological Survey*, Fig. 79, p. 139, ©1968. Aldine Publishing Co: Chicago. Reproduced by permission of Dr. Jen-Hu Chang.)

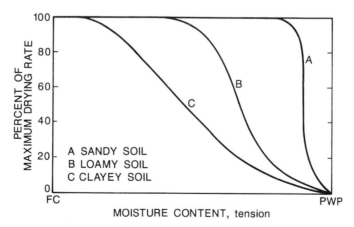

FIG. 26.7 Drying rate of three types of soil. FC = field capacity; PWP = permanent wilting point. (From Chang, J.-H., *Climate and Agriculture: An Ecological Survey*, Fig. 78, p. 138, ©1968. Aldine Publishing Co: Chicago. Reproduced by permission of Dr. Jen-Hu Chang.)

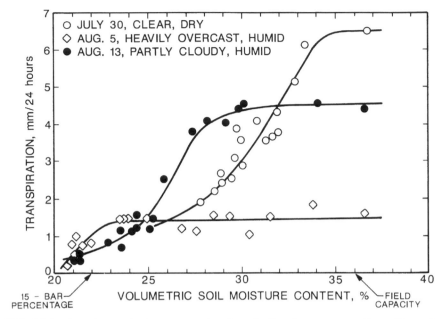

FIG. 26.8   Actual transpiration rate as a function of soil moisture content. (From Chang, J.-H., *Climate and Agriculture: An Ecological Survey*, Fig. 75, p. 136, ©1968. Aldine Publishing Co: Chicago. Reproduced by permission of Dr. Jen-Hu Chang.)

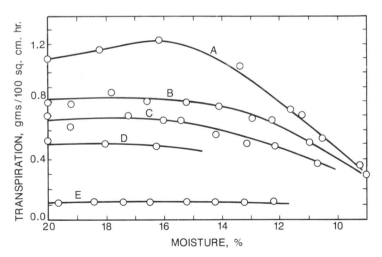

FIG. 26.9   The transpiration of kidney beans in grams 100 cm^{-2} hr^{-1} versus the moisture percentage of the soil at various light intensities at a temperature of 20°C and a relative humidity of 40%. **A:** light intensity 4.5×10^4 ergs cm^{-2} s^{-1}; **B:** 2.4; **C:** 1.4; **D:** 0.66; E = results from Veihmeyer and Hendrickson (undated). Data from Bierhuizen (1958). (From Chang, J.-H., *Climate and Agriculture: An Ecological Survey*, Fig. 74, p. 135, ©1968. Aldine Publishing Co: Chicago. Reproduced by permission of Dr. Jen-Hu Chang.)

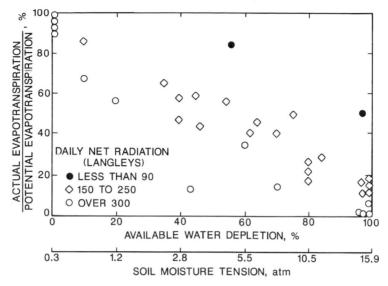

FIG. 26.10 Relative daily actual evapotranspiration rate from short grass as function of moisture depletion from the root zone and soil moisture tension averaged over the total root depth. (From Chang, J.-H., *Climate and Agriculture: An Ecological Survey*, Fig. 76, p. 137, ©1968. Aldine Publishing Co: Chicago. Reproduced by permission of Dr. Jen-Hu Chang.)

was measured with precision weighing lysimeters in an irrigated alfalfa field (Table 26.1). The data in Table 26.1, and other data, show that free water evaporation need not always show the upper limit or the potential evapotranspiration in sub-humid and arid regions as it does, apparently, in humid regions (Rosenberg, 1974, p. 178).

TABLE 26.1 Lysimetrically Measured Alfalfa Evapotranspiration (ET) Compared with Pan Evaporation in Nebraska

Location	May 17–22, 1967 (5-day total) Alfalfa about 25 cm tall
	mm
Alfalfa in lysimeters at Mead, Nebraska	51.91
Alfalfa ET due to advection	19.10
Land exposure pans	45.66
Lake exposure pans	36.08

Data extracted from Rosenberg, 1974, p. 177.

III. ADVECTION

In the preceding paragraph, we used the term *advection*. Advection is defined as the exchange of energy, moisture, or momentum as a result of horizontal heterogeneity (Chang, 1968, p. 140). If an area upwind of an irrigated field is hot and dry, then sensible (measurable) heat will be transferred to the irrigated field and its evapotranspiration rate will be increased. However, if the advected air is colder than the vegetation, then the evapotranspiration rate will be relatively low. Advection is a serious problem in arid and semi-arid climates.

Advected energy involves the *clothesline effect* (Chang, 1968, p. 140). When warm air blows through a small plot with little or no guard area, a severe horizontal heat transfer occurs. The clothesline effect represents either the experimental bias because of the small size of the field or the border conditions unrepresentative of the large field as a whole. The clothesline effect cannot be tolerated in agronomic or climatological investigations. Where advection is important, plant growth may be improved by having larger irrigated fields to minimize the clothesline effect.

Advected energy also involves an *oasis effect* (Chang, 1968, p. 140). Inside a large field, the vertical energy transfer from the air above to the crop is called the oasis effect. The oasis effect must be reckoned with as a climatic characteristic, because it affects the evapotranspiration rates many kilometers into an irrigated field (unlike the clothesline effect).

IV. EXAMPLE CALCULATION TO DETERMINE POTENTIAL EVAPOTRANSPIRATION

Let us now follow an easy method developed by Kanemasu (1977) that we can use to estimate potential evapotranspiration. Farmers who irrigate need to know how much moisture is being used by their crops. By estimating PET, they can tell the amount of water lost by plants. As we have said, PET is evaporation from a wet surface. It is limited by the energy that the surface can absorb. The more energy it absorbs, the higher the evaporation is. So evaporation is much higher on a sunny day than on a cloudy day, and PET depends primarily on the energy from the sun. Various methods of estimating PET require data on solar radiation, temperature, humidity, and wind speed. [We will not discuss the various methods used to measure evapotranspiration, but the interested reader is referred to the following publications for a discussion of methods: Rose (1966, pp. 78–87); Slatyer (1967, pp. 56–64); Tanner (1967); Rosenberg (1974, pp. 159–205); Jury and Tanner (1975); Kanemasu et al. (1979).] To estimate PET, Kanemasu

(1977) chose the Priestley-Taylor method, because it requires relatively easy-to-obtain information (solar radiation and average temperature).

Figures 26.11 and 26.12 show the relationship between the daily solar radiation and daily PET at various mean temperatures. In Kansas, solar radiation on a clear summer day would typically be about 650 cal cm^{-2} day^{-1}; on a cloudy day, it would be about 450 cal cm^{-2} day^{-1}; and, on an overcast day, it would be about 150 cal cm^{-2} day^{-1}.

Example calculation: Suppose that one wanted to know the potential evapotranspiration for corn in Kansas. The solar radiation is 600 cal cm^{-2} day^{-1}; maximum temperature is 30°C (86°F); and minimum temperature is 25°C (77°F). One would calculate the average temperature as (30 + 25)/2 = 27.5°C (81.5°F). One then looks at Fig. 26.11 (for wheat and corn) and selects the appropriate point between the 30 and 20°C lines. The PET value is about 0.24 inch of water per day (0.61 cm per day). If maximum and minimum temperatures are not available, one can use the noon temperature.

Under a full crop cover and when water is not limiting (plants are not severely stressed), actual evapotranspiration (ET) and PET are approximately equal (Kanemasu, 1977). Therefore, under normal cropping conditions, ET would equal PET for an extended period during the summer: for example, from pre-tasselling to blister stages in corn. Under situations of

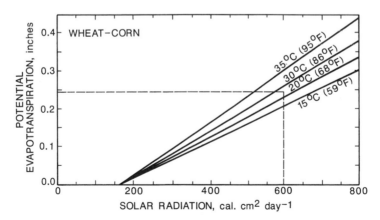

FIG. 26.11 Potential evapotranspiration for winter wheat and corn as a function of solar radiation and mean temperature. 1 inch = 2.54 cm. (From Kanemasu, E.T., An easy method of estimating potential evapotranspiration. *Keeping Up With Research.* No. 30, ©1977. Kansas Agricultural Experiment Station: Manhattan, Kansas. Reproduced by permission of the Publications Coordinator, Department of Communications, College of Agriculture, Kansas State University.)

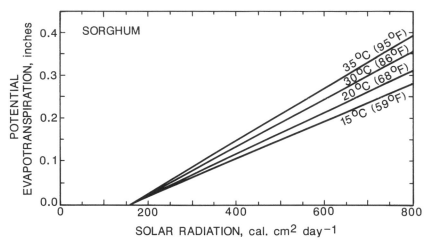

FIG. 26.12 Potential evapotranspiration for sorghum as a function of solar radiation and mean temperature. 1 inch = 2.54 cm. (From Kanemasu, E.T., An easy method of estimating potential evapotranspiration. *Keeping Up With Research*. No. 30, ©1977. Kansas Agricultural Experiment Station: Manhattan, Kansas. Reproduced by permission of the Publications Coordinator, Department of Communications, College of Agriculture, Kansas State University.)

little crop cover (e.g., poor stand development, early and late in the growing season), actual evapotranspiration can be less than PET. Then, evaporation from the soil surface is important.

Although the procedure outlined by Kanemasu (1977) gives only an approximation of daily PET, it allows quick estimates of daily water loss from several crops during a major portion of their growing season when irrigation is often necessary to avoid stress. To maintain the root zone at an optimum soil water content, evapotranspiration losses must be matched by rain or irrigation. In Kansas, for example, a 2-inch (5-cm) irrigation on corn can be used up in eight hot days (8 × 0.25 inches = 2 inches). Thus, it is important to estimate the amount of water lost by evapotranspiration, to know when to irrigate.

V. APPENDIX: BIOGRAPHY OF HOWARD PENMAN

Howard Latimer Penman (1909–1984), an English agricultural physicist, was born in 1909 at Dunston-on-Tyne in County Durham. He was raised in modest circumstances. Because of his outstanding ability and interest in science, he qualified for a first-class honors degree in physics at Armstrong College, now the University of Newcastle-on-Tyne, where he did his earliest research and later was awarded an M.S. in physics. Having earned a Ph.D. following his research on photochemistry at the Shirley Institute in Manchester, he took a position in 1937 in the Physics Department of the Rothamsted Experimental Station, where he remained until his retirement in 1974, with a three-year interruption during the war to work with the Admiralty (van Bavel, 1985). He was head of the physics department at Rothamsted Experimental Station from 1954 to 1974 and was president of the Royal Meteorological Society from 1961 to 1963 (Royal Meteorological Society, 1985). He and his wife had no children.

Internationally known among soil scientists, agricultural meteorologists, and hydrologists for his classical work on evaporation under natural conditions, he did equally innovative research on the movement of gases and vapors in soil, and, in cooperation with R.K. Schofield, on diffusive gas exchange by plant leaves (van Bavel, 1985). This last work is less well known, but it anticipated by decades later work by others that clarified the linkage between plant transpiration and photosynthesis through the stomatal mechanism. Algorithms now used in models of crop growth and water use differ little from the original equations given by Penman and Schofield.

Penman's writings on the relation between evaporation from agricultural lands and atmospheric conditions established his wide reputation, and he had scientific contacts on every continent. He was unremittingly dedicated to the idea that physics had a significant contribution to make in agriculture (van Bavel, 1985). In public, he was a stern lecturer and acrimonious debater, but, on the personal level he was a kind, gentle, and helpful person. His interests included gardening and music. He was passionate about music and was a faithful member of choral groups in London. It was his habit to take the train after work into the city, rehearse all night, and then return around midnight to his home in Harpenden. He died October 13, 1984, at St. Alban's City Hospital after a brief illness (van Bavel, 1985).

REFERENCES

Bierhuizen, J.F. (1958). Some observations on the relation between transpiration and soil moisture. *Neth J Agr Sci* 6; 94–98.

Brun, L.J., Kanemasu, E.T., and Powers, W.L. (1972). Evapotranspiration from soybean and sorghum fields. *Agronomy J* 64; 145–148.

Chang, J.-H. (1968). *Climate and Agriculture. An Ecological Survey.* Aldine: Chicago.

Jury, W.A., and Tanner, C.B. (1975). Advection modification of the Priestley and Taylor evapotranspiration formula. *Agronomy J* 67; 840–842.

Kanemasu, E.T. (1977). An easy method of estimating potential evapotranspiration. *Keeping Up With Research*. No. 30. Kansas Agricultural Experiment Station: Manhattan, Kansas.

Kanemasu, E.T., Wesely, M.L., Hicks, B.B., and Gerber, J.F. (1979). Techniques for calculating energy and mass fluxes. In *Modification of the Aerial Environment of Plants* (Barfield, B.J., and Gerber, J.F., Eds.), pp. 156–182. American Society of Agricultural Engineering: St. Joseph, Michigan.

Monteith, J.L. (1973). *Principles of Environmental Physics*. American Elsevier: New York.

Penman, H.L. (1948). Natural evaporation from open water, bare soil, and grass. *Proc Roy Soc* A193; 120–145.

Ritchie, J.T. (1972). Model for predicting evaporation from a row crop with incomplete cover. *Water Resources Res* 8; 1204–1213.

Rose, C.W. (1966). *Agricultural Physics*. Pergamon Press: Oxford.

Rosenberg, N.J. (1974). *Microclimate: The Biological Environment*. Wiley: New York.

Royal Meteorological Society (1985). Howard Penman, OBE, FRS. *Weather* 40(3): 97.

Slatyer, R.O. (1967). *Plant-Water Relationships*. Academic Press: London.

Tanner, C.B. (1967). Measurement of evapotranspiration. In *Irrigation of Agricultural Lands* (Hagan, R.M., Haise, H.R., and Edminster, T.W., Eds.), pp. 534–574. American Society of Agronomy: Madison, Wisconsin.

Thornthwaite, C.W. (1948). An approach toward a rational classification of climate. *Geogr Rev* 38; 55–94.

van Bavel, C.H.M. (1967). Changes in canopy resistance to water loss from alfalfa induced by soil water depletion. *Agr Meteorol* 4; 165–176.

van Bavel, C.H.M. (1985). In memoriam. H.L. Penman. *Soil Sci* 139; 385–386.

27

Water and Yield

In this chapter we look at water and yield and, in particular, the relationship between evaporation (or transpiration) and yield. If we could develop a reasonably simple relation (equation), we could predict the effect of water deficits on field yields, a desirable goal. To assess the relation between water and yield, Tanner and Sinclair (1983, pp. 7–11) looked at five different analyses done by the following investigators: de Wit (1958); Arkley (1963); Bierhuizen and Slatyer (1965); Stewart (1972); and Hanks (1974). Here we present only de Wit's analysis, the earliest one and basis for subsequent work. (For a biography of de Wit, see the Appendix, Section VI.)

I. DE WIT'S ANALYSIS

De Wit (1958) showed that for dry, high-radiation climates, yield and transpiration were related as

$$Y/T = m/T_{max'} \tag{27.1}$$

where Y = total dry matter mass per area, T = total transpiration per area during growth to harvest, and T_{max} = mean daily free water evaporation for the same period. The constant m is related to the WR/pan used by Briggs and Shantz (1917) ($1/m \approx WR/pan$) where WR = water requirement. De Wit showed that m was governed mainly by species and, for a first approximation, it was independent of soil nutrition and water availability unless there was a serious nutrition deficiency or unless soil water was too high (e.g., due to inadequate aeration).

De Wit proposed that this relation should hold until T approaches a maximum production governed by the growing conditions. The relation in Equation 27.1 could be simplified for humid regions because, when water was not limiting, fluctuations in intercepted radiation, although reflected in transpiration and growth, would not affect appreciably the ratio T/T_{max}. De Wit found under these conditions that

$$Y/T = n, \qquad\qquad (27.2)$$

where n is a constant, gave a better description than does Equation 27.1.

The value of m in Equation 27.1 can be approximated with Equation 27.3 from water use efficiency and mean daily pan evaporation (E_{pan}):

$$m = (Y/T)E_{pan}. \qquad\qquad (27.3)$$

In the Great Plains of the United States, de Wit (1958) found, using data of Briggs and Shantz and a number of other sources, that m was equal to 55, 115, and 207 kg ha^{-1} day^{-1} for Grimm alfalfa, Kubanka wheat, and Red Amber sorghum, respectively (Tanner and Sinclair, 1983, pp. 8–9) [or 5.5, 11.5, and 20.7 grams dry matter per kilogram water per day, respectively (Chang, 1968, p. 128)]. In the Netherlands, the value for n for beets, peas, and oats was 6.1, 3.4, and 2.6 grams per kilogram water per day, respectively (Chang, 1968, p. 128). The value of m and n are more dependent on the climatic conditions than on the nutrient level of the soil and the availability of water, provided that the nutrient level is not too low and the availability of water is not too high. These values are also independent of the degrees of mutual shading, provided that the leaf mass is not too dense. Where these conditions are not fulfilled, the m and n values are larger (Chang, 1968, p. 128).

Table 27.1 compares the values of m derived from the experiments of Briggs and Shantz (1914) and subsequent field observations (Tanner and Sinclair, 1983, p. 8). In Table 27.1, the m's developed from the data of Briggs and Shantz (1914) and Hanks et al. (1969) do not include root dry matter, whereas an estimate of root yield was made for the other data. Also, pan evaporation was used directly with no correction to free water evaporation, as made by de Wit. [According to LeGrand and Myers (1976), readings taken of pan evaporation tell how much water evaporates from lakes, if one applies a pan coefficient of about 0.70.]

Except for the Wisconsin data, corn, sorghum, and millet give the highest values of m, followed by the grain cereals, potatoes, and then the legumes. The data for corn indicate a high level of variability in m, even though the corn crops were subjected to nearly the same experimental treatments. The high m for potato and alfalfa in Wisconsin may indicate that Equation 27.2

TABLE 27.1 Experimental Estimates of m (kg ha^{-1} d^{-1}) From Data of Briggs and Shantz (1914) and More Recent Field Experiments

Crop	Briggs and Shantz	Subsequent field data	Source
Corn	213 ± 14	215 ± 20	UT, Stewart et al., 1977
		258 ± 1	CO, Stewart et al., 1977
		262 ± 46	AZ, Stewart et al., 1977
		314 ± 12	CA, Stewart et al., 1977
Grain sorghum	240 ± 10	141 ± 6	Great Plains, Hanks et al., 1969
Millet	260 ± 35	150 ± 18	Great Plains, Hanks et al., 1969
Wheat	158 ± 10	125 ± 15	Great Plains, Hanks et al., 1969
Potato	160 ± 8	217 ± 24	WI, Tanner, 1976 (unpublished)
Alfalfa	90 ± 11	214 ± 26	WI, Tanner, 1977 (unpublished)
Soybean	102 ± 7	128 ± 34	KS, Teare et al., 1973

From Tanner, C.B., and Sinclair, T.R. Efficient water use in crop production: Research or research?, p. 8. In Taylor, H.M., Jordan, W.R., and Sinclair, T.R. (Eds.), *Limitations to Efficient Water Use in Crop Production*, © 1983, American Society of Agronomy, Crop Science Society of America, and Soil Science Society of America: Madison, Wisconsin. Reprinted by permission of the American Society of Agronomy, Crop Science Society of America, and Soil Science Society of America.

rather than Equation 27.1 is applicable to this humid region. If so, it is difficult to know whether to use Equation 27.1 or Equation 27.2, because gradations in humidity occur not only between locations, but also seasonally at one location. Table 27.1 also shows that there is no consistent improvement in m between the crops grown in 1912–1913 (Briggs and Shantz, 1914) and more recently, excluding the Wisconsin data for the reason discussed above. Thus, to the extent that m is a measure of T efficiency for total biomass production, it appears that there has been no increase in T efficiency since Briggs and Shantz did their work at the beginning of the 1900s.

II. RELATIONSHIP BETWEEN YIELD AND TRANSPIRATION AND YIELD AND EVAPOTRANSPIRATION

Let us look at figures showing the relationship between yield and transpiration and yield and evapotranspiration. Figures 27.1 and 27.2 show the relationship between yield and transpiration as determined by Arkley (1963), who used data from Briggs and Shantz (1913a). The classical work by Briggs and Shantz demonstrated a close relation between transpiration and dry matter production. That is, dry matter is decreased by water deficits. In their experiments, the linear relationship held for different varieties of oats (Fig. 27.1) and barley (Fig. 27.2).

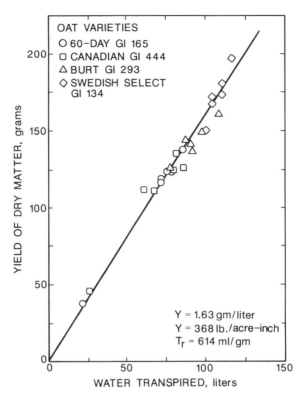

FIG. 27.1   Relationship between yield of dry matter and amount of water transpired by oat varieties. Data obtained by Briggs and Shantz (1913a, 1913b) and shown by Arkley, 1963. (From Chang, J.-H., *Climate and Agriculture: An Ecological Survey*, Fig. 69, p. 126, ©1968. Aldine Publishing Co: Chicago. Reproduced by permission of Dr. Jen-Hu Chang.)

For the same plant species, the efficiency of water use may vary according to the climate (Chang, 1968, p. 220). Stanhill (1960) compared measurements of pasture growth and potential evapotranspiration at seven localities in different parts of the world. In Fig. 27.3, the cumulative measured dry-weight yields are plotted against cumulative measured transpiration. A linear relationship exists at each site, but the slope of the line changes with latitude. In general, the growth rate per unit of water used is higher at high latitudes. This is a result of the increased respiration rate in the tropics.

The relationship between evapotranspiration and dry-matter production may or may not be linear. This is partly because the fraction of evapo-

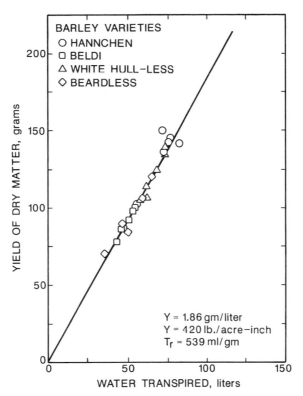

FIG. 27.2 Relationship between yield of dry matter and amount of water transpired by barley varieties. Data obtained by Briggs and Shantz (1913a, 19193b) and shown by Arkley,
1963. (From Chang, J.-H., *Climate and Agriculture: An Ecological Survey*, Fig. 70, p. 127,
©1968. Aldine Publishing Co: Chicago. Reproduced by permission of Dr. Jen-Hu Chang.)

ration that does not contribute to plant growth varies throughout the crop
life cycle. Figures 27.4, 27.5, and 27.6 show the relationship between yield
and evapotranspiration as determined by Allison et al. (1958) and Staple
and Lehane (1954). Even when dry matter production does increase linearly with evapotranspiration, the regression line seldom passes through
the zero point. In other words, evapotranspiration in the field might be
appreciable when the yield is still zero (Chang, 1968, pp. 211–212). Allison
et al. (1958) analyzed the yields of a number of crops grown in a lysimeter
near Columbia, South Carolina, for a period of more than five years. Their
data indicated that the first 18 inches (46 cm) of evapotranspired water
were required to produce only enough for plant survival (Fig. 27.4). The

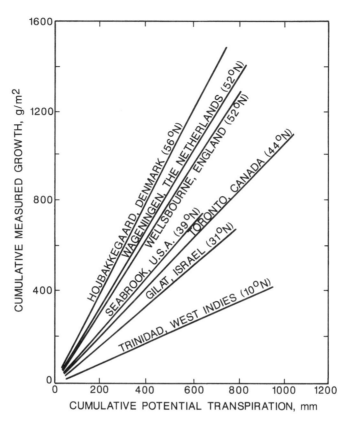

FIG. 27.3 Measurements of potential evapotranspiration and dry matter production from pastures. Data of Stanhill, 1960. (From Chang, J.-H., *Climate and Agriculture: An Ecological Survey*, Fig. 114, p. 222, ©1968. Aldine Publishing Co: Chicago. Reproduced by permission of Dr. Jen-Hu Chang.)

increase in dry matter was almost linear with increasing amounts of water used from 18 to 22 inches (46 to 56 cm). Staple and Lehane (1954) studied the use of water by spring wheat grown in tanks and in the open field in Swift Current, Canada. They reported that 4.9 inches (12 cm) of water for tanks and 5.64 inches (14 cm) for the field were necessary to establish the plants (Figs. 27.5 and 27.6). Beyond this, the yield in the tanks increased nearly linearly. But in the field the yield increased curvilinearly. In either case, the maximum production potential was not realized because of the shortage of water.

Before concluding this section on the relationship between water and yield, let us briefly look at the situation of an individual leaf. Up to now, we

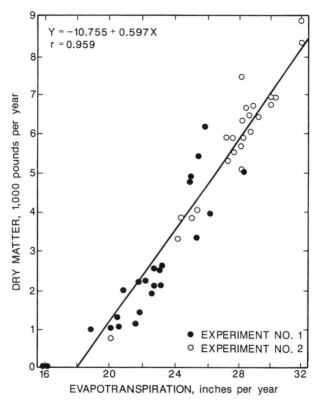

FIG. 27.4 Relationship between crop yields and water use. Data of Allison et al., 1958. (From Chang, J. H., *Climate and Agriculture: An Ecological Survey*, Fig. 105, p. 212, ©1968. Aldine Publishing Co: Chicago. Reproduced by permission of Dr. Jen-Hu Chang.)

have been considering groups of leaves as they might exist under field conditions. For a single leaf, the net assimilation, or net photosynthesis, increases with light intensity to the saturation point and then levels off. The transpiration rate will, however, increase linearly with radiation to a much higher intensity. Thus, the ratio between transpiration and photosynthesis will vary according to the radiation intensity in a manner postulated by de Wit (1958) (Fig. 27.7). This same relationship was later quantitatively presented by Bierhuizen (1959) (Fig. 27.8). The high ratio occurring at extremely low radiation intensity is because transpiration has some value, whereas photosynthesis first has to compensate for the respiration. This high ratio is of little significance because of the low rates of both processes. The lowest ratio is reached at a radiation intensity of 0.1 to 0.2 langleys per minute. Such low radiation intensities are observed only in the early morning and late afternoon.

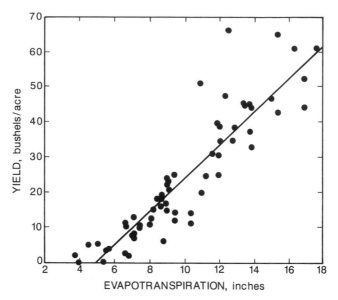

FIG. 27.5   Relationship between wheat yield and evapotranspiration in tanks, 1922–1952. Data from Staple and Lehane, 1954. (From Chang, J.-H., *Climate and Agriculture: An Ecological Survey*, Fig. 106, p. 213, ©1968. Aldine Publishing Co: Chicago. Reproduced by permission of Dr. Jen-Hu Chang.)

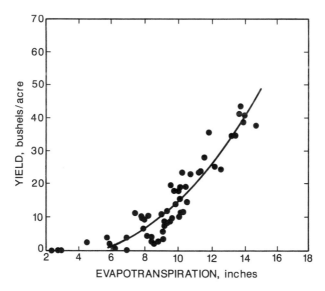

FIG. 27.6   Relationship between wheat yield and evapotranspiration on field plots. Data from Staple and Lehane, 1954. (From Chang, J.-H., *Climate and Agriculture: An Ecological Survey*, Fig. 107, p. 213, ©1968. Aldine Publishing Co: Chicago. Reproduced by permission of Dr. Jen-Hu Chang.)

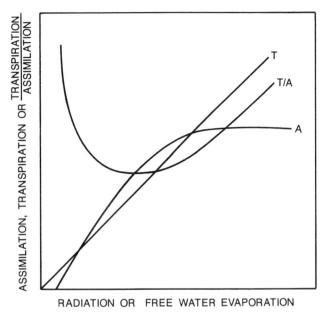

FIG. 27.7 Relationship between net assimilation (*A*), transpiration (*T*), and the transpiration to assimilation ratio (*T/A*) for leaves of plants as a function of the radiation or free water evaporation. Figure from de Wit, 1958. (From Chang, J.-H., *Climate and Agriculture: An Ecological Survey*, Fig. 67, p. 124, ©1968. Aldine Publishing Co: Chicago. Reproduced by permission of Dr. Jen-Hu Chang.)

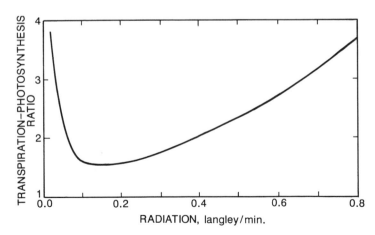

FIG. 27.8 Relationship between radiation and the transpiration-photosynthesis ratio. Figure from Bierhuizen, 1959. (From Chang, J.-H., *Climate and Agriculture: An Ecological Survey*, Fig. 68, p. 124, ©1968. Aldine Publishing Co: Chicago. Reproduced by permission of Dr. Jen-Hu Chang.)

As the radiation intensity increases, beyond 0.2 langleys per minute, the ratio of transpiration to photosynthesis for a single leaf increases nearly linearly. Thus, for a single leaf, the efficiency of water use in the production of dry matter will be lower in areas of high radiation, such as in the arid tropics.

III. WATER AND MARKETABLE YIELD

In many instances, the reductions of yields of grain and other marketable parts of crops are roughly in proportion to the decreases in transpiration induced by water deficits (Tanner and Sinclair, 1983, p. 18). However, often there are stages of development, such as pollination, at which marketable yield may be extraordinarily affected. Figure 27.9 shows a generalized relation between yield and adequacy of water at different stages of growth. The curve was developed for sugar cane in Hawaii, but can be applied, in general, to other crops (Chang, 1968, pp. 214–215). Table 27.2 summarizes moisture-sensitive periods for selected crops during which a water deficit depresses the economic yield much more than at other periods (Chang, 1968, p. 216). Varieties (cultivars) may also respond differently under drought conditions. A drought-resistant variety may follow the upper broken curve in Fig. 27.9, whereas a variety less resistant to drought may follow the lower broken curve.

IV. WATER AND QUALITY

We need to note also that water deficits may be necessary to increase the quality of a crop. So far, we have been concerned only with the relationship between water and dry matter (or marketable yield) production. The quality of an agricultural product, however, is not necessarily related to the yield. In analyzing the relationship between water and crop quality, one must differentiate between natural rainfall and controlled irrigation water. Rainfall usually is accompanied by high cloudiness and low radiation, but the application of irrigation water is not complicated by a change of unfavorable weather conditions.

The effects of irrigation on crop quality are summarized in Table 27.3 from Chang (1968, pp. 223–224). In general, adequate irrigation throughout periods of active vegetative growth results in an improvement in crop quality. However, during the ripening period, moderate moisture stress often has been found to be desirable, especially in the case of certain compounds such as rubber, sugar, and tobacco. For example, the rubber content of guayule is increased by a slight moisture stress. The withdrawal

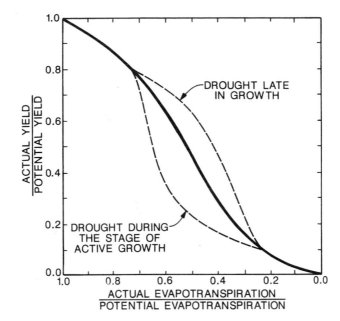

FIG. 27.9 Generalized relationship between yield and adequacy of water application. (From Chang, J.-H, *Climate and Agriculture: An Ecological Survey*, Fig. 108, p. 214, ©1968. Aldine Publishing Co: Chicago. Reproduced by permission of Dr. Jen-Hu Chang.)

of irrigation water several weeks before harvest is a common practice in sugar-cane culture. Late water stress also has been found to increase the sucrose concentration of sugar beets. The aroma of Turkish tobacco is improved by water stress late in the crop cycle. The flavor and taste of most fruits also can be enhanced by the same means (Chang, 1968, p. 224).

V. CROP-WATER-USE EFFICIENCY

Here are a few final comments on crop-water-use efficiency (Tanner and Sinclair, 1983, pp. 18–21). Experimentally, we need to do three things: 1) Be able to distinguish transpiration (T) from evaporation (E) in studies of evapotranspiration (ET); 2) be able to make estimates of vapor pressure deficit (VPD) because data on yield and transpiration are normalized by using VPD to account for differences in years and locations; and 3) improve our understanding of dry-matter partitioning into roots, shoots, and marketable yield.

When increased water-use efficiency is found as a result of improved management, the increases result from increased transpiration as a fraction of the ET. ET efficiency is increased, although T efficiency is changed little,

TABLE 27.2   Moisture-Sensitive Stages (from Chang, 1968, p. 216)

Crop	Critical stage
Cauliflower	No critical moisture-sensitive stage; frequent irrigation required from planting to harvest
Lettuce	Just before harvest when the ground cover is complete
Cabbage	During head formation and enlargement
Broccoli	During head formation and enlargement
Radishes and onions	During the period of root or bulb formation
Snap beans	During flowering and pod development
Peas	At the start of flowering and when the pods are swelling
Turnips	From the time when the size of the edible root increases rapidly until harvest
Potatoes	After the formation of tubers
Potatoes (White Rose)	From stolonization to the beginning of tuberization
Soybeans	Period of major vegetative growth and blooming
Oats	Commencement of ear emergence
Wheat	During heading and filling
Barley	Effects of water stress on grain yield and protein content shown to be greater at the early boot stage than at the soft dough stage, and shown to be greater at the soft dough stage than at the onset of tillering or ripening stages
Corn	Period of silking and ear growth
Cotton	At the beginning of flowering
Apricots	Period of floral bud development
Cherries and peaches	Period of rapid growth prior to maturity
Olives	Later stages of fruit maturity

Table reproduced with the permission of Dr. J.-H. Chang.

if any (Tanner and Sinclair, 1983, p. 20). Conditions such as low fertility, water stress, plant disease, or insects that lower the leaf area so that the canopy is no longer closed will increase soil evaporation and thereby lower Y/ET. Factors such as poor growing temperatures and extreme infertility can lower both Y/T and Y/ET. Nevertheless, a decrease in leaf area index has to be severe before substantial changes in Y/T will be observed. Changes in ET efficiency occur more readily than changes in T efficiency.

Changing plant architecture is not likely to change T efficiency significantly in canopies achieving a leaf area index of about 3. However, canopy structure and population can modify the loss due to evaporation relative to the loss due to transpiration, and, therefore, can affect ET efficiency more than T efficiency. Crop breeding can change rates of maturation to take advantage of seasonal water availability and perhaps change rooting habits to increase soil water supply or change the timing of withdrawal. Such changes may aid in the efficient use of water and ET efficiency without changing the T efficiency.

TABLE 27.3   Effect of Irrigation on Crop Quality

Crop	Effect
Pasture	Irrigation increased the protein and decreased the fat contents of the herbage but had little effect on the crude fiber and ash content.
Vegetables	Maintaining a low moisture stress during the whole growth period generally resulted in the highest yield and quality.
Snap beans	Irrigation decreased the percentage of pods that were badly crooked or severely malformed. Fibrous content of beans was generally reduced.
Sweet corn	Irrigation increased the number of marketable ears per plant, the average weight per ear and the gross yield of unhusked ears, and the percentage of usable corn cut from these ears for canning or freezing.
Soybeans	Irrigated soybeans had slightly lower oil content and slightly higher protein content.
Barley	Irrigation increased the yield of grain and improved malting quality, mainly by increasing extract.
Potatoes	Irrigation that gave good increase in yield of potatoes very seldom reduced the specific gravity and was more likely to increase it.
Tobacco	Irrigated tobacco had lower nicotine and protein, but higher carbohydrate content.
Fruits	Canned peaches that were tough and leathery in texture, pears that remained green and hard a week or more after the ripening season, prunes that were sunburned, and walnuts with partly filled shells were some of the results of a relatively long time without readily available moisture.
Olives	The higher yield obtained by irrigation was due to an increase in fruit size, rather than in the number of fruits. Irrigated groves had a higher oil content than unirrigated ones.

For references for the results, see Chang, 1968. (From Chang, 1968, pp. 223–224.) Table reproduced with the permission of Dr. J.-H. Chang.

The crop can be managed (e.g., population and fertility) to increase or decrease leaf area index, thus changing the partitioning of E and T and ET efficiency. Preventing evaporation from the soil and transpiration from weeds also modifies the partitioning of E and T. However, there is a limit to the improvement in water-use efficiency that such manipulations can provide. The ET efficiency can only approach the T efficiency as the upper limit.

To summarize, it appears that there are two ways to modify significantly the T efficiency based on total dry matter of crops (Tanner and Sinclair, 1983, p. 20). First, crops can be grown in humid climates where the vapor pressure deficit is small and advection is minimal. However, in these regions, sunlight is usually less and total yields may be smaller.

Second, the partitioning of total dry matter can be changed to create more marketable products. This would increase the T efficiency of the marketable yield. This option means changing the chemistry of the plant. The changes would have to be large, and, consequently, are unlikely. Therefore, changing the T efficiency of the marketable yield seems improbable. Tanner and Sinclair (1983, p. 25) conclude that transpiration efficiency is a relatively difficult to manipulate variable. Transpiration efficiencies of different crops have changed little since Briggs and Shantz did their work at the beginning of the 1900s. Even though the likelihood of large improvements in T efficiency is small, crop water-use efficiency can be improved, as was noted in the preceding section (e.g., changing rooting habits, increasing leaf area index, minimizing soil evaporation, and preventing transpiration from weeds).

VI. APPENDIX: BIOGRAPHY OF CORNELIUS DE WIT

Cornelius ("Kees") Teunis de Wit (1924–1993), professor at the Agricultural University, Wageningen, The Netherlands, was born east of Arnhem. His introduction to agriculture was while he worked as a farm laborer during World War II (Rabbinge, 1995). His thesis at Wageningen was notable for its theoretical nature and foreshadowed his founding of the department of theoretical production. Early in his career, he worked in Burma. Later, he developed strong ties with Mali, Israel, and the United States. In the 1950s, after writing his dissertation on fertilizer placement, he wrote classic monographs on competition, on the relation between transpiration and crop yields, and on the photosynthesis of leaf canopies. He calculated the population that the earth's photosynthesis could feed. In the 1960s and 1970s, de Wit and his colleagues at Wageningen took up the dynamic simulation of crop growth, incorporating biochemistry, development from seeds to grain, the soil and atmosphere around the crop, and its pests (American Society of Agronomy, 1994). His countrymen elected him a senator in the parliament of Gelderland, and, in The Hague during the 1980s, he served on the Netherlands Scientific Council for Governmental Policy. Afflicted by diabetes, he retired in February, 1989, honored by ceremonies attended by scientists from many countries. Undaunted, he continued to work and advised the Consultative Group on International Agricultural Research (CGIAR) and suggested research that should be carried out on crops in different regions.

He was a Knight of the Order of The Netherlands Lion and Foreign Associate of the National Academy of Sciences of the United States. In

1984, he was co-winner, with Don Kirkham, of the Wolf Prize in agriculture; Kirkham was recognized for theoretical work and de Wit for development of numerical models. The citation read, "for their innovative contributions to the quantitative understanding of soil water and othe environmental interactions influencing crop growth and yield."

De Wit and his wife had two children. He died at age 69 on December 8, 1993 (American Society of Agronomy, 1994).

REFERENCES

Allison, F.E., Roller, E.M., and Raney, W.A. (1958). Relationship between evapotranspiration and yields of crops grown in lysimeters receiving natural rainfall. *Agronomy J* 50; 506–511.

American Society of Agronomy (1994). Deaths. Cornelius T. de Wit. *Agronomy News*, February issue, p. 25. American Society of Agronomy: Madison, Wisconsin.

Arkley, R.J. (1963). Relationships between plant growth and transpiration. *Hilgardia* 34; 559–584.

Bierhuizen, J.F. (1958). Some observations on the relation between transpiration and soil moisture. *Neth J Agr Sci* 6; 94–98.

Bierhuizen, J.F. (1959). Plant growth and soil moisture relationships. In *Plant-Water Relationships in Arid and Semiarid Conditions*. United Nations Educational, Scientific and Cultural Organization: Paris. (Cited by Chang, 1968, p. 251)

Bierhuizen, J.F., and Slatyer, R.O. (1965). Effect of atmospheric concentration of water vapor and CO_2 in determining transpiration-photosynthesis relationships of cotton leaves. *Agr Meteorol* 2; 259–270.

Briggs, L.J., and Shantz, H.L. (1913a). The water requirement of plants. I. Investigations in the Great Plains in 1910 and 1911. *US Dep Agr Bur Plant Industry*, Bull. No. 284. United States Department of Agriculture: Washington, DC.

Briggs, L.J., and Shantz, H.L. (1913b). The water requirement of plants. II. A review of the literature. *US Dep Agr Bur Plant Industry*, Bull. No. 285. United States Department of Agriculture: Washington, DC.

Briggs, L.J., and Shantz, H.L. (1914). Relative water requirement of plants. *J Agr Res* 3; 1–63.

Briggs, L.J., and Shantz, H.L. (1917). The water requirement of plants as influenced by environment. *Proc 2nd Pan-Am Sci Congr* 3; 95–107.

Chang, J.-H. (1968). *Climate and Agriculture: An Ecological Survey*. Aldine: Chicago.

de Wit, C.T. (1958). Transpiration and crop yields. *Versl Landbouwk Onderz* No. 64.6. Institute for Biological and Chemical Research on Field Crops and Herbage: Wageningen, The Netherlands.

Hanks, R.J. (1974). Model for predicting plant yield as influenced by water use. *Agronomy J* 66; 660–665.

Hanks, R.J., Gardner, H.R., and Florian, R.L. (1969). Plant growth-evapotranspiration relations for several crops in the central Great Plains. *Agronomy J* 61; 30–34.

LeGrand, F.E., and Myers, H.R. (1976). Weather observation and use. *OSU Extension Facts* No. 9410. Oklahoma State University: Stillwater, Oklahoma.

Rabbinge, R. (1995). Prof. Dr. Ir. Cornelius Teunis de Wit (1924–93). *J Ecology* 83; 345–346. (The same obituary by R. Rabbinge appears in *Agr Systems* 47; I–III, 1995.)

Stanhill, G. (1960). The relationship between climate and the transpiration and growth of pastures. In *Proceedings, Eighth International Grassland Congress*. (Cited by Chang, 1968, p. 281.)

Staple, W.J., and Lehane, J.J. (1954). Weather conditions influencing wheat yields in tanks and field plots. *Can J Agr Sci* 34; 552–564.

Stewart, J.I. (1972). Prediction of Water Production Functions and Associated Irrigation Programs to Minimize Crop Yield and Profit Losses due to Limited Water. Ph.D. Thesis, University of California: Davis, California. University Microfilms #73-16, 934. (Cited by Tanner and Sinclair, 1983, p. 27.)

Stewart, J.I., Hagan, R.M., Pruitt, W.O., Hanks, R.J., Riley, J.P., Danielson, R.E., Franklin, W.T., and Jackson, E.B. (1977). Optimizing crop production through control of water and salinity levels in the soil. *Pub. no. PRWG151-1*. Utah Water Research Lab, Utah State University: Logan, Utah. (Cited by Tanner and Sinclair, 1983, p. 27.)

Tanner, C.B., and Sinclair, T.R. (1983). Efficient water use in crop production: Research or research? In *Limitations to Efficient Water Use in Crop Production* (Taylor, H.M., Jordan, W.R., and Sinclair, T.R., Eds.), pp. 1–27. American Society of Agronomy, Crop Science Society of America, Soil Science Society of America: Madison, Wisconsin.

Teare, I.D., Kanemasu, E.T., Powers, W.L., and Jacobs, H.S. (1973). Water-use efficiency and its relation to crop canopy area, stomatal regulation, and root distribution. *Agronomy J* 65; 207–211.

INDEX

489